Bruno Maresca   Susan Lindquist   (Eds.)

Heat Shock

With 83 Figures

Springer-Verlag
Berlin Heidelberg New York
London Paris Tokyo
Hong Kong Barcelona
Budapest

Ph. D. Bruno Maresca
International Institute of Genetics and Biophysics
Via Marconi 12
80125 Naples, Italy

Ph. D. Susan Lindquist
Howard Hughes Medical Institute
Department of Molecular Genetics and Cell Biology
The University of Chicago
1103 E 57th Street
Chicago, IL 60637, USA

Book design by Donatella Capone, Secretary IIGB Press, Naples, Italy
Cover design by Leonardo Coen Cagli

ISBN 3-540-54111-X Springer-Verlag Berlin Heidelberg New York
ISBN 0-387-54111-X Springer-Verlag New York Berlin Heidelberg

Typesetting: Camera ready by author
31/3145-543210 - Printed on acid-free paper

Preface

In the last few years the study of the Heat Shock Response has moved to the center stage in cell biology. Control mechanisms regulating the synthesis of heat shock proteins (HSPs) have for several years provided important general insight on the regulation of gene expression and continue to do so. But the major revelation of the last few years, which has sparked interest from all quarters of biology, is the discovery that HSPs (and closely related proteins produced at normal temperatures) play major roles in an extraordinary variety of normal cellular processes. In the past few years HSPs have been the focus of investigations in many areas of cell biology, including protein trafficking, signal transduction, DNA replication, RNA transcription, protein synthesis, and in the assembly of diverse protein structures. They have been the focus of biomedical investigations in immunology, infectious diseases, chronic degeneration, hyperthermia, cancer research, to name a few. The explosion of interest in heat shock proteins is evidenced by a doubling in the number of publications on varies aspects of heat shock response in 1990 compared to 1989.

A common interest in the heat shock response brought together 225 scientists - biochemists, cell biologists, geneticists, immunologists and physicians - from around the world to discuss their new findings at the HEAT SHOCK MEETING, held at Ravello (Italy), September 17-20, 1990. This was the first major meeting on the response to be held in Europe: previous meetings were held at Cold Spring Harbor, New York or at Keystone, Colorado.

The central phenomenon of the HEAT SHOCK response is the rapid and specific induction of a small set of proteins when cells or whole organism are abruptly exposed to temperatures that are about 5° to 10°C above their normal physiological growth temperature. The same set of proteins is induced by many other type of stresses, for example ethanol, anoxia, and heavy metal ions, and their induction correlates with the development of tolerance to more severe stress (high temperatures, higher concentrations of ethanol, etc.). HSPs are induced in both prokaryotic and eukaryotic cells and are among the most highly conserved proteins known. In same cases the proteins are also produced at a substantial level at normal temperatures; in others, closely related proteins encoded by separate genes are produced at normal temperatures.

In the last few years, several laboratories world-wide have elucidated some of the mechanisms of regulation of gene expression both at transcriptional (by identifying a specific consensus sequence, HSE) and translational level (e.g., the characterization in different systems of a DNA binding protein, HSF).

The work presented at the HEAT SHOCK MEETING in Ravello can roughly be divided into three general areas: Regulation, Function, and Medical Importance. Presentations concerning the regulation of the response focussed on aspects of transcriptional control. Heat shock genes carry consensus elements which direct transcription under stressful conditions. In eukaryotic cells, this consensus element (HSE) binds a specific transcription factor (HSF). The changes in HSF that are triggered by stress were explored, including changes in its modification state, its oligomeric structure, and its DNA binding capacity in vivo and in vitro. Other presentations concerned the involvement of other proteins in the transcriptional activation of heat shock genes.

After years of frustration concerning possible functions of HSPs, enormous progress has recently been made. Examples of co-operation between HSPs in performing fundamental cellular tasks formed a major topic of discussion. For example, it was demonstrated that HSPs are involved in several steps of mitochondrial protein import and assembly. Members of the HSP70 family act in the cytosol to keep proteins destined for mitochondrial import in an unfolded, translocational-competent state and act inside the mitochondria to accept the unfolded proteins as they are being transported. HSP60 functions inside the mitochondria, facilitating the proper folding and assembly of the newly transported proteins. The DnaK, GrpE and DnaJ proteins function together in the replication of bacteriophage and plasmid DNAs, in the renaturation of denatured proteins, and in the regulation of the response. GroEL and GroES proteins act together to facilitate the assembly of oligomeric proteins.

In other systems there were evidence for individual HSPs serving important functions. For example, HSP47 appears to be involved in collagen assembly; one member of the HSP70 family helps to target specific proteins to lysosomes for degradation; ubiquitin may play a similar role; HSP82 is required for steroid hormone receptors to achieve an activation-competent state and may help to tether some receptors, such as the glucocorticoid receptor to the cytoskeleton; ubiquitin and ubiquitin-conjugating enzymes help cells to cope with high temperatures by removing denatured proteins and protein complexes from the cell. Whether these diverse functions also involve the cooperation of other HSPs is not yet clear.

A full session of the meeting was devoted to the relationship between HSPs and the immuno response. HSPs have important roles in immunity as well as in auto-immunity. Stress proteins are among the dominant antigens in the immune response to a large spectrum of pathogens:

Mycobacterium tubercolosis, M.leprae, Plasmodium falciparum, Schistosoma mansoni *etc. Indeed, HSPs are primary targets for the mysterious γδT-cells, which are located in skin and gut and may provide a first line of defense against diverse pathogens. Another connection between HSPs and the immune response is the key role that is apparently played by HSP70 family members in antigen presentation by MHC-class II molecules. Furthermore, during the process of inflammation, oxygen free radicals are produced and these, in turn, induce the synthesis of heat shock proteins, suggesting HSPs may play a protective role in inflammatory responses.*

Other presentations focused on the function of HSPs on a broader, biological scale. The induction of HSPs is correlated with the induction of tolerance to extreme temperatures, protecting organisms both from lethality and from heat-induced developmental defects. HSPs play a role in protecting mitochondrial ATPase in dimorphic fungi during the heat-induced phase transition, and presumably play a similar role in other organisms. The induction of HSPs also helps to prevent the denaturation of proteins in mammalian and bacterial cells. A mutation in the yeast hsp104 gene demonstrate that HSP104, together with other as yet unidentified factors, is responsible for induced tolerance to extreme temperatures in yeast cells. One of these factors is likely to be HSP70. Experiments in mammalian cells strongly suggest a key role for this protein in thermotolerance.

In conclusion, recent developments emphasize the central role that heat shock proteins play not only in cellular homeostasis, but also in an extraordinary variety of other fundamental cellular processes.

We are particularly grateful for the support provided to the HEAT SHOCK MEETING by the National Research Council (Comitato Biotecnologie), The Regione Campania, the Institute of Philosophical Studies. In addition, we would like to thank the "viamarconidieci Foundation" and Ms Donatella Capone for the flawless organization of the Meeting.

Bruno Maresca

Susan Lindquist

Contents

Analysis of Heat Shock Regulation

Heat Shock Protein Functions in *E. coli* and Yeast

Analysis of Heat Shock Protein Functions

Medical Applications of Heat Shock Response

Contributors

The numbers in brackets are the opening page numbers of the contributors' articles.

Olivier Bensaude, Institute Pasteur, 28, Rue du Dr. Roux, 75724 Paris-Cedex 15, France **(97)**

Willi K.H. Born, National Jewish Center for Immunology and Respiratory Medicine, 1400 Jackson Street, Denver, CO, 80206; USA **(227)**

Thomas Bosch, Zoologisches Institut der Universität München, Luisenstraße 14, 8 München 2, Germany **(133)**

Jan R. Brown, Scarborough Campus, University of Toronto, 1265 Military Trail, Scarborough Ontario, MIC IA4, Canada **(291)**

Bernd Bukau, Zentrum für Molekulare Biologie der Universität Heidelberg, Im Neuenheimer Feld, 282, D-6900 Heidelberg, Germany **(55)**

Elizabeth A. Craig, Department of Physiology and Chemistry, University of Wisconsin, 1300 University Ave., Madison, WI, 53706, USA **(77)**

Fred J. Dice, Department of Physiology, Tufts University School of Medicine, 139 Harrison Avenue, Boston, MA, 02111, USA **(181)**

Costa Georgopoulos, University of Utah School of Medicine, 50 N. Medical dr., Salt Lake City, UT, 84132, USA **(45)**

George M. Hahn, Department of Radiation Oncology, Stanford University, Stanford, CA, 94305-5468, USA **(249)**

Arthur Horwich, Yale University School of Medicine, Department of Human Genetics, 333 Cedar St., New Haven, CT, 06510-8005, USA **(165)**

Stefan Jentsch, Friederich-Miescher-Laboratorium der Max-Planck-Gesellschaft, Spemannstr. 37-39, Tübingen 7400, Germany **(85)**

Gloria C. Li, Department of Medical Physics, Memorial Sloan-Kettering Cancer Center, 1275 York Ave., New York, NY, 10021, USA **(257)**

Susan L. Lindquist, Department of Molecular Genetics and Cell Biology, University of Chicago, 1103 E 57th St., Chicago, IL, 60637, USA **(123)**

John T. Lis, Department of Biochemistry, Molecular and Cell Biology, Cornell University, Wing Hall, Ithaca, NY, 14853, USA **(3)**

Bruno Maresca, International Institute of Genetics and Biophysics, Via Marconi, 12, 80125 Naples, Italy **(143)**

R. John Mayer, Biochemistry Department, University of Nottingham Medical School, Queen's Medical Centre, Nottingham NG7 2UH, UK **(299)**

Richard I. Morimoto, Department of Biochemistry, Molecular and Cell Biology, Northwestern University, 2153 Sheridan Rd., Evanston, IL, 60201, USA **(17)**

Martin E. Munk, Department of Medical Microbiology and Immunology, University of Ulm, Ulm D-7900, Germany **(209)**

Kazuhiro Nagata, Chest Disease Research Institute, Kyoto University, Sokyo-ku, Kyoto 606, Japan **(105)**

Nancy S. Petersen, Department of Molecular Biology, University of Wyoming, University Station Box 394, Laramie, WY, 82071, USA **(155)**

Nikolaus Pfanner, Institut für Physiologische Chemie, Physikalische Biochemie und Zell Biologie der Ludwig-Maximilians Universität München, Goethestraße, 33, D-8000 München 2, Germany **(175)**

Susan K. Pierce, Northwestern University, Department of Biochemistry Molecular Biology and Cell Biology, 2153 Sheridan Road, Evanston, IL, 60208-3500, USA **(215)**

Barbara Polla, Unité d'Allergologie, Hôpital Cantonal Universitaire, 24 Rue Micheli du Crest, Geneva 12114, Switzerland **(279)**

Milton J. Schlesinger, Department of Microbiology and Immunology, Washington University School of Medicine, Box 8053, St. Louis, MO, 63110, USA **(111)**

Stephen J. Ullrich, Department of Health and Human Services, National Institutes of Health, NCI, Bethesda, MD, 20892, USA **(269)**

Richard Voellmy, Department of Biochemistry and Molecular Biology, University of Miami School of Medicine, 1600 NW 10th Ave., Miami, FL, 33136-1015, USA **(35)**

Sue H. Wickner, Laboratory of Molecular Biology, NCI, National Institutes of Health, Bldg 37 2E14, Bethesda, MD, 20892, USA **(67)**

John B. Winfield, Division of Rheumatology and Immunology, 932 FLOB 231 H - CB # 7280, University of North Carolina, CH Chapel Hill, NC, 27514, USA **(235)**

Carl Wu, Laboratory of Biochemistry, National Institutes of Health, Bldg 37 Rm 4C09, Bethesda, MD, 20892, USA **(9)**

Ichiro Yahara, Department of Cell Biology, The Tokyo Metropolitan Institute of Medical Science, Honkomagome 3-18-22, Bunkyo-ku, Tokyo 113, Japan **(119)**

Douglas B. Young, MRC Tuberculosis/Infections, Hammersmith Hospital, Du Cane Rd., London, W12 OHS, UK **(203)**

Richard A. Young, Whitehead Institute for Biomedical Research, Nine Cambridge Center, Cambridge, MA, 02142, USA **(193)**

Analysis of
Heat Shock Regulation

Transcriptional Activation of *Drosophila* Heat Shock Genes

J.T. Lis
Section of Biochemistry, Molecular
and Cell Biology Biotechnology Building
Cornell University
Ithaca, NY 14853
USA

Introduction

In this chapter, I consider two distinct, but related, arenas in the transcriptional regulation of heat shock genes. The first is the interaction of the common component of the regulatory regions of heat shock genes, the heat shock element (HSE) and the regulatory protein, heat shock factor (HSF). The second is the association of an RNA polymerase II with the 5' ends of heat shock genes, an association that occurs even in the absence of heat shock. This RNA polymerase II molecule is transcriptionally engaged, but arrested, at the start of heat shock genes. The potential interplay of the HSF-HSE complex and this arrested RNA polymerase II molecule will be discussed in the context of a mechanism for transcriptional activation during the heat shock response.

The HSE and its interaction with multimeric HSF

The HSEs are a conserved feature of the regulatory regions of heat shock genes of species from yeast to man. Heat shock genes possess one or more HSEs that are required for the stress-induced transcription of these genes.

The analyses of regulatory regions of a variety of *Drosophila* heat shock genes revealed that HSEs are composed of a simple contiguous array of the alternately-oriented, 5 bp modular unit, -GAA- (→) (Amin et al., 1988; Xiao and Lis, 1988). The number of 5 bp units in an HSE can vary, as can the number of HSEs associated with a particular heat shock gene.

The regulatory protein, HSF, binds tightly to the region of DNA containing the HSEs. We found that purified heat shock factor (HSF) can bind *in vitro* with dyad symmetry to an HSE containing as few as two inverted 5 bp units in either head-to-head (→ ←) or tail-to-tail (← →) arrangements (Perisic et al., 1989). Although HSF binding is not detected to a single 5 bp unit, it binds with equal affinity to either the head-to-head or tail-to-tail arrangements of pairs of 5 bp units. This cooperative binding to two different configurations of HSF is puzzling. It suggests that if HSF binds to these dyad sequences as a dimer then it must possess remarkable flexiblity in the region between the domain of the polypeptide responsible for cooperative interactions and the domain responsible for contacting DNA sequences of the 5 bp unit. Alternatively, HSF may contain more than two subunits and different pairs of subunits of this multimer might be used to contact the two different permutations of the pairs of 5 bp units.

A glimpse into the multimeric nature of the *Drosophila* HSF was provided by binding studies of HSF to DNA fragments containing different arrays of alternately-oriented 5 bp units (→ ← , → ← → , → ← → ← , etc.) (Perisic et al., 1989). This series of HSEs contains arrays of two to nine of these 5 bp units. HSF binds to these different arrays and protects all of the 5 bp units from attack by either DNaseI or methidiumpropyl-EDTA Fe•(II). Thus, the size of the footprint increase by 5 bp (on average) with the addition of each 5 bp unit. In contrast, the size of the HSF•HSE complexes, as assessed by their mobility during agarose gel electrophoresis, increases most distinctly with the addition of every three 5 bp units. This apparent discrepancy could be explained if HSF is a trimer and if two subunits of HSF contacting DNA are sufficient to generate a stable protein/DNA complex. HSEs with two or three 5 bp units produce complexes with the same apparent size on native gels. We propose that each of these complexes has a single HSF multimer bound, but in the case of an HSE containing two 5 bp units only two HSF subunits interact with DNA. The addition of the third 5 bp unit would allow the third subunit of HSF to interact with DNA and therefore the region of DNA covered by HSF would increase without a concomitant increase in the size of the HSF/DNA complex. An array with four 5 bp units would potentially bind two HSF trimers to give a larger complex in which two subunits of each trimer interact with DNA. Arrays with five and six subunits would also have two trimers bound and show no increase in the size-of the complex, but the number of HSF subunits in contact with DNA would increase with each 5 bp unit.

Additional support for the trimeric nature of *Drosophila* HSF came from chemical crosslinking studies of purified HSF (Perisic et al., 1989). Crosslinking generates dimer and trimer size HSF complexes both in the presence and in the absence of DNA. At a high crosslinker concentration, the only detectable form of HSF is a species with an apparent molecular weight of 350 kd, approximately the size expected for an HSF trimer.

Yeast HSF was also shown to form trimers (Sorger and Nelson, 1989). In this independent study, two truncated HSF subunits of different size associated to form four different size multimers in a ratio of 1:3:3:1 - a pattern consistent with that of trimers. Moreover, crosslinking of yeast HSF generated trimer size products. In addition, Sorger and Nelson (1989) identified a domain required for multimerization that has similarity to the α-helicies that cause trimerization of haemoglutinin of Influenza virus. They suggested that HSF, like haemoglutinin, is a trimer held together *via* a three stranded coiled-coil.

Wu and colleagues recently observed that the cloned Drosophila HSF produced in *E. coli* has a mobility during electrophoresis on native gels expected of a hexamer of a 110 kd protein (Clos et al., 1990). Perhaps HSF subunits first trimerize and these trimers further associate to give a still larger multimer. Chemical crosslinking may be efficient between subunits of the trimer and less efficient between interacting trimers. Indeed, with larger amount of HSF available from the *E.coli* produced *Drosophila* HSF, crosslinked complexes larger than trimer size are detected (Clos et al., 1990). The binding properties of these preparations are much like the preparations of native HSF from *Drosophila* cells (Perisic et al., 1989), in that each multimer seems to occupy preferentially sets of three 5 bp units.

Quantitative assays of the binding of HSF to different numbers and arrangements of 5 bp units provided further support for the trimeric nature of the standard HSF binding site (Xiao et al., 1990). First, an HSE with three alternately-oriented 5 bp units (→ ← → , a standard binding site for HSF) would be expected to make contact to DNA with three subunits of the HSF and thus it should bind HSF more tightly than an HSE with two 5 bp units. Indeed, it does bind HSF 15-30 fold more tightly than either the head-to-head (→ ←) or tail-to-tail (← →) dimers of 5 bp units. Second, three tandemly-oriented 5 bp units (→ → →) could potentially bind HSF, since this tandem array contains two 5 bp units (the first and the third) in positions that are identical to the standard trimeric binding site. Indeed, an array of three tandemly-oriented 5 bp units has an affinity similar to that of two inverted 5 bp units. Moreover, methylation interference assays indicate that the contacts important for this binding are in the first and third repeats.

HSF binds to heat shock elements with striking cooperativity. At one level, the 5 bp units of a single timeric binding site (→ ← →) act cooperatively to bind a single HSF multimer. This site binds HSF 15-30 fold more tightly than an HSE containing two 5 bp units and over 1000 fold more tightly than a single 5 bp unit. At a higher level, multimers of HSF interact cooperatively to bind to HSEs containing longer arrays of 5 bp units with remarkable avidity. An HSE with six 5 bp units binds HSF very tightly, and the half life of such complexes is greater than 48 hours. The HSE containing six 5 bp units has in effect two trimeric binding sites. The binding of HSF to one of this pair of trimeric binding sites facilitates HSF binding to the second site by over 2000-fold at 25°C. This cooperativity appears even more important at the heat shock temperature of 37°C. Perhaps this strong cooperativity in binding accounts for the requirement for multiple HSF binding sites for in vivo expression (Amin et al., 1985; Dudler and Travers, 1984; Simon et al., 1985).

6

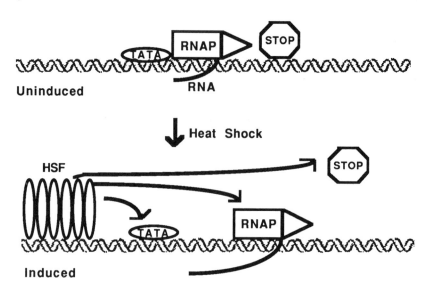

Figure 1. A highly schematic model outlining the associations of proteins with the *hsp70* gene promoter both prior and following heat shock activation.

Post-initiation control of heat shock gene transcription

It has been assumed that transcription factors accelerate the rate of transcription by facilitating recruitment of RNA polymerase to the promoter or the rate of initiation of tne recruited polymerase. We have observed by UV-crosslinking experiments that a molecule of RNA polymerase II is already present at the transcription start site of the *hsp70* gene in cells that have not been heat shocked (Gilmour and Lis, 1986). This *in vivo* crosslinking assay was done using irradiation times as short as a 60 microsecond pulse, which was made possible with a xenon flash lamp. The presence of this molecule of RNA polymerase on the 5' end of the HSP70 gene demonstrates that in vivo RNA polymerase has access and is associated with the promoter/start region of the gene even before heat shock. What appears to be rate limiting in transcription is not polymerase access to the promoter but a step subsequent to this association.

The crosslinking experiment does not distinguish an RNA polymerase that is simply bound to the promoter from one that is transcriptionally engaged. To address this, the association of RNA polymerase with the 5' end of the *hsp70* gene was further investigated by nuclear run-on assays (Rougvie and Lis, 1988). These assays have shown that this RNA polymerase is transcriptionally engaged but arrested approximately 25 nucleotides from the transcription start. Thus, this RNA polymerase has initiated transcription and has synthesized an approximately 25 nucleotide RNA before being blocked from further elongation.

The rate-limiting step that is influenced by transcription factor(s) could conceivably be any of several steps in the pathway leading to the production of a complete transcript. In eukaryotes, however, it has generally been assumed that the rate-limiting step influenced by transcription factors is a step in the initiation process, either the recruitment of RNA polymerase to a gene, or the formation of the first phosphodiester bond. In contrast to this view, we have found that RNA polymerase II is already transcriptionally engaged on the *hsp70* promoter in the absence of heat-induction. Why does this transcribing RNA polymerase stop? There are several possible models. Perhaps RNA polymerase II cannot efficiently escape the promoter on its own because of its affinity for the promoter sequences and the general transcription factors such as TFIID that reside near the start site. Alternatively, RNA polymerase may be blocked by chromatin structure of the uninduced *hsp70* gene. Finally, the "stop" signal may represent the binding of a specific protein that blocks the progress of RNA polymerase II.

In response to heat shock the rate with which RNA polymerase transcribes the gene increases dramatically. What is responsible for releasing these prebound polymerase molecules into productive transcription? An obvious candidate is HSF. HSF could act through interactions with general transcription factors such as TFIID, or directly on RNA polymerase, or through interactions that effectively remove a "stop" signal. Further mutagenesis of the *hsp70* promoter and characterization of arrested polymerase will be required to determine which, if any, of these models is correct.

To assess the generality of a rate-limiting step occurring early in the transcript elongation process, we have examined the effects of sarkosyl on the nuclear run-on transcription of several *Drosophila* genes (Rougvie and Lis, 1990). An arrested (or paused) RNA polymerase exists on a second major heat shock gene, *hsp26*. Intriguingly, the presence of a paused RNA polymerase is not limited to heat shock inducible genes. Analyses of RNA polymerase distribution on several genes, which are constitutively expressed in *Drosophila* cell culture, reveals that many, but not all, appear to have paused RNA polymerase at their 5' ends. The rates of RNA polymerase binding to, and initiating transcription from, the promoters of β1-tubulin, α-tubulin, glyceraldehyde-3-phosphate dehydrogenase-1 and -2, and poly-ubiquitin genes appear to be higher than the rate at which the RNA polymerase gains access to the body of these genes. Perhaps modulation of early elongation is a generally-used mode of transcriptional regulation.

References

Amin, J, Ananthan, J and Voellmy, R, (1988) Key Features of Heat Shock Regulatory Elements. Mol. Cell. Biol., 8: 3761-3769.

8

Amin, J, Mestril, R, Lawson, R, Klapper, H and Voellmy, R, (1985) The Heat Shock Consensus Sequence is not Sufficient for hsp70 Gene Expression on *Drosophila melanogaster*. Mol. Cell. Biol., 5: 197203.

Dudler, R and Travers, AA, (1984) Upstream elements necessary for optimal function of the hsp70 promoter in transformed flies. Cell, 38: 391-398.

Gilmour, DS and Lis, JT, (1986) RNA polymerase II interacts with the promoter region of the noninduced hsp70 gene in *Drosophila melanogaster* cells. Mol. and Cell. Biol., 6: 3984-3989.

Perisic, 0, Xiao, H and Lis, JT, (1989) Stable binding of Drosophila heat shock factor to head-to-head and tail-to-tail repeats of a conserved 5 bp recognition unit. Cell, 59: 797-806.

Rougvie, AE and Lis, JT, (1988) The RNA polymerase II molecule at the 5' end of the uninduced hsp70 gene of *D. melanogaster* is transcriptionally engaged. Cell, 54: 795-804.

Rougvie, AE and Lis, JT, (1990) Post-initiation transcriptonal control in *Drosophila melanogaster*. Mol. Cell. Biol., 10: 6041-6045.

Simon, JA, Sutton, CA, Lobell, RB, Glaser, RL and Lis, JT, (1985) Determinants of heat shock-induced chromosome puffing. Cell, 40: 805-817.

Sorger, PK and Nelson, HCM, (1989) Trimerization of a yeast transcriptional activator via a coiled-coil motif. Cell, 59: 807-813.

Xiao, H and Lis, JT, (1988) Germline transformation used to define key features of heat-shock response elements. Science, 239: 1139-1142.

Structure and Function of *Drosophila* Heat Shock Factor

C. Wu, J. Clos, J.T. Westwood, V. Zimarino[#], P.B. Becker, S. Wilson[*]
Laboratory of Biochemistry, National Cancer Institute,
National Institutes of Health
Bethesda, MD 20892
USA

Introduction

In eukaryotic cells, the synthesis of heat shock proteins is subject to transcriptional and post-transcriptional control in eukaryotic cells (reviewed by Craig, 1985; Lindquist, 1986). Heat shock-inducible transcription is mediated by a positive control element, the heat shock element (HSE), defined as three repeats of a 5-nucleotide [-GAA-] module, arranged in alternating orientation (Pelham, 1982; Amin et al., 1988; Xiao and Lis, 1988). Multiple copies of the HSE are found upstream of all heat shock genes. A heat shock transcriptional activator, termed heat shock factor (HSF), binds to HSEs and activates transcription of heat shock genes *in vitro* (Wu, 1984a; Wu, 1984b; Parker and Topol, 1984; Topol et al., 1985). Although the sequence of the HSE has been highly conserved in evolution, HSF purified from yeast, *Drosophila*, and human cells differ in molecular size (150 kD, 110 kD and 83 kD, respectively; Sorger and Pelham, 1987; Wu et al., 1987; Goldenberg et al., 1988). Yeast and higher eukaryotes also differ in the regulation of HSF activity. In yeast, HSF bound constitutively to the HSE apparently stimulates transcription when phosphorylated under heat shock conditions. In *Drosophila* and vertebrate cells, HSF is unable to bind to the HSE unless the cells are heat shocked (for a review, see Wu et al., 1990). The heat-inducible binding of HSF appears to be a major regulatory step in the pathway to heat shock gene activation in higher eukaryotes.

The induction and reversal of HSF binding activity *in vivo* does not require new protein synthesis (Zimarino and Wu, 1987; Kingston et al., 1987; Zimarino et al., 1990a). In addition, HSF extracted from nonshocked cell cytosol can be activated *in vitro* by heat (Larson et al., 1988), low pH (Mosser et al., 1990), and by interaction with antibodies

---

[*] Genex Corp., Gaithersburg, Maryland, USA.
[#] Institute of Microbiology, University of Copenhagen, Oster Farimagsgade 2A, DK-1353 Copenhagen K, Denmark.

raised to the active form of HSF (Zimarino et al., 1990b). These results suggest that the pre-existent, inactive form of HSF can assume the active conformation without an enzymatic modification of protein structure.

We have cloned the *Drosophila* HSF gene and found that cloned HSF synthesized in *E. coli*, or translated *in vitro* in a reticulocyte lysate at non-heat shock temperatures binds to DNA with maximal affinity. In contrast, cloned HSF expressed in *Xenopus* oocytes binds to DNA with maximal affinity only after heat shock induction. These results suggest that HSF is under repression in nonshocked *Drosophila* cells, which is relieved upon heat shock.

Isolation of the *Drosophila* HSF gene

We cloned the *Drosophila* HSF gene by microsequencing tryptic peptides of purified HSF protein and designing oligonucleotide probes (Clos et al., 1990). The 2.8 kb of HSF cDNA sequence reconstructed from six overlapping cDNA clones reveals a single open reading frame of 691 amino acids (2073 nt). The molecular mass of *Drosophila* HSF, calculated from the deduced amino acid sequence, is 77,300 daltons, significantly lower than the apparent mass of 110,000 daltons measured by SDS gel electrophoresis (Wu el al., 1987).

Evidently, *Drosophila* HSF has an anomalous mobility on SDS gels; a similar anomaly was observed with yeast HSF (Sorger and Pelham, 1988; Wiederrecht et al., 1988). For purposes of discussion, we continue to use the molecular size of HSF protein as measured by SDS gel electrophoresis. The *Drosophila* HSF protein sequence predicts an acidic protein (pI = 4.7). The overall distribution of charged residues along the length of the protein sequence is nonuniform: the N-terminal one-third of HSF (amino acids 1-240) is relatively basic (predicted pI = 10.25), while the C-terminal two-thirds (amino acids 240-691) is relatively acidic (predicted pI = 4.1). In addition, there is an unusual N-terminal cluster of 9 acidic residues in a row (amino acids 18 to 26).

Cloned HSF is an active, DNA binding transcription factor in the absence of heat shock

Naturally occurring HSF extracted from the cytosol of nonshocked *Drosophila* cells shows a basal affinity for DNA, which can be significantly increased by a direct heat treatment *in vitro*, or by reaction with polyclonal antibodies raised to the *in vivo*-activated form of HSF (Zimarino et al., 1990b). When cloned HSF is synthesized by *in vitro* translation in a rabbit reticulocyte lysate at 25°C, or at 30°C, neither heat treatment (34°C) nor reaction with anti- HSF increased HSF affinity for DNA (Clos et al., 1990). The constitutive DNA binding activity of HSF synthesized *in vitro* could be due to an activating substance in the reticulocyte lysate. However, reticulocyte lysates do not activate HSF when incubated with cytosol from unshocked *Drosophila* cells.

Cloned HSF expressed at a high or a low level in *E. coli* at 18°C using the T7 RNA polymerase-dependent expression system (Studier and Moffatt, 1986) also showed maximal binding affinity without heat or anti-HSF treatment (Clos et al., 1990). The data suggest that cloned HSF protein synthesized outside the environment of a higher eukaryotic cell has an intrinsic affinity for DNA. In addition, HSF produced in *E. coli* is able to function as a transcription factor in an *in vitro* transcription system derived from *Drosophila* embryos (Soeller et al, 1988; Biggin and Tjian, 1988). Addition of the cloned protein to the transcription extract resulted in a 7-fold increase of transcription from a promoter carrying two HSEs, relative to the transcription from the same promoter lacking HSEs (Clos et al., 1990). Hence, recombinant HSF protein is capable of functioning as a transcription factor in a binding site-dependent manner, apparently without further modification by a heat shock-induced enzymatic activity.

Transcriptional activity and phosphorylation

On the basis of sensitivity to phosphatases and labeling experiments with 32P, it has been suggested that increased phosphorylation of yeast and human HSF are necessary for transcriptional competence of the protein (Sorger and Pelham, 1987; Sorger, 1990; Larson et al., 1988). However, a direct role for phosphorylation in activating the transcription function of HSF remains remains open for further analysis. On the other hand, the ability of cloned *Drosophila* HSF protein produced in *E. coli* to stimulate transcription from a heat shock promoter *in vitro* suggests that cloned HSF is able to fold to a transcriptionally active conformation in the absence of a heat shock-induced phosphorylation (Clos et al., 1990). A complication with such studies of transcription in cell-free extracts is that a need for phosphorylation may become evident only when transcriptional activity is measured by assays that are closer to *in vivo* conditions in the context of assembled chromatin.

Oligomeric state of HSF in solution

The native size of *Drosophila* HSF as estimated by pore exclusion limit electrophoresis suggests that a significant fraction of cloned HSF protein forms a hexamer free in solution. A similar size estimation of cloned or natural HSF protein bound to the HSE suggests that a HSF hexamer binds to DNA with high affinity (Clos et al., 1990). Chemical cross-linking of a dilute HSF solution shows that cloned HSF protein is composed of trimers, hexamers, and very large complexes beyond the limit of gel analysis (Clos et al., 1990). Taken together, these

Dros HSF   PAFLAKLWRLVDDADTNRLICWTKDGQSFVIQNQAQFAKELLPLNYKHNNMASFIRQLNMYGFHKI – 112
Yeast HSF   PAFVNKLWSMLNDDSNTKLIQWAEDGKSFIVTNREEFVHQILPKYFKHSNFASFVRQLNMYGWHKV – 238

Sigma 32   LQELADRYGVSAERVRQLEFNAMKKL – 278
Sigma 70   LEEVGKQFDVTRERIRQIEAKALRKL – 598
                                            |←—α—→| t |←——α——→|

1a

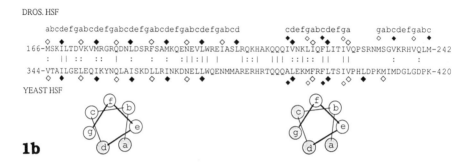

Figure 1. Conserved structural motifs.
(A) Alignment of protein sequences conserved between *Drosophila* HSF and yeast HSF. Similar or identical residues are boxed. Also boxed are the two conserved pentapeptides in the putative recognition helix of the helix-turn-helix motif (α - t - α) of σ32 and σ70.
(B) Comparison of the heptad repeats of hydrophobic amino acids found in *Drosophila* and yeast HSF sequences. The repeats are made up of hydrophobic residues at positions a (open diamonds) and d (filled diamonds), in the

results indicate that HSF probably exists in equilibrium as trimers, hexamers, and even larger oligomers free in solution, but the oligomeric state that binds to a HSE (three alternating [-GAA-] modules) is primarily hexameric. The question thus arises whether one or two subunits of a hexamer participate in the recognition of each [-GAA-] module. If one HSF subunit interacts with one [-GAA-] module, three subunits of the hexamer remain free, in principle, to bind to another HSE. It will be important to determine the stoichiometry of HSF binding to HSE.

Previous reports have shown that the oligomerization state of HSF in *Drosophila* and yeast is primarily trimeric (Perisic et al., 1989; Sorger and Nelson, 1989). Evidence for trimerization of *Drosophila* HSF was based the observation of a 350 kD cross-linked HSF product visualized by protein blot (Western) analysis. It is possible that the full oligomerization potential of HSF is absent in some preparations of natural HSF, or that the cloned protein associates more readily as a hexamer. The oligomerization state of full-length yeast HSF is as yet undetermined.

Conserved sequences between *Drosophila* and yeast HSF

Despite the high degree of homology among heat shock proteins between species as diverse as *E. coli* and *Drosophila* (about 50% identity, for HSP70; Bardwell and Craig, 1984), the sequences of *Drosophila* and yeast HSF have diverged over a large portion of the proteins (Clos et al., 1990; Wiederrecht et al., 1988; Sorger and Pelham, 1988). A search for similarities revealed two major regions of local conservation. Deletion analyses show that these conserved regions function in specific and high affinity binding to DNA. The most conserved region between *Drosophila* and yeast HSF (50% identical over 66 amino acids) is included within the DNA binding domains of *Drosophila* and yeast HSF (Fig. 1 A).

Within this region, two pentapeptides show an intriguing similarity to the putative recognition helix of bacterial sigma factors (Gribskov and Burgess, 1986; Helmann and Chamberlin, 1988). The similarity to sigma factors may define an α-helical element of the HSF DNA binding domain that is important for DNA interaction (Clos et al., 1990). On the whole, however, the DNA binding domains of yeast and *Drosophila* HSF do not fall into any known category of DNA binding motifs; hence a solution of the structure of this domain by physical and genetic studies will introduce a new member to the growing family of protein structures that recognize specific DNA sequences.

The second region conserved between *Drosophila* and yeast HSF is implicated in the oligomerization of HSF (Sorger and Nelson, 1989; Clos et al., 1990). This region contains an unusual arrangement of hydrophobic heptad repeats (Fig. 1 B). Three arrays of such repeats are found, one of which is positioned one residue out of phase with another array. When viewed in a backbone model of an α-helix, such a helix has hydrophobic residues juxtaposed at four positions on one helical face. Such a helix would have the potential to associate simultaneously with two neighboring helices of the same type by hydrophobic interactions characteristic of leucine zipper coiled-coils (Landschulz et al., 1988; O'Shea et al., 1989). Although it is unclear at present how the three arrays of hydrophobic heptad repeats might direct the oligomerization of HSF, their remarkable degree of conservation suggests that they all have functional roles.

The conserved amino acids in the oligomerization domain are not limited to hydrophobic residues. Identical residues include polar amino acids (three glutamines in a row [QQQ]), hydrophobic [W,F,I,L], basic [R,K] and acidic [E] amino acids. Although hydrophobic interactions are the major stabilizing force between coiled-coils, additional specificity may be conferred by charged or polar interactions, mediated by residues outside the heptad repeat (Cohen and Parry, 1990).

Negative regulation of HSF binding activity *in vivo*

The native form of HSF in crude extracts of unshocked *Drosophila*, *Xenopus*, and vertebrate cells shows a basal affinity for DNA by *in vitro* assays, which is increased about 10-fold when cells are induced by heat shock (Zimarino et al., 1990a). When recombinant HSF is synthesized after microinjection of *Xenopus* oocytes with HSF RNA transcribed *in vitro*, the DNA binding activity shows significant inducibility after heat shock (Clos et al., 1990). Thus, in contrast to the full DNA binding capacity of HSF synthesized in *E. coli* or in a reticulocyte lysate, the intrinsic affinity of HSF for DNA is suppressed in nonshocked *Xenopus* oocytes. The results suggest that the naturally occurring form of HSF in unshocked *Drosophila* cells is under negative control, which is relieved upon heat shock.

The suppression *in vivo* could be due to folding of HSF to a conformation that is unable to oligomerize, or to an association of HSF with a specific inhibitory substance. It is intriguing to speculate that (constitutively synthesized) heat shock proteins themselves may participate in the suppression of HSF activity. Heat shock proteins have been shown to act as molecular detergents or chaperones in protein folding and unfolding (for reviews see Pelham, 1990; Rothman, 1989). A clear demonstration of heat shock protein involvement in HSF regulation would be a satisfying finding in view of suggestive evidence from many sources that heat shock proteins autoregulate their synthesis.

Conclusions

From the earliest studies of the heat shock response, a bewildering multiplicity of stress inducers have presented a challenge to the search for a common stress signal transduction pathway. Besides heat, inducers of the stress response include drugs affecting energy metabolism, oxidizing agents, sulfhydryl reagents, chelating agents, heavy metals, ionophores, amino acid analogues, etc. (Ashburner and Bonner, 1979; Nover, 1984). We now suggest that these unrelated inducers may all be acting by altering the inactive conformation of HSF. Assuming that HSF protein is maintained in an inactive state by an essential combination of hydrophobic, charged, and polar interactions, it is conceivable that the disruption of a subset of these forces by any one inducer of the stress response could be sufficient to trigger a change of state (Clos et al., 1990). In this view, a solution to the enigma of stress signal transduction may be found in the molecular architecture of HSF protein itself, and in the interactions with its negative regulators.

References

Amin, J, Ananthan, J and Voellmy, R, (1988) Key features of heat shock regulatory elements. Mol. Cell. Biol., 8: 3761-3769.

Ashburner, M and Bonner, JJ, (1979) The induction of gene activity in Drosophila by heat shock. Cell, 17: 241-254.

Bardwell, JC and Craig, EA, (1984) Major heat shock gene of Drosophila and Escherichia coli heat inducible dnaK gene are homologous. Proc. Natl. Acad. Sci. USA, 81: 848-852.

Biggin, MD and Tjian, R, (1988) Transcription factors that activate the ultrabithorax promoter in developmentally staged extracts. Cell, 53: 699-711.

Clos, J, Westwood, JT, Becker, PB, Wilson, S, Lambert, K and Wu, C, (1990) Molecular cloning and expression of a hexameric heat shock factor subject to negative regulation. Cell, in press.

Cohen, C, and Parry, DAD, (1990) α-helical coiled coils and bundles: How to design an α-helical protein. Proteins, 7: 1-15.

Craig, EA, (1985) The heat shock response. Crit. Rev. Biochem., 18: 239-280.

Goldenberg, CJ, Luo, Y, Fenna, M, Baler, R, Weinmann, R and Voellmy, R. (1988) Purified human factor activates heat shock promoter in a HeLa cell-free transcription system. J. Biol. Chem., 263: 19734-19739.

Gribskov, M and Burgess, RR, (1986) Sigma factors from E. coli, B. subtilis, phage SPO1, and phage T4 are homologous proteins. Nucl. Acids Res., 14: 6745-6763.

Helmann, JD and Chamberlin, MJ (1988) Structure and function of bacterial sigma factors. Ann. Rev. Biochem., 57: 839-872.

Kingston, RE, Schuetz, TJ and Larin, Z, (1987) Heat inducible human factor that binds to a human hsp 70 promoter. Mol. Cell. Biol., 7: 1530-1534.

Landschulz, WH, Johnson, PF and McKnight, SL, (1988) The leucine zipper: a hypothetical structure common to a new class of DNA binding proteins. Science, 240: 1759-1764.

Larson, JS, Schuetz, TJ and Kingston, RE, (1988) Activation in vitro of sequence specific DNA binding by a human regulatory factor. Nature, 335: 372-375.

Lindquist, S, (1986) The heat shock response. Ann. Rev. Biochem., 55: 1151-1191.

Mosser, DD, Kotzbauer, PT, Sarge, KD and Morimoto, R, (1990) In vitro activation of heat shock transcription factor DNA-binding by calcium and biochemical conditions that affect protein conformation. Proc. Nat. Acad. Sci. USA, 87: 3748-3752.

Nover, L, Hellmund, D, Neumann, D, Scharf, K-D and Serfling, E, (1984) The heat shock response of eukaryotic cells. Biol. Zentr., 103: 357-435.

O'Shea, EK, Rutkowski, R, Stafford, WFIII and Kim, PS, (1989) Preferential heterodimer formation by isolated leucine zippers from Fos and Jun. Science, 245: 646-648.

Parker, CS and Topol, J, (1984) A Drosophila RNA polymerase II transcription factor contains a promoter-region-specific DNA binding activity. Cell, 36: 357-369.

Pelham, HRB, (1982) A regulatory upstream element in the Drosophila hsp 70 heat shock gene. Cell, 30: 517-528.

Pelham, HRB, (1990) Functions of the hsp70 protein family: an overview. In: Stress Proteins in Biology and Medicine, Morimoto, RI, Tissieres, A and Georgopolous, C (eds.). Cold Spring Harbor Laboratory Press, 287-299.

Perisic, O, Xiao, H and Lis, JT, (1989) Stable binding of Drosophila heat shock factor to head-to-head and tail-to-tail repeats of a conserved 5 bp recognition unit. Cell, 59: 797-806.

Rothman, JE, (1989) Polypeptide chain binding proteins: Catalysts of protein folding and related processes in cells. Cell, 59: 591-601.

Soeller, WC, Poole, SJ and Kornberg, T, (1988) In vitro transcription of the Drosophila engrailed gene. Genes Dev., 2: 68-81.

Sorger, PK and Pelham, HRB (1987) Purification and characterization of a heat shock element binding protein from yeast. EMBO J., 6: 3035-3041.

Sorger, PK and Pelham, HRB, (1988) Yeast heat shock factor is an essential DNA-binding protein that exhibits temperature-dependent phosphorylation. Cell, 54: 855-864.

Sorger, PK and Nelson, HCM (1989) Trimerization of a yeast transcriptional activator via a coiled-coil motif. Cell, 59: 807-813.

Sorger, PK, (1990) Yeast heat shock factor contains separable transient and sustained response transcriptional activators. Cell, 62: 793-805.

Studier, FW and Moffatt, BA, (1986) Use of bacteriophage T7 RNA polymerase to direct selective high-level expression of cloned genes. J. Mol. Biol., 189: 113-130.

Topol, J, Ruden, DM and Parker, CS (1985) Sequences required for *in vitro* transcriptional activation of a *Drosophila* hsp70 gene. Cell, 42: 527-537.

Wiederrecht, G, Seto, D and Parker, CS, (1988) Isolation of the gene encoding the S. cerevisiae heat shock transcription factor. Cell, 54: 841-853.

Wu, C, (1984a) Two protein-binding sites in chromatin implicated in the activation of heat shock genes. Nature, 309: 229-234.

Wu, C, (1984b) Activating protein factor binds *in vitro* to upstream control sequences in heat shock gene chromatin. Nature, 311: 81-84.

Wu, C, Wilson, S, Walker, B, Dawid, I, Paisley, T, Zimarino, V, and Ueda, H, (1987) Purification and properties of *Drosophila* heat shock activator protein. Science, 238: 1247-1253.

Wu, C, Zimarino, V, Tsai, C, Walker, B and Wilson, S, (1990) Transcriptional regulation of heat shock genes. In: Stress Proteins in Biology and Medicine, Morimoto, RI, Tissieres, A and Georgopoulos, C, (eds). Cold Spring Harbor Laboratory Press, 429-442.

Xiao, H and Lis, JT (1988) Germline transformation used to define key features of the heat shock response element. Science, 239: 1139-1142.

Zimarino, V and Wu, C (1987) Induction of sequence specific binding of *Drosophila* heat shock activator protein without protein synthesis. Nature, 327: 727-730.

Zimarino, V, Tsai, C and Wu, C (1990a) Complex modes of heat shock factor activation. Mol. Cell. Biol., 10: 752-759.

Zimarino, V, Wilson, S and Wu, C, (1990b) Antibody-mediated activation of *Drosophila* heat shock factor *in vitro*. Science, 249: 546-549.

In vivo and in vitro Studies on the Activation and Binding of Human Heat-shock Transcription Factor

K. Abravaya, K.D. Sarge, B. Phillips, V. Zimarino, R.I. Morimoto
Department of Biochemistry, Molecular Biology and Cell Biology
Northwestern University
Evanston, Illinois 60208
USA

Introduction

The rapid transcriptional induction of heat shock genes upon physiological stress such as heat shock is mediated by heat shock transcription factor (HSF) which recognizes a target sequence, the heat shock element (HSE) (Goldenberg et al., 1988; Parker and Topol, 1984; Pelham, 1982; Wu et al., 1987). Heat shock elements consist of an array of inverted repeats of the sequence NGAAN, although the arrangement and number of these NGAAN units can vary (Amin et al., 1988, Xhiao and Lis, 1988). The dynamics of HSF-HSE interactions differ among organisms. In *Drosophila*, HSF binds to HSE only upon heat shock (Thomas and Elgin, 1988; Wu, 1984), while in yeast, HSF is constitutively bound to HSE (Jakobson and Pelham, 1988; McDaniel et al., 1989; Sorger et al., 1987) and its transcriptional activity is induced by heat shock (Sorger and Pelham, 1988).
We have previously identified two distinct HSE binding activities in HeLa cells (Mosser et al., 1988). Non-heat shocked cells contain a constitutive HSE binding activity (CHBA); while a distinct HSE-binding activity (HSF) distinguished by electrophoretic mobility on the gel shift assay, is present in heat shocked cells and is only detected during the activation of the stress response. The increase in the levels of HSF-HSE complex detected during activation of the stress response is independent of new protein synthesis, indicating that HSF exists in a non-DNA binding form at normal temperatures and is converted to the DNA-binding, transcriptionally active form in response to heat shock (Zimarino et al., 1987). The stress-induced form of HSF has been purified from several species and corresponds in size to polypeptides of 150kd in *S. cerevisiae* (Sorger and Pelham, 1987), 110kd in *Drosophila* (Wu et al., 1987), and 83kd in humans (Goldenberg et al., 1988).
Activation of HSF DNA-binding ability can also be examined *in vitro* using cytosolic extracts from human and insect cells (Larson et al., 1988; Zimarino et al., 1990). In these systems induction of HSF DNA-

a

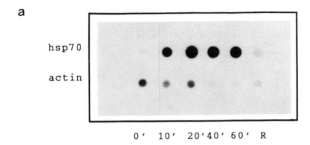

hsp70

actin

0' 10' 20'40' 60' R

b

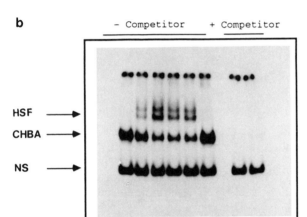

− Competitor   + Competitor

HSF ⟶

CHBA ⟶

NS ⟶

0' 10'20'40'60' R   0' 40'

c

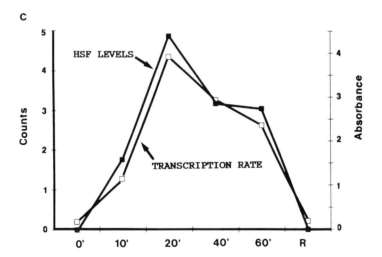

HSF LEVELS

Counts

Absorbance

TRANSCRIPTION RATE

0'   10'   20'   40'   60'   R

Figure 1. Analysis of transcription rates and HSF levels in heat-shocked HeLa cells. (A) Run-on analysis of *hsp70* and ß-actin transcription in HeLa cells at the indicated times of heat shock. R (recovery), 2 hr heat-shocked cells were returned to 37°C. Labeled *hsp70* and ß-actin transcripts were hybridized to immobilized plasmid DNA containing the coding region of the human *hsp70* gene (Hunt and Morimoto, 1985) and human ß-actin (Gunning et al., 1983) cDNA respectively. A decrease in the transcription rate of the β-actin gene during heat shock is consistently observed in our laboratory. (B) Gel retardation assay of an HSE-containing oligomer by whole cell extracts (WCE) prepared from HeLa cells at the indicated times of heat shock. The oligomer, derived from the HSE-containing region of the human *hsp70* promoter, was described previously (Mosser et al., 1988). Complexes due to non-specific (NS) DNA-binding proteins, constitutive HSE binding activity (CHBA), and heat shock factor (HSF) are indicated by arrows. WCE's from samples taken at 0 min and 40 min of heat shock were also used in incubations which included a 100-fold excess of unlabeled oligomer (+ competitor). (C) Graphic comparison of HSF levels and transcription rates at the indicated times of heat shock. Filters from the transcription assay, the autoradiogram of which is shown in A, were scanned and quantified using a Molecular Dynamics 400A Phosphoimager. Values for transcription rates were machine counts of radioactivity bound to immobilized *hsp70* DNA, plotted on an arbitrary scale. HSF levels were determined by laser densitometry of the autoradiogram shown in B. (From Abravaya et al., 1991)

binding activity occurs at temperatures similar to those which induce binding *in vivo* during heat shock (Larson et al., 1988). *In vitro* activation of HSF DNA binding can also be accomplished by a variety of biochemical conditions which are known to affect protein conformation, including low pH, non-ionic detergents, urea, and calcium (Mosser et al., 1990). Activation of DNA binding can also be induced by incubating *Drosophila* cell extracts in the presence of specific anti-HSF antibodies (Zimarino et al., 1990). Together, these observations strongly suggest that activation of HSF DNA binding involves protein conformational changes and perhaps protein-protein interactions. Recently a trimeric and hexameric structure for yeast (Sorger and Nelson, 1989) and *Drosophila* HSF (Clos et al., 1990; Perisic et al., 1990) has been described. Although the native structure of *Drosophila* HSF prior to heat shock has not been examined, a mechanism for activation of DNA binding has been suggested to involve oligomerization.

In order to elucidate the mechanism of heat shock activation in human cells, we have performed genomic footprinting to examine, *in vivo*, interactions of proteins with the human heat shock, *hsp70* promoter before, during and after the activation of the stress response. These studies have provided a detailed view of HSF-HSE interactions and revealed the kinetics of this interaction. In addition, we have applied biochemical techniques to study the properties of human HSF before and during heat shock.

Genomic footprinting of the *hsp70* promoter

Interactions of proteins with the *hsp70* promoter during heat shock were examined by genomic footprinting (Abravaya et al., 1991; Mueller and Wold, 1989). To compare changes in DNA:protein interactions with changes in HSF levels and transcription rate of the *hsp70* gene, during the activation of the stress response; the transcription rate was measured by nuclear run-on analysis (Mosser et al., 1988), and the levels of HSF were determined by gel shift assays (Mosser et al., 1988).

hsp70 transcription rates and HSE-binding factors: The transcription rate of the *hsp70* gene increased rapidly upon transfer of the cells to 42°C and was enhanced 20-fold after 20 min of heat shock, then declined slowly from this maximal level. Transcription returned to basal levels after 2 hr recovery at 37°C (Fig. 1A). By gel shift analysis, using an HSE-containing oligomer derived from the human *hsp70* promoter, two HSE-specific complexes can be detected (Fig. 1B) (Mosser et al., 1988). Non-heat shocked cells contain an HSE-binding activity (CHBA) which gives rise to a faster migrating complex, while a distinct HSE binding activity (HSF), which forms a slower migrating complex on native gels is only present in heat shocked cells and its level is correlated with the increase in transcription rate of the *hsp70* gene (Fig. 1C). Concomitant with an increase in HSF levels during heat shock, CHBA levels decline (Fig. 1B and 16). Since an HSE binding activity was observed both in non-heat shocked cells and heat shocked cells, it was not clear whether the HSE region of the human *hsp70* promoter is constitutively occupied *in vivo* or is occupied only during heat shock, and this could only be addressed by genomic footprinting experiments.

Genomic footprinting of the hsp70 basal promoter: Genomic footprinting was performed using a ligation mediated polymerase chain reaction (Mueller and Wold, 1989). To determine interactions of proteins with the human *hsp70* promoter prior to heat shock, the methylation pattern of DNA isolated from non-heat shocked cells was compared to the pattern of deproteinized DNA methylated *in vitro*, i.e., naked DNA (Fig. 2B and C, lanes 0' and N). Changes in dimethylsulfate (DMS) reactivity mapped to G's at positions -71, -65, -64, -48, -47, -45, -44, and -21 on the coding strand and at positions -41 and -38 on the non-coding strand (Fig. 2B and C, lanes 0' and N; open arrows and open stars). These alterations in DMS reactivity map to the CCAAT and Sp1 sites in the human *hsp70* basal promoter (Fig. 2A) and suggest binding of these factors. Protection of a G residue immediately downstream of the TATA element (G-21) suggests binding to this site as well.
Basal expression of the *hsp70* gene in non-heat shocked HeLa cells requires the proximal 75 nucleotides of the promoter, which contain consensus binding sites for several transcription factors including CCAAT, Sp1 and TFIID (Morimoto et al., 1990; Williams et al., 1989; Wu et al., 1986). Other factors which activate expression of the human *hsp70* gene, such as serum stimulation and the adenovirus E1A protein, appear to act through the same proximal region of the *hsp70* promoter (Morimoto et al., 1990; Williams et al., 1989). The CCAAT and TATA elements within the proximal region of this promoter have been shown to be essential for basal expression of the gene, while the Sp1 site also contributes to basal transcription (Morimoto et al., 1990; Williams et al., 1989; Wu et al., 1986). The binding of purified or partially purified CCAAT transcription factor (CTF), TFIID, and Sp1 to

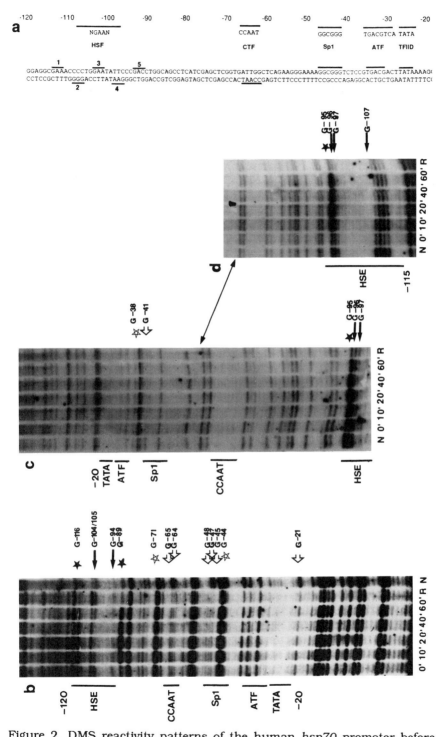

Figure 2. DMS reactivity patterns of the human *hsp70* promoter before, during, and after heat shock.
a. The sequence of the promoter region (Hunt and Morimoto, 1985) analyzed

22

by genomic footprinting. Sites which are perfect or imperfect matches to consensus sites for known transcription factors are underlined. The consensus sequence and transcription factors which bind to these sites *in vitro* are indicated above each underlined site. NGAAN sequences, comprising an array of repeated inverted units characteristic of HSE's, are underlined and numbered.
b. Coding strand methylation patterns in genomic DNA isolated at the indicated times during heat shock. R (recovery), genomic DNA isolated from cells after a 2 hr recovery at 37°C. N (naked), protein-free DNA, methylated *in vitro*. Arrows denote guanine residues protected from methylation; stars denote guanines hypersensitive to methylation. Open arrows and open stars indicate differences in methylation pattern between DNA isolated from non-heat shocked cells and naked DNA; closed arrows and closed stars denote differences in methylation pattern between DNA isolated during heat shock (lanes 10', 20', 40', 60') and DNA isolated prior to heat shock (lane 0'). Band intensities were determined by densitometric scanning. Because of interlane variations in overall band intensities and variations in band intensities between different regions of the same lane, guanines were denoted as hypersensitive or protected from methylation based on normalization to neighbouring bands which did not show any altered reactivity.
c. Same as b except that the non-coding strand is shown.
d. Same as c except electrophoresis was carried out for a shorter period of time than in c to analyze shorter fragments. (From Abravaya et al., 1991)

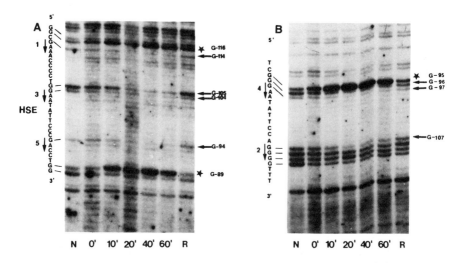

Figure 3. DMS reactivity patterns of the HSE region. Both coding (A) and non-coding (B) strands are shown, with the sequence of the HSE region indicated to the left of each gel and each G aligned with its corresponding band on the gel. The NGAAN sites are numbered as in Fig. 1A. Arrows indicate G's protected from methylation and stars indicate guanines hypersensitive to methylation. (From Abravaya et al., 1991)

their recognition sequences on the human *hsp70* promoter has been reported (Morgan et al., 1987; Morgan, 1989; Nakajima et al., 1988). A comparison of the methylation patterns of DNA isolated from heat shocked and non-heat shocked cells revealed no changes in the basal promoter elements (Fig. 2B and 2C). We thus conclude that interactions of protein factors with basal promoter elements are not altered during heat shock.

Genomic footprinting of the heat shock element before and during heat shock and following recovery: A comparison of DMS reactivity patterns in the HSE region of the promoter (-115 to -95), which contains five inverted repeats of the consensus NGAAN sequence, revealed no differences between naked DNA and DNA isolated from non-heat shocked cells (Fig. 3, lanes 0' and N). This result suggests that factors are not bound to the HSE region prior to heat shock in HeLa cells. During heat shock, protection of the consensus G in each of the five NGAAN sites was observed (G-114, -107, -104, -97, and -94) (Fig. 3), protections at sites 3 and 4 being the strongest. The guanine residues adjacent to the conserved G's were also protected at sites 3 and 4 (G-105 and G-96). Striking hypersensitivities (G-95 and G-89) downstream of the HSE were observed on both strands. Upstream of the HSE, a slight (less than two fold) hypersensitivity at G-116 was also detected. The temporal pattern of changes in reactivity is generally consistent among all of the affected G's, i.e., protections and hypersensitivities appear at 10 min and are maximal at 20 or 40 min, and start to weaken by 60 min, paralleling changes in both the transcription rate and levels of heat induced HSE binding activity. The methylation pattern of DNA isolated from cells after 2 hr of recovery at 37°C was similar to that of DNA from non-heat shocked cells (Fig. 3, lanes 0' and R), suggesting

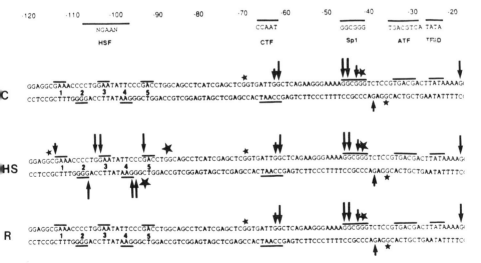

Figure 4. Summary of DMS reactivity patterns of the *hsp70* promoter before, during and after heat shock. The sequence of the promoter region analyzed by genomic footprinting is shown as in Fig. 1. The arrows and stars denote protections and hypersensitivities respectively. The small arrows and stars represent less than two fold changes. C: control, non-heat shocked cells, HS: heat shocked cells, R: recovered cells.

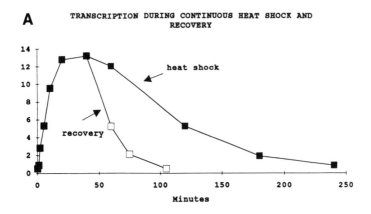

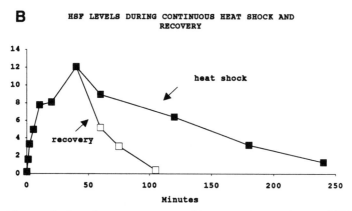

Figure 5. Graphic comparison of transcription rates and HSF levels during continuous heat shock and recovery.
(A) Transcription rate. Cells were heat shocked at 42°C for indicated times (closed squares) or were returned to 37°C for 15, 30 and 60 min (open squares) following a 40 min of heat shock. Transcription rate was measured and quantitated as in Figure 1.
(B) HSF levels. HSF levels were measured and quantitated as in Figure 1.

that HSF is no longer bound to the HSE at this time. The methylation pattern of the *hsp70* promoter before, during heat shock, and after recovery is shown in Figure 4.

Throughout the entire period of heat shock, the guanine reactivity pattern exhibits marked hypersensitivities. These enhanced methylations may signify a distortion in chromatin structure which is created by HSF binding. Bending of DNA upon HSF binding to the *Drosophila hsp70* promoter has been previously reported (Shuey and Parker, 1986a; Shuey and Parker, 1986b). Alternatively, enhanced methylation has been postulated to result from locally increased DMS concentration in hydrophobic pockets created by protein-DNA interactions (Becker and Schutz, 1988; Shuey and Parker, 1986a).

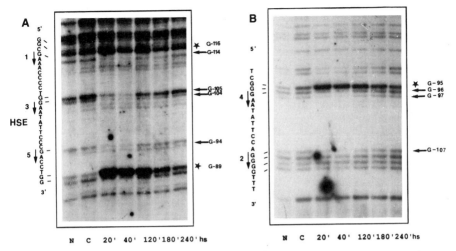

Figure 6. DMS reactivity pattern of the HSE region during continuous heat shock (A) coding strand, (B) non-coding strand. DNA was isolated from non-heat shocked, control (C) cells, and cells heat shocked for indicated times. N: naked DNA. For details see Figure 3.

During exposure of HeLa cells to continuous 4 hr of heat shock, the stress response attenuates. After two hours of heat treatment, the rate of *hsp70* gene transcription is reduced by 50%, and by 4 hr, it is completely abolished (Fig. 5A). Changes in the levels of heat induced HSE binding activity measured by gel mobility shift assay parallel changes in the transcription rate of the *hsp70* gene (Fig. 5B). Attenuation of the stress response is faster when the cells are allowed to recover at 37°C. Within 15' of recovery at 37°C, both the transcription rates and HSF levels are reduced by 60% (Fig. 5A and B). Genomic footprinting of samples isolated during the last 2 hr of 4 hr continuous heat treatment revealed that protections of the conserved G residues are considerably weakened after two hours of heat shock, and are no longer detectable by 3 hr of heat treatment (Fig. 6). Genomic footprinting of samples isolated during recovery revealed that after 15' of recovery at 37°C, interactions at all five NGAAN sites are barely detectable, and are completely abolished by 30' of recovery (Fig. 7). Thus the *in vivo* occupancy of the 5 NGAAN sites that comprise the human HSE parallels both the levels of stress induced HSF, and the transcription rate of the *hsp70* gene during the attenuation of the stress response.

Biochemical evidence for multiple forms of heat shock transcription factor

We have investigated the biochemical features of three forms of HSF: (1) the non-DNA-binding form of HSF which exists in non-heat shocked cells, (2) *in vitro* activated HSF, and (3) HSF from *in vivo* heat shocked cells. Our ability to *in vitro* activate the non-DNA-binding form of HSF

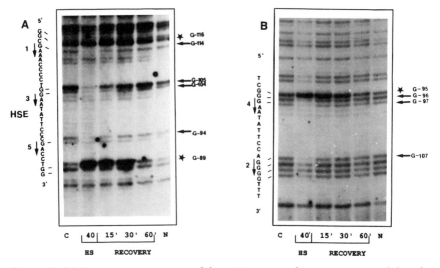

Figure 7. DMS reactivity pattern of the HSE region during recovery. (A) coding strand, (B) non-coding strand. DNA was isolated from non-heat shocked (C), 40 min heat-shocked cells (HS), and cells which were returned to 37°C for the indicated times following the 40 min of heat shock (**recovery**).

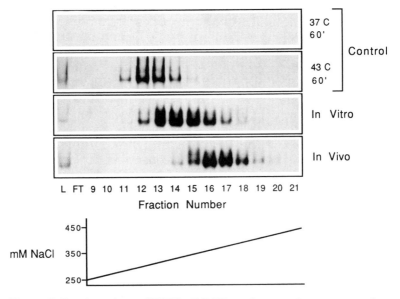

Figure 8. Fractionation of HSF by DEAE-sepharose chromatography. Extracts were prepared from cells grown at 37°C or heat shocked at 43°C for one hour. Extracts from control cells kept on ice or activated *in vitro* by heating to 43°C for one hour and extracts from heat shocked cells were loaded on a 0.5 ml DEAE-sepharose column in 20 mM HEPES (pH 7.9), 10% glycerol, 50 mM NaCl. After washing with loading buffer, bound proteins were eluted with a 100-550 mM linear NaCl gradient. Aliquots of the load, flowthrough, and fractions 9-21 of each sample were analyzed by gel shift assay either as is (*in vitro* and *in vivo* activated samples) or after incubation either at 37° or 43°C (control extracts). The approximate salt gradient relative to fractions shown is indicated.

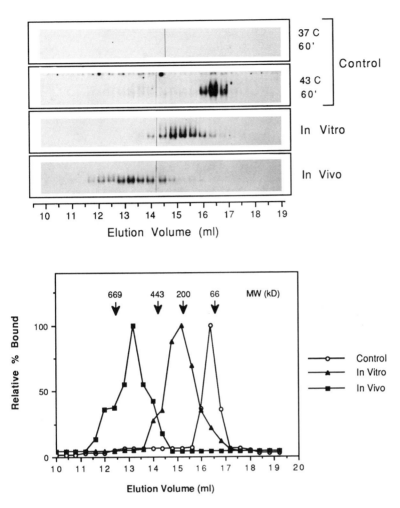

Figure 9. Fractionation of DEAE-sepharose fractionated HSF by superose 6 chromatography. Peak fractions from the DEAE-sepharose chromatography of the non-DNA-binding control form of HSF (DEAE fraction 12), *in vitro* activated HSF (DEAE fraction 15), and *in vivo* activated HSF (DEAE fraction 17) were run on a Pharmacia superose 6 gel filtration column in 20mM HEPES (pH 7.9), 10% glycerol, 100mM NaCl, and 0.2% Triton x-100. Fractions were assayed as described in the legend for Figure 1. The elution of each form of HSF relative to protein molecular weight standards is shown in the lower panel.

even after chromatographic fractionation has enabled us to analyze the biochemical properties of this heretofore undetectable form of HSF.

We have used DEAE-sepharose chromatography to examine for net charge differences between the three forms of HSF. When extracts from cells grown at 37°C are chromatographed on DEAE and an aliquot of each fraction is activated *in vitro* by heating to 43°C for 1 hr,

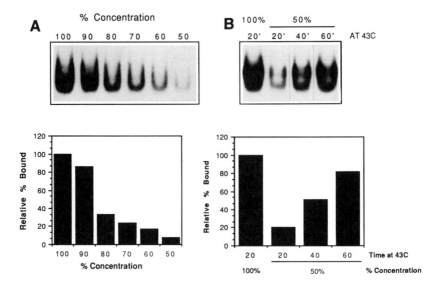

Figure 10. Activation kinetics of DEAE-fractionated control HSF. (A). An aliquot of the peak DEAE fraction 12 of the non-DNA-binding control form of HSF was exchanged to 20mM HEPES (pH 7.9), 10% glycerol, and 100mM NaCl (0.1 M HCN), by several washes on a centricon 30 ultrafiltration cartridge. Samples were activated in vitro (43°C, 1 hr) either at full strength (undiluted, 100%) or after dilution with 0.1M HGN to 90, 80, 70, 60, and 50% concentration of the undiluted sample. Aliquots of each sample containing the same proportional amount of original control HSF activity were assayed by gel shift. (B). The undiluted (100%) peak DEAE fraction for control HSF was in vitro activated (43°C) for 20 min and a 50% dilution was activated at 43°C for 20, 40, or 60 min and then assayed by gel shift. Quantitation of the gel shifts is shown below each gel shift lane.

an HSE-binding activity is observed which elutes from the column with a peak at fractions 12 and 13 (Fig. 8). However, if the control cell extract is heated for 1 hr at 43°C before fractionation, the HSE-binding activity elutes at a higher salt concentration and peaks in fractions 14 and 15. HSF from cells heat shocked in vivo elutes at an even higher salt concentration, peaking in fractions 16 and 17. These results reveal that the three forms of HSF have distinct biochemical features with respect to their net negative charge, with the non-DNA-binding control form of HSF being the least negatively charged, the in vitro activated form of HSF having more negative charge, and the in vivo activated form of HSF being the most negatively charged. This is not surprising in light of the fact that HSF isolated from heat shocked cells is known to be more phosphorylated than HSF activated in vitro (Larson et al., 1988); however the basis of the charge difference between the in vitro and in vivo activated forms of HSF and the non-DNA-binding control form of HSF is unclear.

We determined the native size of each form of HSF by superose 6 gel filtration chromatography using an aliquot of the peak fraction of each form eluted from the DEAE-sepharose column. As can be seen in Figure 9, the non-DNA-binding control form of HSF (DEAE fraction 12) was fractionated on superose 6 chromatography and each fraction was then *in vitro* activated (43°C, 1 hr) to reveal an apparent native molecular weight of 80-100 kd. The *in vitro* activated form of HSF (DEAE fraction 15), has an apparent size of 200-300 KD and the *in vivo* activated form of HSF (DEAE fraction 17) elutes in the range of 500-600 KD. Identical results were obtained using crude extracts on superose 6 chromatography.

These results reveal that the three forms of HSF, in addition to having different net negative charges, also have dramatically different native sizes. Although the estimations of native molecular weight are only approximations given that they are based on spherical protein standards, it is clear that the ability of HSF to bind DNA is associated with an increase in native size.

The ability to *in vitro* activate the DNA binding of the partially purified form of control HSF provided an opportunity to further investigate features of the activation reaction. We have observed during experiments on *in vitro* activation of HSF that the rate of activation displays a marked concentration dependence. Figure 10A shows that if DEAE-fractionated control HSF (DEAE fraction 12) is *in vitro* activated by heating for 20 min at 43°C (undiluted fraction 12) or at several dilutions down to 50% of the original concentration, a dramatic decrease in the levels of activated HSF is observed. This effect is most likely due to a change in the kinetics of the activation reaction and not to irreversible inactivation of HSF DNA-binding activity because nearly all of the activity present in the original undiluted sample can be obtained from the 50% dilution simply by increasing the period of incubation at 43°C (Fig. 10B). These results show that the reaction velocity of HSF DNA binding activation *in vitro* is directly proportional to its concentration in the reaction sample. This suggests that the interaction of HSF either with itself or other proteins, or both, are involved in the acquisition of DNA binding by HSF. These findings are entirely consistent with our observations on the increase in native size of HSF upon activation of DNA binding.

Conclusions

We have examined by a combination of *in vivo* and *in vitro* methodologies, events associated with the transcriptional induction of the human *hsp70* gene in response to heat shock. Several promoter elements involved in basal expression of the *hsp70* gene are apparently occupied by factors in non-heat shocked cells. The interactions with the Sp1, CCAAT and TATA elements in the basal promoter may be important for growth regulation and adenovirus Ela trans-activation (Williams et al., 1989; Wu and Morimoto, 1985). We find no evidence that these interactions are perturbed during heat shock, implying that the binding of factors to these sites is affected neither by binding of HSF to the HSE element nor by the elevated temperature *per se.*

Our data further reveals that prior to heat shock, the HSE region of the human *hsp70* promoter is not occupied by protein. Thus the mechanism of heat shock transcriptional regulation in human cells resembles that of Drosophila rather than yeast, the only other two species for which *in vivo* analyses of HSF-HSE interactions have previously been carried out (McDaniel et al., 1989; Thomas and Elgin, 1988; Wu et al., 1987). The observation that an HSE-specific binding activity is detected in non-heat shocked cells *in vitro* but not *in vivo* may indicate that this factor is non-nuclear or is incapable of binding *in vivo* to the promoter. The relationship of the constitutive HSE binding activity to HSF and the significance of the inverse correlation between changes in their levels during heat shock remains to be determined. If both binding activities correspond to interactions with the same factor, a post-translational modification of the factor might be necessary for its interaction with DNA *in vivo* but not *in vitro.*

The interaction of stress-induced HSF with the HSE affects a region encompassing at least 25 base pairs, which includes five adjacent and inverted NGAAN sites. Our results are in agreement with *in vitro* studies which demonstrated that a region including HSE sites 2, 3, 4, and 5 was protected from DNase digestion when a human *hsp70* promoter fragment was incubated with nuclear extract from heat shocked HeLa cells (Kingston et al., 1987). To define the sequences of the human *hsp70* promoter which confer heat shock responsiveness, our laboratory has transfected human cells with promoter fusion constructs. These studies have defined sequences between -107 and -91, which contain HSE sites 3, 4, and 5, as the region of the human *hsp70* promoter which confers heat shock responsiveness (Williams and Morimoto, 1990; Wu et al., 1986). One cannot conclude, however, that sites 1 and 2 are not required, because the linkers used for construction of the 5' deletion mutants inadvertently recreated two appropriately spaced NGAAN units upstream of site 3. Further transfection studies will therefore be required to determine whether sites 1 and 2 are necessary for or contribute to heat inducible expression of the *hsp70* gene. Analysis of linker scanner and additional 5' deletion mutants of the promoter revealed that sites 4 and 5 alone do not confer heat inducibility and that loss of site 5, when the remaining four sites are present, reduces but does not abolish responsiveness to heat (Williams and Morimoto, 1990). These results are consistent with our *in vivo* genomic footprinting data which indicates the involvement of multiple NGAAN sites during heat shock induction of the human *hsp70* gene.

The *in vivo* footprint studies provide a dynamic and detailed view of HSF-HSE interactions which accompany the heat induced transcriptional activation of the human *hsp70* gene. The data showing protection of guanine residues in each of five adjacent NGAAN units provides essential *in vivo* evidence supportive of the view that these units constitute the binding sites for HSF. A combination of factors may underlie the observation that all five sites are not equivalently protected. Although HSE sites 3 and 4, which are perfect matches to the consensus are the strongest binding sites, our data reveals that the match to the consensus is not the sole determinant of strength of binding, since HSE site 1 shows a weak protection. The nucleotides which flank the GAA's may also influence the strength of interactions as well as the relative position of each NGAAN binding unit within the array.

Recent reports have indicated that in both *Drosophila* (Clos et al., 1990; Perisic et al., 1990) and yeast (Sorger and Nelson, 1989), HSF exists *in vitro* as a trimer or hexamer, and evidence presented here also supports oligomerization of human HSF during heat shock. A model has been proposed wherein HSF binds to arrays of repeated NGAAN sites as a trimer, with each NGAAN unit binding a monomeric subunit (Perisic et al., 1990). Data on which this model is based indicate that two contiguous NGAAN units suffice to bind a trimer: the third subunit of the trimer apparently need not be associated with DNA if a binding site is not present. An array of 5-6 units is proposed to bind two trimers. Our data indicating occupancy of all five binding sites in the human HSE *in vivo*, would be consistent with HSF binding as a hexamer or two trimers or alternatively as a single trimer. In the model that posits binding of a single trimer, binding would occur in a combination of several coexisting alternative configurations.

The most dramatic feature of HSF binding are the assymmetrically distributed hypersensitivities to DMS. The hypersensitivities are usually observed prior to detectable HSF binding and persists well after protections disappear. It is possible that the presence and positions of these hypersensitivities are an intrinsic feature of HSF-HSE interactions and would be present regardless of the location of the HSE in the *hsp70* promoter. Alternatively, these hypersensitivities may result from interactions of HSF with factors bound to adjacent sequences, in this case with proteins bound to basal promoter elements. Footprinting studies of other heat shock promoters might provide useful information with which to address these questions.

The process by which activated HSF loses DNA binding activity during recovery and more interestingly during continuous heat shock is as intriguing as the activation process. Assuming that HSF activation is a rapidly reversible process, then a decrease in the activation signal during continuous heat shock would result in the observed deactivation. Denaturation and misfolding of proteins has been suggested to provide the signal for the stress response and HSF activation (Mosser et al., 1990; Parsel and Sauer, 1989; Pelham, 1989). During recovery at 37°C, misfolding of proteins would cease once the cells are returned to this temperature, and in the absence of an activating signal, the stress response would attenuate readily. During continuous heat shock, a renaturation or prevention of denaturation could be postulated to occur at later times, diminishing the activation signal. Such events could be facilitated by heat shock proteins (Pelham, 1989), which accumulate during heat shock, thus providing an autoregulatory loop. It is possible that activation of HSF results from a conformational change in the factor itself, allowing the protein to bind DNA; at later times of heat shock, such a conformational change may be prevented or return to a non-DNA-binding conformation may be facilitated, thus leading to the decline of the stress response.

Acknowledgments

This work was supported by grants from the National Institutes of Health, American Cancer Society, and March of Dimes to R.I.M. and NIH postdoctoral fellowships to K.A., B.P. and K.D.S.

References

Abravaya, X., Phillips, B and Morimoto, RI, (1991) Heat shock induced interactions of HSF and the human HSP70 promoter examined by *in vivo* footprinting. Mol. Cell. Biol., 11: 586-592.

Amin, J, Ananthan, J and Voellmy, R, (1988) Key features of heat shock regulatory elements. Mol. Cell. Biol., 8: 3761-3769.

Becker, PB and Schutz, G, (1988) Genomic footprinting. Genetic Engineering, Principles and Methods J. K. Setlow, ed.,Plenum press, New York, 10: 1-19.

Clos, J, Westwood, JT, Becker, PB, Wilson, S, Lambert, K and Wu, C, (1990) Molecular cloning and expression of a hexameric *Drosophila* heat shock factor subject to negative regulation. Cell, in press.

Goldenberg, CJ, Luo, Y, Fenna, M, Baler, R, Weinmann, R and Voellmy, R, (1988) Purified human factor activates heatshock promoter in a HeLa cell-free transcription system. J. Biol. Chem., 263: 19734-19739.

Gunning, P, Ponte, P, Okayama, H, Engel, J, Blau, H and Kedes, L, (1983) Isolation and characterization of full length c-DNA clones for human α-β- and -actin mRNAs: Skeletal but not cytoplasmic actins have an amino-terminal cysteine that is subsequently removed. Mol. Cell. Biol., 3: 787-795.

Hunt, C and Morimoto, RI (1985) Conserved features of eukaryotic HSP70 genes revealed by comparison with the nucleotide sequence of human HSP70. Proc. Natl. Acad. Sci. USA, 82: 6455-6459.

Jakobson, BR and Pelham, HRB, (1988) Constitutive binding of yeast heat-shock factor to DNA *in vivo*. Mol. Cell. Biol., 8: 5040-5042.

Kingston, RE, Schuetz, TJ and Larin, Z, (1987) Heat inducible human factor that binds to a human hsp70 promoter. Mol. Cell. Biol., 7: 1530-1534

Larson, JS, Schuetz, TJ and Kingston, RE, (1988) Activation *in vitro* of sequence specific DNA binding by a human regulatory factor. Nature (London), 335: 372-375.

McDaniel, D, Caplan, AJ, Lee, MS, Adams, CC, Fishel, BR, Gross, D, and Garrard WT, (1989) Basal-level expression of the yeast HSP82 gene requires a heat shock regulatory element. Mol. Cell. Biol., 9: 4789-4798.

Morgan, WD, Williams, GT, Morimoto, RI, Greene, J, Kingston, RE, and R. Tijan (1987) Two transcriptional activators, CCAAT-box binding transcription factor and heat shock transcription factor, interact with a human HSP70 gene promoter. Mol. Cell. Biol., 7: 1129-1138.

Morgan, WD (1989) Transcription factor Sp1 binds to and activates a human hsp70 gene promoter. Mol. Cell. Biol., 9: 4099-4104.

Morimoto, RI, Abravaya, K, Mosser, D, and Williams, GT, (1990) Transcription of the human hsp70 gene: cis-acting elements and trans-acting factors involved in basal, adenovirus E1A and stress-induced expression. Stress Proteins, Elsevier Press, Amsterdam, in press.

Mosser, DD, Theodorakis, NG and Morimoto, RI, (1988) Coordinate changes in heat shock element binding activity and hsp70 gene transcription rates in human cells. Mol. Cell. Biol., 8: 4736-4744.

Mosser, DD, Kotzbauer, PT, Sarge, KD and Morimoto, RI, (1990) *In vitro* activation of heat shock transcription factor DNA-binding by calcium and biochemical conditions that affect protein conformation. Proc. Natl. Acad. Sci. USA, 87: 3748-3752.

Mueller, PR and Wold, B, (1989) *In vivo* footprinting of a muscle specific enhancer by ligation mediated PCR. Science, 246: 780-785

Nakajima, N, Horikoshi, N and Roeder, RG, (1988) Factors involved in specific transcription by mammalian RNA polymerase II: purification, genetic specificity, and TATA box promoter interactions of TFIID. Mol. Cell. Biol., 8: 4028-4040.

Parker, CS and Topol, J, (1984) A Drosophila RNA polymerase II transcription factor specific for the heat shock gene binds to the regulatory site of an hsp70 gene. Cell, 37: 273-283.

Parsell, DA and Sauer, RT, (1989) Induction of a heat shock like response by unfolded protein in *Escherichia coli*: dependence on protein level not protein degradation. Genes & Devel., 3: 1226-1232.

Pelham, HRB, (1982) A regulatory upstream promoter element in the *Drosophila* hsp70 heat-shock gene. Cell, 30: 517-528.

Pelham, HRB, (1989) Heat shock and the sorting of luminal ER proteins. EMBO J., 8: 3171-3176.

Perisic, O, Xiao, H and Lis, JT, (1989) Stable binding of *Drosophila* heat shock factor to head-to-head and tail-to-tail repeats of a conserved 5 bp recognition unit. Cell, 59: 797-806.

Shuey, DJ and Parker, CS, (1986a) Binding of *Drosophila* heat shock gene transcription factor to the hsp70 promoter. J. Biol. Chem., 261: 7934-7940.

Shuey, DJ and Parker, CS, (1986b) Bending of promoter DNA on binding of heat shock transcription factor. Nature, 323: 459-461.

Sorger, PK and Pelham, HRB, (1987) Purification and characterization of a heat-shock element binding protein from yeast. EMBO J., 6: 3035-3041.

Sorger, PK, Lewis, MJ, and Pelham, MHB (1987) Heat shock factor is regulated differently in yeast and HeLa cells. Nature (London), 329: 81-84.

Sorger, PK and Pelham, HRB, (1988) Yeast heat shock factor is an essential DNA-binding protein that exhibits temperature dependent phosphorylation. Cell, 54: 855-864.

Sorger, PK and Nelson, HCM, (1989) Trimerization of a yeast transcriptional activator via a coiled-coil motif. Cell, 59: 807-813.

Thomas, GH and Elgin, SCR, (1988) Protein/DNA architecture of the DNAse I hypersensitive region of the *Drosophila* hsp26 promoter. EMBO J., 7: 2191-2201.

Williams, GT and Morimoto, RI, (1990) Maximal stress-induced transcription from the human hsp70 promoter requires interactions with the basal promoter elements independent of rotational alignment. Mol. Cell. Biol., 10: 3125-3136.

Williams, GT, McClanahan, TR and Morimoto, RI, (1989) Ela transactivation of the human HSP70 promoter is mediated through the basal transcriptional complex. Mol. Cell. Biol., 9: 2574-2587.

Wu, B, and Morimoto, RI, (1985) Transcription of the human HSP70 gene is induced by serum stimulation. Proc. Natl. Acad. Sci. USA, 82: 6070-6074

Wu, B, Kingston, R and Morimoto, RI, (1986) Human HSP70 promoter contains at least two distinct regulatory domains. Proc. Natl. Acad. Sci. USA, 83: 629-633.

Wu, C, (1984) Two protein-binding sites in chromatin implicated in the activation of heat shock genes. Nature (London), 311: 8184.

Wu, C, Wilson, S, Walker, S, Dawid, I, Paisley, T, Zimarino, V and Ueda, H, (1987) Purification and properties of Drosophila heat shock activator protein. Science, 238: 1247-1253.

Xiao, H and Lis, JT, (1988) Germline transformation used to define key features of the heat shock response element. Science, 239: 1139-1142.

Zimarino, V and Wu, C, (1987) Induction of sequence specific binding of *Drosophila* heat shock activator protein without protein synthesis. Nature (London), 327: 727-730.

Zimarino, V, Wilson, S and Wu, C, (1990) Antibody-mediated activation of *Drosophila* heat shock factor *in vitro*. Science, 249: 546-549.

Mechanisms of Regulation of Small Heat Shock Protein Genes in *Drosophila*

R. Voellmy, Y. Luo[*], R. Mestril[#], J. Amin, J. Ananthan
Department of Biochemistry and Molecular Biology
University of Miami,
Miami, Florida 33101
USA

Introduction

Small heat shock protein (hsp) genes of *Drosophila* are not only activated by heat but also by the steroid hormone ecdysterone. Promoter deletion analysis has revealed that the two stimuli regulate gene expression by distinct mechanisms. Sequence elements that are critically important for ecdysterone activation of the *hsp27* and *hsp23* genes have been identified and have been used to purify ecdysterone receptor by specific DNA affinity chromatography. Purified receptor enhances transcription *in vitro* from promoters containing the above sequence elements. Thus, ecdysterone receptor is a *bona fide* transcription factor and is essential for the hormonal activation of small hsp genes. Recent experiments suggest that an additional factor modifies this regulation.

Regulation of small heat shock protein genes

Of the seven heat shock genes clustered in region 67B of the *D.melanogaster* genome, four code for small heat shock proteins with molecular weights of 27, 26, 23 and 22 kDa (HSP27-HSP22). We have been concerned with the regulation of two of these genes, coding for HSP27 and HSP23. Both genes are not only heat-regulated but are also active at different stages of normal development (Sirotkin and Davidson, 1982; Ireland et al., 1982). They are expressed at high levels in late third instar larvae and early pupae in which titers of the molting hormone ecdysterone are high (Table 1). This developmental regulation is not shared by all hsps. Most notably, inducible HSP70 genes are totally silent at these stages. This suggests that the developmental

---

[*] Laboratory of Biochemistry and Molecular Biology, The Rockefeller University, New York, N.Y. 10021, USA.
[#] Department of Medicine, UC San Diego, California 92103, USA.

Table 1

Developmental regulation of the *D.melanogaster hsp27* and *hsp23* gene promoters			
Stage/condition	***hsp27** activity**	***hsp23** activity**	***hsp70**activity**
Embryo	+	-	-
Early 3ʳᵈ instar larvae	+/-	-	-
Late 3ʳᵈ instar larvae	+	+	-
Early pupae	+	+	-
Adult	+	-	-
S3 cells	-	-	-
S3 cells + ecdysterone	+	+	-
Transfected promoters in S3 cells	-	-	-
Ditto + ecdysterone	+	+	-

* Positive and negative promoter activity is listed. Specific nervous system expression of some promoters is not being considered here (Pauli et al., 1990, and results presented at this conference by Haas, Klein and Kloetzel).

activity of the *hsp27* and 23 genes cannot simply be a consequence of stressful events occurring in the course of development. Ireland et al. (Ireland et al., 1982) have demonstrated that this regulation can be reproduced in cultured *Drosophila* (S3 and Kc) cells by simple addition of ecdysterone to the medium, and that it represents most likely a primary hormone response since transcription of (some) small hsp genes can be hormone-activated in the presence of the protein synthesis inhibitor cycloheximide.

To identify cis-acting elements involved in the regulation of these genes, we have developed a DEAE-dextran-based procedure for the transfection of *D.melanogaster* S3 cells (Lawson et al., 1984). The question whether transiently introduced heat shock promoters are correctly regulated by ecdysterone in *D.melanogaster* cells has been answered by examining the expression of transfected hsp:ß-galactosidase hybrid genes: a transfected *hsp23* hybrid gene is hormone-regulated while *hsp84* and *hsp70* hybrid genes are not (Lawson et al., 1985; see also Table 1). Morganelli et al. (Morganelli et al., 1985) have confirmed our findings and have shown that all small hsp gene promoters are ecdysterone-regulated in transfected cells. Using the above transient expression system, an extensive deletion

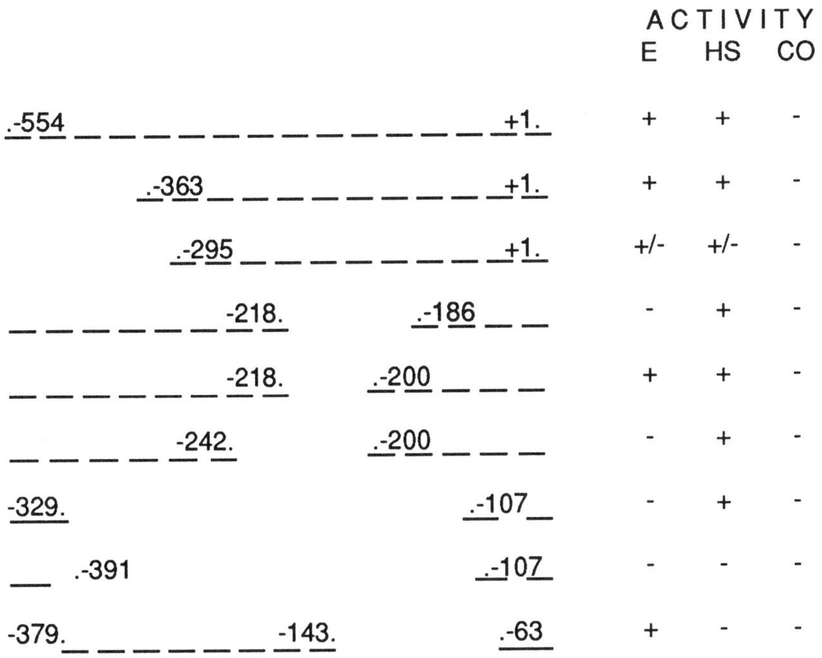

```
                                                        A C T I V I T Y
                                                        E    HS   CO

.-554 _ _ _ _ _ _ _ _ _ _ _ _ _ _ +1.                   +    +    -

        .-363 _ _ _ _ _ _ _ _ _ +1.                     +    +    -

            .-295 _ _ _ _ _ _ _ _ +1.                   +/-  +/-  -

_ _ _ _ _ _ -218.         .-186 _ _                     -    +    -

_ _ _ _ _ _ -218.      .-200 _ _ _                      +    +    -

_ _ _ _ -242.          .-200 _ _ _                      -    +    -

-329.                              .-107_               -    +    -

___  .-391                         .-107                -    -    -

-379. _ _ _ _ _ _ -143.            .-63                 +    -    -
```

Figure 1. Analysis of the *hsp23* promoter. Expression of mutant promoters in transfected S3 cells.
Numbers refer to distances in nt from the start of transcription site of the *hsp23* gene. For original data see Mestril et al. (Mestril et al., 1986).

analysis of the *hsp23* promoter has been carried out in our laboratory (Mestril et al., 1986). Selective mutants are shown schematically in Figure 1, and their respective activities and regulation by ecdysterone and heat shock are indicated. These experiments have revealed that essential elements for ecdysterone as well as for heat regulation are included in the proximal 350-400 nt of promoter sequence. Two elements of critical importance for ecdysterone activation are immediately downstream from position -200 (relative to the transcription start site) and upstream from -218, respectively. Deletion of either element does not affect the heat-induced expression of the *hsp23* promoter. The element downstream from position -200 will be referred to here as *hsp23* HERE (for hydroxyecdysone-responsive element).

The nucleotide sequence of the *hsp23* gene promoter reveals consensus HSE (heat shock element = heat shock transcription factor recognition site; for detailed descriptions of the element see Amin et al., (Amin et al., 1988) and Xiao and Lis (Xiao and Lis, 1988) sequences at -413 and at -145.

Mutant analysis suggests that one of the two HSEs and an additional element(s) from the region between the HSEs, that may include a weak binding site(s) for heat shock transcription factor, are minimally

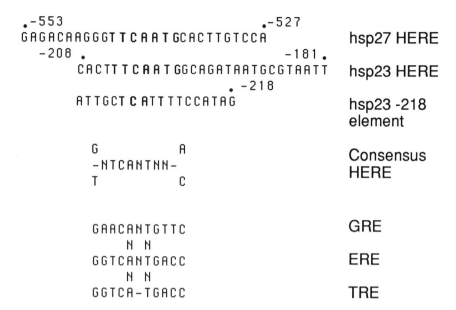

Figure 2. Sequences involved in ecdysterone regulation. Sequence comparisons. GRE: glucocorticoid response element. ERE: estrogen response element. TRE: thyroid response element.

required for promoter activity in heat-treated cells. Deletion of both HSEs eliminates heat regulation without affecting ecdysterone-stimulated promoter activity (Fig. 1). This indicates strongly that heat shock and ecdysterone regulation involve totally independent mechanisms.

Riddihough and Pelham (Riddihough and Pelham, 1986; Riddihough and Pelham, 1987) have mapped a sequence element critically important for ecdysterone activation at position -553 in the promoter of the *hsp27* gene. This element is referred to here as *hsp27* HERE, and has been shown to be a binding site for a cellular factor (Riddihough and Pelham, 1987). Assuming that ecdysterone regulation of small hsp genes is a primary hormone response, elements responsible for this regulation are likely to represent binding sites for an activated ecdysterone receptor. The existence of such a receptor has been previously established (see Cherbas et al., 1984, for a review). Surprisingly, however, the nucleotide sequences of the two *hsp23* elements and the *hsp27* HERE are only relatively weakly homologous (Fig. 2). Note that the *Drosophila* sequences appear to resemble consensus binding sites of vertebrate steroid and thyroid receptors. Exonuclease III protection experiments with nuclear extract from ecdysterone-treated *Drosophila* S3 cells have indicated the presence of several protein binding sites between positions -300 and -180 in the *hsp23* promoter. No evidence for binding sites further upstream has been obtained. These findings have been confirmed by gel retardation

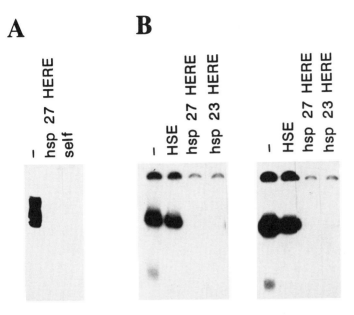

Figure 3. Gel retardation assays with crude extract proteins and purified ecdysterone receptor.

A: A fragment containing sequences from -242 to -181 of the *hsp23* promoter has been endlabeled by polynucleotide kinase and used as probe in binding assays with nuclear extract. Competitors have been at 50 fold molar excess over probe.

B: Binding assays with purified receptor using short synthetic *hsp27* (left panel) or *hsp23* (right panel) HEREs, endlabeled as before, as probes. Competitors have been at 500 fold molar excess. Experimental details will be described elsewhere.

assays using fragments of different lengths with sequences from the same region. They also have indicated that the different sites are bound by a common factor. The same factor also binds to the *hsp27* HERE sequence as shown by competiton binding assays (Fig. 3). Presumably, this factor is identical with ecdysterone receptor.

Putative ecdysterone receptor has been purified, using specific DNA affinity chromatography on columns containing *hsp27* HERE sequences, to apparent homogeneity (30,000 fold purification). Filter binding assays with radio-iodinated ponasterone (Cherbas et al., 1988), a high affinity analog of ecdysterone, have shown that hormone binding activity co-purifies with *hsp27* HERE DNA binding activity and, thus, have provided strong evidence for the identity of the purified protein with ecdysterone receptor. Purified receptor binds to *hsp27* and to *hsp23* HERE sequences as well as to other related sequences (Fig. 3). A "consensus" binding site has been derived and is shown in Figure 2.

Constructs B23-6 containing six copies of an *hsp23* HERE sequence (lanes 1 and 2) and B27-4 containing four copies of an *hsp27* HERE

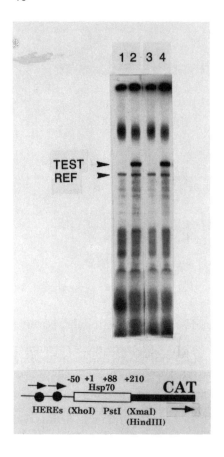

Figure 4. Accurate transcription from promoters containing HERE sequences is stimulated by ecdysterone receptor.

sequence (lanes 3 and 4) have been used as templates (Test). Transcriptions *in vitro* in HeLa extract have been carried out in the absence (lanes 1 and 3) or presence of added purified ecdysterone receptor. As a positive specificity control, mLP (major late promoter) cat DNA has also been included in the reactions. Analysis of transcripts by reverse transcription. Experimental details will be described elsewhere.

Synthetic *hsp27* and *hsp23* HERE elements have been inserted, in multiple copies, upstream from the basal *hsp70* promoter in construct D50-cat (Goldenberg et al., 1988). D50-cat has been constructed by replacing sequences between the NdeI and the HindIII sites in pSV2-cat with a *D.melanogaster hsp70* gene segment that contains sequences from positions 50 to +210 (relative to the start of transcription) but lacks HSEs. The above constructs, that are shown schematically in Figure 4, have been used as templates in *in vitro* transcription experiments. Reactions have been carried out with HeLa cell extract (Dignam et al., 1983) and with or without added purified ecdysterone receptor, and transcripts have been analyzed by reverse transcription using a radio-labeled cat primer (containing sequences from the beginning of the coding region of the chloramphenicol acetyltransferase gene), electrophoresis on an acrylamide-urea gel and autoradiography. As shown in Figure 4, both an *hsp27* and an *hsp23* HERE-containing promoter are activated by ecdysterone receptor and, in its presence, are transcribed correctly and efficiently. Further experimentation has confirmed that the enhancement of transcription by the receptor requires its specific binding to target sequences in the promoter. Thus, ecdysterone receptor is a bona fide transcription factor that can activate *in vitro* promoters of both the *hsp27* and the *hsp23* types.

Conclusions

Recent experiments suggest that the regulation of the small hsp genes may be more complex than has been suggested by the *in vitro* transcription experiments. While ecdysterone receptor is critically involved in the larval-ecdysterone activation of the *hsp27* and *hsp23* genes, there appears to be a second factor that also plays a role. Time course experiments as well as studies with protein synthesis inhibitors suggest that *hsp27* gene transcription responds rapidly to an increase in the concentration of ecdysterone, while expression of the *hsp23* gene only changes dramatically hours later, and only after some other factor has been newly synthesized. Interestingly, this "early" and "late" regulation is conferred exclusively by the HERE sequences of the two promoters. As has been mentioned earlier, these HERE sequences appear to contain only a binding site for a single factor, i.e., ecdysterone receptor, since 1) they produce single complexes in gel retardation assays with crude nuclear extract from cells exposed to ecdysterone for 24 hr, 2) binding to one type of HERE sequence can be competed by the other, and 3) purified receptor binds to HERE sequences and produces complexes indistinguishable from those formed by nuclear extract proteins. Thus, high level expression from the *hsp23* but not the *hsp27* promoter may require that ecdysterone receptor interacts with a second factor, and that this interaction (or modification) enables it to stimulate effectively *hsp23* promoter activity. The postulated second factor would have to be encoded by a gene of the *hsp27* type that is activated rapidly by ecdysterone. That transcription *in vitro* of promoters containing both *hsp27* and *hsp23* HERE sequences is stimulated similarly by added receptor may mean that the purified receptor has been modified prior to its isolation and is now able to transcribe both types of promoters or, alternatively, that *in vitro* transcription from the *hsp23* promoter is less sensitive to the presence or absence of the postulated second factor than transcription within cells for reasons that are not known.

Acknowledgments

This work has been supported by an NIH grant.

42

References

Amin, J, Ananthan, J and Voellmy, R, (1988) Key features of heat shock regulatory elements. Mol. Cell. Biol., 8: 3761-3769.

Cherbas, L, Fristrom, JW and O'Connor, JD, (1984) The action of ecdysone in imaginal discs and Kc cells of *Drosophila melanogaster*. Biosynthesis, metabolism and mode of action of invertebrate hormones. Springer Berlin Heidelberg New York.

Cherbas, P, Cherbas, L, Lee, SS and Nakanishi, K, (1988) 26-^{125}I-iodoponasterone A is a potent ecdysone and a sensitive radioligand for ecdysone receptors. Proc. Natl. Acad. Sci. USA, 85: 2096-2100.

Dignam, JD, Lebowitz, RM and Roeder, RG, (1983) Accurate transcription initiations by RNA polymerase II in a soluble extract from isolated mammalian cells. Nucl. Acids Res., 11: 1475-1489.

Goldenberg, CJ, Luo, Y, Fenna, M, Baler, R, Weinmann, R and Voellmy, R, (1988) Purified human factor activates heat shock promoter in a HeLa cell-free transcription system. J. Biol. Chem., 263: 19734-19737.

Ireland, R, Berger, E, Sirotkin, K, Yund, M, Osterbur, D and Fristrom, J, (1982) Ecdysterone induces the transcription of four heat shock genes in *Drosophila* S3 cells and imaginal discs. Dev. Biol., 93: 498-507.

Lawson, R, Mestril, R, Schiller, P and Voellmy, R, (1984) Expression of heat shock-ß-galactosidase hybrid genes in cultured *Drosophila* cells. Mol. Cell. Biol., 198: 116-124.

Lawson, R, Mestril, R, Luo, Y and Voellmy, R, (1985) Ecdysterone selectively stimulates the expression of a 23,000-Da heat-shock protein-ß-galactosidase hybrid gene in cultured *Drosophila* cells. Dev. Biol., 110: 321-330.

Mestril, R, Schiller, P, Amin, J, Klapper, H, Ananthan, J and Voellmy, R, (1986) Heat shock and ecdysterone activation of the *Drosophila melanogaster* hsp23 gene; a sequence element implied in developmental regulation. EMBO J., 5: 1667-1673.

Morganelli, CM, Berger, EM and Pelham, HRB, (1985) Transcription of *Drosophila* small *hsp-tk* hybrid genes is induced by heat shock and by ecdysterone in transfected *Drosophila* cells. Proc. Natl. Acad. Sci. USA, 82: 5865-5869.

Pauli, D, Tonka, CH, Tissieres, A and Arrigo, AP, (1990) Tissue-specifc expression of the heat shock protein hsp27 during *Drosophila melanogaster* development. J. Cell Biol., 111: 817-828.

Riddihough, G and Pelham, HRB, (1986) Activation of the *Drosophila* hsp27 promoter by heat shock and by ecdysone involves independent and remote regulatory sequences. EMBO J., 5: 1653-1658.

Riddihough, G and Pelham, HRB, (1987) An ecdysone response element in the *Drosophila* hsp27 promoter. EMBO J., 6: 3729-3734.

Sirotkin, K and Davidson, N, (1982) Developmentally regulated transcription from *Drosophila melanogaster* chromosomal site 67B. Dev. Biol., 89: 196-210.

Xiao, H and Lis, JT, (1988) Germline transformation used to define key features of heat-shock response elements. Science, 239: 1139-1142.

Heat Shock Protein Functions in E.coli and Yeast

The Biological Role of the Universally Conserved *E. coli* Heat Shock Proteins

D. Ang, T. Ziegelhoffer, A. Maddock, J. Zeilstra-Ryalls,
C. Georgopoulos, O. Fayet*, K. Liberek#, D. Skowyra#,
J. Marszalek#, J. Osipiuk#, Sz. Wojtkowiak#, M. Zylicz#
Department of Cellular, Viral and Molecular Biology
University of Utah Medical Center
Salt Lake City, Utah 84132
USA

Introduction

The heat shock or stress response has been universally conserved among organisms (reviewed in Morimoto et al., 1990; Georgopoulos et al., 1990; Gross et al., 1990). When the gram-negative bacterium, *E. coli*, is treated with a heat shock, the cell responds by transiently accelerating the rate of transcription of heat shock genes. This increased transcription is promoted by the $E\sigma^{32}$ RNA polymerase holoenzyme. The genes of approximately half of the 20 or so observed heat shock proteins in *E. coli* have been cloned and their corresponding proteins purified and characterized. Surprisingly, five of the major heat shock genes, namely *dnaK*, *dnaJ*, *grpE*, *groES*, and *groEL*, were previously identified because certain mutations in them block bacteriophage λ growth (reviewed in Friedman et al., 1984; Georgopoulos et al., 1990). The *dnaK*, *dnaJ*, and *grpE* gene products are required for bacteriophage λ growth specifically at the level of DNA replication, whereas the *groES* and *groEL* gene products are required specifically at the level of prohead assembly. In this paper, we summarize our previous and current studies on these five genes and suggest models for the mechanism of action of their gene products.

The dnaK, dnaJ, and grpE proteins

The *dnaK* and *dnaJ* genes form an operon at 0.3 min on the *E. coli* genetic map, the order being promoter-*dnaK*-*dnaJ*. The *grpE* gene is monocistronic and maps at 56 min (reviewed in Friedman et al., 1984).

---

* Centre de Recherche de Biochimie et de Genetique Cellulaires du Centre National de la Recherche Scientifique, 31062 Toulouse, France
# Division of Biophysics, Department of Molecular Biology, University of Gdansk, 24 Kladki, 80-822 Gdansk, Poland

The dnaK protein is highly conserved in nature, being approximately 50% identical at the amino acid sequence level to the major eukaryotic heat shock protein, hsp70 (Bardwell and Craig, 1984). It possesses weak 5'-nucleotidase and autophosphorylating activities (Zylicz et al., 1983; Bochner et al., 1986). The dnaJ protein is also conserved, being homologous to the sec63, sis1, and ydj1 proteins of yeast (Sadler et al., 1989; K. Arndt, personal communication; A. Caplan, personal communication). The grpE protein is similarly conserved in nature, as suggested by detection of a protein in yeast that crossreacts to anti-grpE antibody (D. Ang, unpublished data).

Mutations in the *dnaK*, *dnaJ*, or *grpE* genes, selected on the basis of blocking bacteriophage λ DNA replication, can be bypassed by compensating mutations in the λP replication gene (Table 1). This result suggests that the dnaK, dnaJ, and grpE proteins of *E. coli* interact with the λP protein. This genetic evidence has been reinforced with direct biochemical experiments using purified proteins, on a one to one basis, demonstrating that λP indeed interacts with dnaK, dnaJ, and grpE (reviewed in Georgopoulos et al., 1990).

All of the *E. coli* and λ proteins defined by genetic experiments as being necessary for *in vivo* λ DNA replication have been purified to homogeneity using *in vitro* complementation assays. Our laboratories, as well as that of R. McMacken (Johns Hopkins University, Baltimore, Maryland), have succeeded in reconstituting an *in vitro* λ DNA replication

Table 1

Mutant	λ[a]	λP'	λdnaK⁺	λdnaJ⁺	λgrpE⁺	Bacterial growth at 43°[b]
wild type	1.0	1.0	1.0	1.0	1.0	1.0
dnaK756	<10^{-5}	1.0	1.0	<10^{-5}	<10^{-5}	<10^{-5}
dnaJ259	<10^{-5}	1.0	<10^{-5}	1.0	<10^{-5}	<10^{-5}
grpE280	<10^{-5}	1.0	<10^{-5}	<10^{-5}	1.0	<10^{-5}

Plating properties of *E. coli* mutants blocked in λ DNA replication

[a] The efficiency of plating (e.o.p.) of bacteriophage λ on the various mutant *E. coli* hosts at 37°C relative to the wild type host is indicated.
[b] The efficiency of colony formation (c.f.u.) at 43°C relative to that at 30°C is indicated.

system composed entirely of purified proteins, and which mimics well the *in vivo* situation (Zylicz et al., 1989; Mensa-Wilmot et al., 1989). This system has been used to determine the steps leading to initiation of λ DNA replication.

Initiation of λ DNA replication includes the following steps (see Figure 1 for details): (a) the λO initiator protein binds specifically to oriλ, (b)

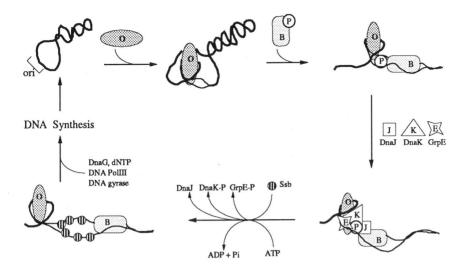

Figure 1. The steps leading to one DNA replication. For details, see text and Zylicz et al., 1989.

the λP initiator protein complexes with *E. coli* dnaB helicase, (c) the λP-dnaB complex is localized to *oriλ*, through a demonstrable λO-λP interaction (The *oriλ*-λO-λP-dnaB complex is very stable. However, the helicase activity of dnaB is inhibited by λP in this stable complex.), (d) in the absence of ATP, the dnaJ, dnaK and grpE proteins bind to the DNA-protein structure at *oriλ*, giving rise to an even larger complex (Dodson et al., 1989), (e) upon addition of ATP, the λP protein dissociates from dnaB and is released from the protein-DNA complex (Liberek et al., 1988; Alfano and McMacken, 1989), thus allowing dnaB to unwind the duplex DNA, (f) the dnaG primase locates the dnaB-ssDNA complex, subsequently synthesizing RNA primers, and (g) the DNA polymerase III holoenzyme extends these RNA primers into DNA. From our work and that of others, it appears that the role of the dnaK, dnaJ and grpE heat shock proteins in λ DNA replication is the *dissociation* of the protein complex assembled at *oriλ*, and the sequestering of the λP protein. This releases dnaB from its complex with λP, activating dnaB as a helicase, and allowing the initiation of λ DNA replication.

Aside from its role in λ DNA replication, the dnaK protein of *E. coli* has been shown to (a) substitute for the yeast HSC70 proteins in the transport of pre-pro-α-factor into membranes (Waters et al., 1989; reviewed by Rothman, 1989), (b) facilitate the transport of a pre-lamB-lacZ fusion protein in *E. coli* (Phillips and Silhavy, 1990), (c) protect *E. coli* RNA polymerase from heat inactivation (Skowyra et al., 1990), and (d) in the presence of ATP, disaggregate heat-inactivated *E. coli* RNA polymerase (Skowyra et al., 1990). Furthermore, purified dnaK protein binds preferentially to the denatured form of BPTI (bovine pancreatic trypsin inhibitor) compared to the native form, and quantitatively dissociates from this complex in the presence of ATP (unpublished data). Its roles in λ DNA replication, protein translocation and protection of proteins under denaturing conditions such as heat shock suggest that dnaK is required for the "shielding" of segments of protein which, when unfolded, may be damaging to the cell and/or to the protein itself. In light of its ability to dissociate protein aggregates, dnaK may be viewed as a "molecular crowbar". On the basis of these results and those of Flynn et al. (Flynn et al., 1989) using the eukaryotic BiP protein, we propose the following model for the mechanism of action and recycling of dnaK. The dnaK protein recognizes and binds to specific sequences or "unstructured" regions of many proteins (some as they emerge from the ribosome?), "shielding" these regions from undesirable intra- or intermolecular interactions. Depending on the protein segment bound by dnaK, such binding may be weak, so that the polypeptide is released quickly. In other situations, hydrolysis of ATP by dnaK may be necessary for release of the polypeptide from dnaK. Still other polypeptides could be bound so tightly to dnaK that the participation of both the dnaJ and grpE proteins is required for the release and recycling of dnaK (exactly analogous to the step catalyzed by groES in Figure 2). In support of this model, it has been shown recently that these three heat shock proteins physically interact, and that this interaction leads to a stimulation of dnaK's ATPase activity (Liberek et al., submitted).

The groE proteins

The *E. coli groES* and *groEL* heat shock genes form an operon, the order being promoter-*groES*-*groEL*, which maps at 94 min (reviewed by Friedman et al., 1984). Both proteins possess seven-fold symmetry, groES being composed of seven 97-residue subunits, and groEL of fourteen 548-residue subunits. The groEL amino acid sequence and oligomeric structure are highly conserved in nature, being approximately 50% identical at the amino acid sequence level to both the chloroplast Rubisco-subunit binding protein (Hemmingsen et al., 1988), and the mitochondrial HSP60 protein (Reading et al., 1989). Mutations in either *groES* or *groEL* exert pleiotropic effects on either phage head or tail assembly at all temperatures, and on *E. coli* growth at high temperature (Georgopoulos and Eisen, 1974). Specifically, mutations in either *groES* or *groEL* interfere with the proper assembly of bacteriophage λ's head-tail connector (reviewed in Georgopoulos et al., 1990). Mutations in either the λB or λE genes have been shown to overcome the block to head-tail connector assembly (Table 2). Recent data suggests that the λB protein forms a complex with groEL and that this complex dissociates in the presence of ATP (Ziegelhoffer, unpublished data). Both of the *groE* functions are essential for *E. coli* viability at all temperatures, since neither gene can be deleted (Fayet et al., 1989).

Both genetic and biochemical data has accumulated, demonstrating that the groES and groEL proteins functionally interact. The genetic data consist of the isolation of mutations in the *groEL* gene that specifically suppress the temperature-sensitive phenotype of certain *groES* alleles (Tilly and Georgopoulos, 1982). The biochemical data consist of the demonstration that the groES and groEL proteins physically interact since (a) they cosediment in glycerol gradients (Chandrasekhar et al., 1986) or co-elute during size chromatography (Viitanen et al., 1990) in the presence of Mg^{2+} and ATP, (b) groES helps the release of pre-β-lactamase (Lecker et al., 1989) or cyanobacterial Rubisco (Goloubinoff et al., 1989) bound to groEL in an ATP hydrolysis-dependent manner, (c) groES inhibits the uncoupled ATPase activity of groEL (Chandrasekhar et al., 1986; Viitanen et al., 1990), and (d) groES binds specifically to a groEL-affinity column in the presence of Mg^{2+} and ATP (Chandrasekhar et al., 1986). The elucidation at the molecular level of the *groES* temperature-sensitive mutations and their *groEL* suppressors may allow the identification of specific amino acid interactions between the two proteins. The recent work of Martel et al. (Martel et al., in press) points to the presence of a conserved amino acid sequence motif in both groES- and groEL-like sequences. These amino acid sequence motifs could be used to nucleate groES-groEL protein interactions.

Overproduction of both the groES and groEL proteins has been shown to correct the temperature-sensitivity of a plethora of missense mutations in a diverse group of genes (Fayet et al., 1986; Jenkins et al., 1986; van Dyk et al., 1989). This result suggests that the groE proteins participate in the correct folding of various polypeptides *in vivo*. Recently, it has been shown that the groE proteins facilitate the *in vivo* export of pre-β-lactamase (Kusukawa et al., 1989) and a mutant pre-lamB-lacZ fusion protein (Phillips and Silhavy, 1990). *In vitro* experiments with purified proteins have shown that groEL protein interacts with the unfolded, but not the folded, forms of pro-ompA and pre-phoE (Lecker et al., 1989), pre-β-lactamase (Laminet et al., 1990), and cyanobacterial Rubisco (Goloubinoff et al., 1989). In the

Table 2

Plating properties of *E. coli* mutants blocked in λ morphogenesis					
Mutant	λ[a]	λB' or lE'	λgroES+	λgroEL+	Bacterial growth at 43°[b]
wild type	1.0	1.0	1.0	1.0	1.0
*groES*30	<10⁻⁵	1.0	1.0	<10⁻⁵	<10⁻⁵
*groEL*140	<10⁻⁵	1.0	<10⁻⁵	1.0	<10⁻⁵

[a] E.o.p. is as defined in the legend of Table 1.
[b] C.f.u. is as defined in the legend of Table 1.

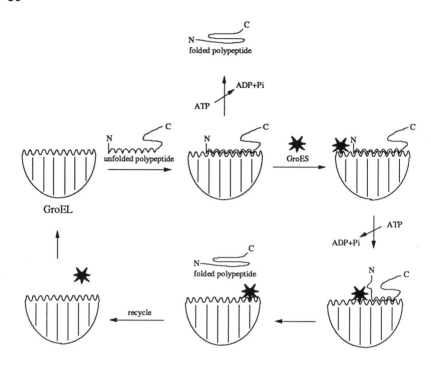

Figure 2. The "cogwheel" model of groEL and groES protein action.

case of pre-β-lactamase, it has been shown that addition of ATP allows the release of pre-β-lactamase (in a form capable of folding). The addition of purified groES accelerates the rate of release of pre-β-lactamase from groEL. In the case of cyanobacterial Rubisco, it has been shown that both groES protein and ATP are absolutely needed for its release (in a folded state) from groEL (Goloubinoff et al., 1989). Surprisingly, the groEL eukaryotic analogues, the mitochondrial hsp60 and chloroplast Rubisco-subunit binding proteins, can substitute for groEL if the *E. coli* groES protein and ATP are present. This last result demonstrates a dramatic conservation in the function of the groEL chaperonin family in evolution. The absolute need for the *E. coli* groES protein for the dissociation of cyanobacterial Rubisco complexed with the eukaryotic HSP60 protein prompted Lubber et al. (Lubben et al., 1990) to search for and identify a beef liver mitochondrial groES analogue, termed mt-cpn10.

The groEL protein recognizes and binds to "unstructured" peptide regions exposed in nascent polypeptides. Although the nature of the "unstructured" peptide features recognized by groEL protein are not known, probably both hydrophilic and hydrophobic regions are recognized, as in the case of the HSP70 family of proteins mentioned above (Flynn et al., 1989). The bound polypeptide may or may not be spontaneously released, depending on the quality and strength of its interaction with groEL. The model envisions that some of the stably bound polypeptides may be released in conjunction with ATP hydrolysis, whereas others need the simultaneous assistance of the groES protein. The groES protein is proposed to displace tightly bound polypeptides through the "cogwheel" action (Fig. 2) and/or a conformational change in the groEL protein. The released polypeptide may assume a different conformation, leading to intra- or even intermolecular folding. In the absence of groES, the release of some of the groEL-bound polypeptide may occur too slowly, or even not at all, essentially preventing the recycling of the groEL protein. The scheme shown in Figure 2 helps explain why mutations in either the *groES* or *groEL* genes will lead to the same phenotype, i.e., unavailability of active and free groEL polypeptides. Insufficient levels of wild type groEL protein or the presence of mutant groEL protein may result in misfolding of nascent polypeptide chains because of reactive peptide sequences, e.g., extremely hydrophilic or hydrophobic sequences, in the nascent chain which bind prematurely to other sections of the polypeptide.

References

Alfano, C and McMacken, R, (1989) Heat shock protein-mediated disassembly of nucleoprotein structures is required for the initiation of bacteriophage lambda DNA replication. J. Biol. Chem., 264: 10709-10718.

Bardwell, JCA and Craig, EA, (1984) Major heat shock gene of *Drosophila* and the *Escherichia coli* heat-inducible dnaK gene are homologous. Proc. Natl. Acad. Sci. USA, 81: 848-852.

Bochner, BR, Zylicz, M and Georgopoulos, C, (1986) *Escherichia coli* DnaK protein possesses a 5'-nucleotidase activity that is inhibited by ApppppA. J. Bacteriol., 168: 931-935.

Chandrasekhar, GN, Tilly, K, Woolford, C, Hendrix, R and Georgopoulos, C, (1986) Purification and properties of the GroES morphogenetic protein of E.coli. J. Biol. Chem., 261: 12414-12419.

Dodson, M, McMacken, R and Echols, H, (1989) Specialized nucleoprotein structures at the origin of replication of bacteriophage lambda. Protein association and disassociation reactions responsible for localized initiation of replication. J. Biol. Chem., 264: 10719-10725.

Fayet, O, Louarn, J-M and Georgopoulos, C, (1986) Suppression of the *E.coli* dnaA46 mutation by amplification of the groES and groEL genes. Mol. Gen. Genet., 202: 435-445.

Fayet, O, Ziegelhoffer, T and Georgopoulos, C, (1989) The groES and groEL heat shock gene products of *Escherichia coli* are essential for bacterial growth at all temperatures. J. Bacteriol., 171: 1379-1385.

Flynn, GC, Chappell, TG, Rothman, JE, (1989) Peptide binding and release by proteins implicated as catalysts of protein assembly. Science, 245: 385-390.

Friedman, DE, Olson, ER, Georgopoulos, C, Tilly, K, Herskowitz, I and Banuett, F, (1984) Interactions of bacteriophage and host macromolecules in the growth of bacteriophage lambda. Microbiol. Rev., 48: 299-325.

Georgopoulos, CP and Eisen, H, (1974) J. Supramol. Struct., 2: 349-359.

Georgopoulos, C, Ang, D, Liberek, K and Zylicz, M, (1990) In Morimoto, R, Tissieres, A and Georgopoulos, C, eds, Cold Spring Harbor Laboratory, Cold Spring Harbor, N.Y., pp. 191-221.

Goloubinoff, P, Christeller, JT, Gatenby, AA and Lorimer, GH, (1989) Reconstitution of active dimeric ribulose bisphosphate carboxylase from an

52

unfolded state depends on two chaperonin proteins and Mg-ATP. Nature, 342: 884-889.

Gross, CA, Straus, DB, Erickson, JW and Yura, T, (1990) In Morimoto, R, Tissieres, A and Georgopoulos, C, eds, Cold Spring Harbor Laboratory, Cold Spring Harbor, N.Y., pp. 167-189.

Hemmingsen, SM, Woolford, C, van der Vies, SM, Tilly, K, Dennis, DT, Georgopoulos, CP, Hendrix, RW and Ellis, RJ, (1988) Homologous plant and bacterial proteins chaperon oligomeric protein asembly. Nature, 333: 330-334.

Jenkins, AJ, Marsh, JB, Oliver, IR and Master, M, (1986) A DNA fragment containing the groE genes can suppress mutations in the E.coli dnaA gene. Mol. Gen. Genet., 202: 446-454.

Kusukawa, N, Yura, T, Ueguchi, C, Akiyama, Y and Ito, K, (1989) Effects of mutations in heat-shock genes groES and groEL on protein export in Escherichia coli. EMBO J., 8: 3517-3521.

Laminet, A, Ziegelhoffer, T, Georgopoulos, C and Plückthun, A, (1990) The Escherichia coli heat shock proteins GroEL and GroES modulate the folding of the beta-lactamase precursor. EMBO J., 9: 2315-2319.

Lecker, S, Lill, R, Ziegelhoffer, T, Georgopoulos, C, Bassford, PJ Jr., Kumamoto, CA and Wickner, W, (1989) Three pure chaperon proteins of E.coli - SecB, trigger factor, and GroEL - form soluble complexes with precursor proteins in vitro. EMBO J., 8: 2703-2709.

Liberek, K, Marszalek, J, Ang, D, Georgopoulos, C and Zylicz, V, submitted.

Lubben, TH, Gatenby, AA, Donaldson, GK, Lorimer, GH and Viitanen, PV, (1990) Identification of a groES-like chaperonin in mitochondria that facilitates protein folding. Proc. Natl. Acad. Sci. USA, 87: 7683-7687.

Martel, R, Cloney, LP, Pelcher, LE and Hemmingsen, SM, Gene, in press.

Mensa-Wilmot, K, Seaby, R, Alfano, C, Wold, MS, Gomes, B and McMacken, R, (1989) Reconstitution of a nine-protein system that initiates bacteriophage lambda DNA replication. J. Biol. Chem., 264: 2853-2861.

Morimoto, R, Tissieres, A and Georgopoulos, C, eds, (1990) "Stress proteins in biology and medicine." Cold Spring Harbor Laboratory, Cold Spring Harbor, N.Y.

Phillips, GJ and Silhavy, TJ, (1990) Heat-shock proteins DnaK and GroEL facilitate export of LacZ hybrid proteins in E. coli. Nature, 344: 882-884.

Reading, DS, Hallberg, RL and Myers, AM, (1989) Characterization of the yeast HSP60 gene coding for a mitochondrial assembly factor. Nature, 337: 655-659.

Rothman, JE, (1989) Polypeptide chain binding proteins: catalysts of protein folding and related processes in cells. Cell, 59: 591-601.

Sadler, I, Chiang, A, Kurihara, T, Rothblatt, J, Way, J and Silver, P, (1989) A yeast gene important for protein assembly into the endoplasmic reticulum and the nucleus has homology to DnaJ, an Escherichia coli heat shock protein. J. Cell Biol., 109: 2665-2675.

Skowyra, D, Georgopoulos, C and Zylicz, M, (1990) The E. coli dnaK gene product, the hsp70 homolog, can reactivate heat-inactivated RNA polymerase in an ATP hydrolysis-dependent manner. Cell, 62: 939-944.

Tilly, K and Georgopoulos, C, (1982) Evidence that the two E.coli groE morphogenetic gene products interact in vivo. J. Bacteriol., 149: 1082-1088.

van Dyk, TK, Gatenby, AA and LaRossa, RA, (1989) Demonstration by genetic suppression of interaction of GroE products with many proteins. Nature, 342: 451-453.

Viitanen, PV, Lubben, TH, Reed, J, Goloubinoff, P, O'Keefe, DP and Lorimer, GH, (1990) Chaperonin-facilitated refolding of ribulosebisphosphate carboxylase and ATP hydrolisis by chaperonin 60 (groEL) are K+ dependent. Biochemistry, 29: 5665-5671.

Waters, MG, Chirico, WJ, Henriquez, R and Blobel, G, (1989) Purification of yeast stress proteins based on their ability to facilitate secretory protein translocation. UCLA Symposia on Molecular and Cellular Biology. Pardue, ML, Feramisco, J and Lindquist, S, eds., Alan R. Liss Inc., N.Y., vol. 96, pp 163-174.

Zylicz, M, LeBowitz, J, McMacken, R and Georgopoulos, C, (1983) The dnaK protein of *E.coli* possesses an ATPase and autophosphorylating activity and is essential in an *in vitro* DNA replication system. Proc. Natl. Acad. Sci. USA, 80: 6431-6435.

Zylicz, M, Ang, D, Liberek, K and Georgopoulos, C, (1989) Initiation of lambda DNA replication with purified host- and bacteriophage-encoded proteins: the role of the dnaK, dnaJ and grpE heat shock proteins. EMBO J., 8: 1601-1608.

E.coli Mutants Lacking the *dnaK* Heat Shock Gene: Identification of Cellular Defects and Analysis of Suppressor Mutations

B. Bukau, G.C. Walker*
Zentrum für Molekulare Biologie
der Universität Heidelberg
INF 282
6900 Heidelberg
FRG

Introduction

We investigated the role of the DnaK (HSP70) chaperone in metabolism of *Escherichia coli* by analyzing cellular defects caused by deletion of the *dnaK* gene (Δ*dnaK52*). Δ*dnaK52* mutants are cold sensitive as well as temperature sensitive and thus possess a very narrow temperature range for growth. At intermediate (30°C) temperature, Δ*dnaK52* mutants display severe defects in major cellular processes such as cell division, chromosome segregation, replication of low copy number plasmids and regulation of heat shock gene expression that lead to poor growth and genetic instability of the cells. These results indicate important functions of DnaK in cellular metabolism of *E. coli* at a wide range of growth temperatures.

In an effort to learn more about the roles of DnaK in normal metabolism we analyzed secondary mutations (*sidB*) that suppress the growth defects of Δ*dnaK52* mutants at 30°C and also permit growth at low temperature. The four members of this class of suppressor mutations that we analyzed map within the *rpoH* gene, which encodes the heat shock-specific sigma subunit (σ^{32}) of RNA polymerase, and cause downregulation of expression of heat shock genes. These findings suggest that the physiologically most significant function of DnaK in metabolism of unstressed cells is its function in heat shock gene regulation.

HSP70 proteins are molecular chaperones that are involved in protein folding as well as dissassembly of protein complexes (Lindquist and Craig, 1988; Morimoto et al., 1990). So far there is little known about

---

* Department of Biology, Massachusetts Institute of Technology, Cambridge, MA 02139, USA.

the cellular processes that require chaperone functions of HSP70 proteins. In eucaryotes, HSP70 proteins and their cognates are required e.g., for secretion of proteins, uncoating of clathrin-coated vesicles, and degradation of proteins in lysosomes (Chiang et al., 1989; Lindquist and Craig, 1988; Morimoto et al., 1990). In E. coli, a single HSP70 member exists, the DnaK heat shock protein (Bardwell and Craig, 1984). Analysis of mutants that carry missense mutations in the dnaK gene revealed that DnaK is essential for initiation of λ DNA replication at all temperatures (Echols, 1986; Libereck et al., 1988) as well as for cell growth at high temperatures (Itikawa and Ryu, 1979; Morimoto et al., 1990). A shift of dnaK missense mutants to nonpermissive temperatures causes rapid cessation of synthesis of RNA and DNA (Itikawa and Ryu, 1979; Paek and Walker, 1987; Sakakibara, 1988). Furthermore, dnaK mutants are defective in degradation of abnormal proteins (Strauss et al., 1988) and of the highly unstable σ^{32} protein (Grossman et al., 1987; Strauss et al., 1987; Tilly et al., 1989), the rpoH gene product, which is the heat shock-specific sigma subunit of RNA polymerase. Partial stabilization of σ^{32} in dnaK mutants contributes to increased expression of heat shock genes at 30°C (Paek and Walker, 1987) and to defects in recovery from the heat shock response (Paek and Walker, 1987; Tilly et al., 1983). The phenotypes resulting from dnaK missense mutations are similar to those resulting from mutations in dnaJ and grpE heat shock genes (Morimoto et al., 1990) and there is increasing evidence that DnaK, DnaJ and GrpE heat shock proteins interact with each other to form functionally active complexes in vivo (Morimoto et al., 1990).

We were interested in analyzing further the cellular functions of DnaK in metabolism of unstressed as well as of stressed E. coli cells. Our approach was to delete the dnaK gene (ΔdnaK52), to analyze the cellular defects of ΔdnaK52 mutants at a variety of temperatures, and then to isolate and characterize secondary mutations that suppress these cellular defects. The results of these studies were published recently (Bukau and Walker, 1989a; Bukau and Walker, 1989b; Bukau and Walker, in press; Bukau et al., 1989) and are reviewed here.

Analysis of cellular defects of ΔdnaK52 mutants

DnaK is not absolutely required for growth at 30°C: the ΔdnaK52::Cm[r] allele was constructed in vitro by deleting the entire promoter region as well as 933 bp of the coding sequence of the dnaK gene and replacing it with a chloramphenicol resistance gene (Paek and Walker, 1987). The ΔdnaK52::Cm[r] allele could be crossed at normal frequency into the chromosome of wild type cells at 30°C thereby replacing the dnaK[+] gene. The resulting ΔdnaK52 mutants completely lack DnaK protein and also contain low cellular concentrations of DnaJ protein (Sell et al., 1990) that is encoded for by the promoter-distal dnaJ gene of the dnaKJ operon. Thus, the DnaK protein is not absolutely required

for growth of *E. coli* at 30°C. However, preliminary experiments
indicated that growth conditions have a significant impact on the
cellular requirements for DnaK at 30°C which are manifested as a
difficulty in transducing the Δ*dnaK52*::Cm^r allele.

Δ*dnaK52 mutants possess multiple cellular defects at 30°C:* although
at 30°C the Δ*dnaK52*::Cm^r allele could be transduced at high
frequency into wild type cells, the transductants grew slowly and
possessed a low viability which decreased further with prolonged
incubation (Bukau and Walker, 1989a). When kept on LB agar plates
at 30°C for about 7 days, Δ*dnaK52* transductants completely lost their
viability; they could be kept alive only if they were constantly transferred
to fresh agar plates, indicating that Δ*dnaK52* mutants are sensitive to
nutrient starvation or drying conditions. Furthermore, prolonged
incubation of the transduction plates led to the frequent appearance

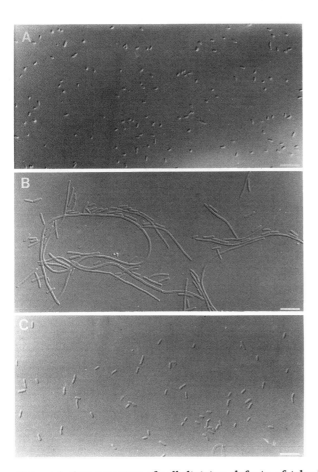

Figure 1. Suppression of cell division defects of Δdnak52 mutants by sidB1
mutation. Wild type cells (MC4100) (A), ΔdnaK52::Cm^r transductants (B), and
ΔdnaK52 sidB1 mutants (C) after growth at 30°C. Bars 10 μm (Bukau and
Walker, 1991)

58

of papillae in the ΔdnaK52 transductant colonies indicating genetic instability of the cells (see below).

Several severe cellular defects of ΔdnaK52 mutants at 30°C were identified (Bukau and Walker, 1989a; Bukau and Walker, 1989b). Microscopic observation of the cells revealed defects in cell division at the level of septation which led to formation of unseptated cell filaments of various lengths (Fig. 1). Interestingly, cell division defects can efficiently be suppressed by overproduction of the FtsZ protein, a key component of the regular cell division machinery required for initiation of septation. Defects in cell division at intermediate temperatures is a general phenotype of ΔdnaK52 mutants since it occurred in all genetic backgrounds tested. The role for DnaK in cell division remains unclear.

Previous reports had indicated involvement of DnaK in DNA replication after heat shock (Itikawa and Ryu, 1979; Paek and Walker, 1987; Sakakibara, 1988). We therefore tested the hypothesis that DnaK also plays a role in DNA metabolism in unstressed cells at 30°C. In fact, ΔdnaK52 mutants possess defects in segregation of chromosomes (Bukau and Walker, 1989b). Fluorescence microscopic analysis revealed that chromosomes were frequently lacking at peripheries of cell filaments of ΔdnaK52 mutants and clustered at other locations (Fig. 2). In other parts of the cell filaments, chromosomes were apparently normally distributed and they were also present in most of the small

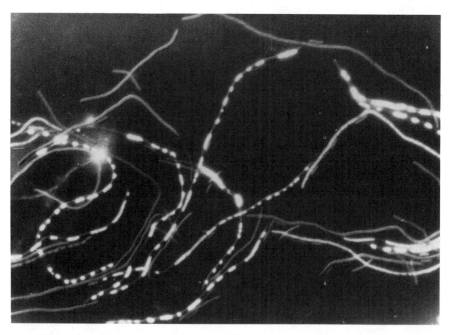

Figure 2. Chromosome segregation defects of ΔdnaK52 mutants at 30°C. Logarithmically growing ΔdnaK52 mutants were stained with DNA binding dye 33342 (Hoechst) and subjected to fluorescence microscopic analysis.

cells found in populations of ΔdnaK52 mutants. Chromosome segregation defects result from defects in replication and/or proper segregation of DNA (Hirota et al., 1968) and therefore these phenotypes of ΔdnaK52 mutants indicate important, although not absolutely essential, direct or indirect functions for DnaK in replication and/or partitioning of chromosomes at 30°C. ΔdnaK52 mutants were found to have a threefold lower rate of DNA synthesis than wild type cells which is consistant with a role of DnaK in DNA replication. However, presence of multiple cellular defects renders identification of defects in chromosomal DNA synthesis of ΔdnaK52 mutants difficult.

As an alternative approach to identifying a role for DnaK in DNA synthesis, we analyzed the stability of low-copy-number plasmids that are experimentally easily accessible and that are widely used as model systems to study replication as well as partitioning of DNA to daughter cells during cell division. We used a derivative of bacteriophage P1, λ-mini-P1 (Funnel, 1988; Sternberg and Austin, 1983), which replicates as a plasmid with a copy number similar to that of the host chromosome. Replication and partitioning of λ-mini-P1 depend on the replication (rep) and partitioning (par) functions of P1. We introduced the ΔdnaK52 allele at 30°C into wild type cells carrying λ-mini-P1 as lysogen and determined plasmid stability (Bukau and Walker, 1989b). λ-mini-P1 was rapidly lost in ΔdnaK52 mutant cells at a rate about 18 times higher than in dnaK⁺ cells. We also found that dnaJ259 mutants were unable to stably maintain λ-mini-P1 plasmids as well. These results indicate functions for DnaK and DnaJ proteins in replication and/or partitioning of this plasmid. To distinguish between these two possibilities, we determined the effect of the ΔdnaK52 mutation on stability of Δpar mini-P1 plasmids (pSP102) (Tilly and Yarmolinsky, 1989) which carry the replication functions but lack the partitioning functions of P1. Introduction of the ΔdnaK52 allele into cells that carry pSP102 plasmids caused a greater-than-100-fold increase in the rate of loss of this plasmid (Bukau and Walker, 1989b). Thus, DnaK is involved at the level of replication of these plasmids. Similar data have recently been obtained by Tilly and Yarmolinsky (Tilly and Yarmolinsky, 1989).

ΔdnaK52 mutants are cold sensitive as well as temperature sensitive for growth: to determine the permissive growth temperature spectrum of *E. coli* cells lacking the DnaK function we tested whether it is possible to transduce the chromosomal ΔdnaK52::Cmʳ allele into wild type cells at a variety of temperatures. At high temperatures above 37°C, the *dnaK* gene could not be deleted, as expected from the previously known temperature sensitivity of *dnaK* missense mutations. Unexpectedly, however, we were also unable to delete *dnaK* at temperatures below 20°C. Furthermore, ΔdnaK52::Cmʳ transductants isolated at 30°C were unable to grow at both, high and low temperatures (Fig. 3). These results indicate essential functions of DnaK at the extreme ends of the growth temperature spectrum.

Taken together, we have shown that ΔdnaK52 mutants possess a

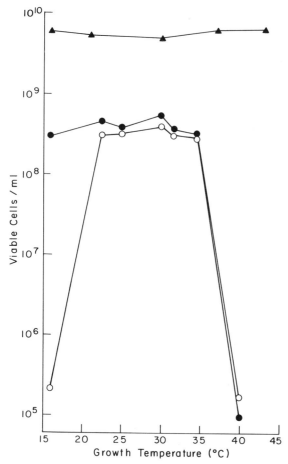

Figure 3. Effects of the sidB1 mutation on growth temperature spectrum of ΔdnaK52 mutants. Wild type cells (Δ), ΔdnaK52 transductants (O) and ΔdnaK52 sidB1 mutants (O) were grown at 30°C, dilutions plated on LB agar plates that were incubated at the indicated temperatures, and the number of viable cells were determined (Bukau and Walker, 1991).

narrow temperature range for growth and have multiple severe cellular defects at permissive intermediate temperature. These defects are likely to be responsible for low viability and genetic instability of ΔdnaK52 mutants at 30°C.

We considered the possibility that the phenotypes of ΔdnaK52 mutants described here were due to reduction in synthesis of DnaJ. Recent work of C. Georgopoulos and coworkers (Sell et al., 1990) revealed that dnaJ null mutations cause cell division defects and growth defects at 30°C as well as temperature sensitivity of growth, although these defects are less severe than in ΔdnaK52 mutants. The fact that ΔdnaK52 mutants contain very little DnaJ might therefore contribute to the observed cellular defects of these mutants. However, lysogenization of ΔdnaK52 mutants with λ phages carrying dnaK⁺ but not *dnaJ*⁺ complemented major cellular defects such as cell division defects, slow growth and plasmid instability indicating that it is lack of DnaK rather than reduction of DnaJ that is primarily responsible for the observed defects of Δ*dnaK52* mutants.

In summary, DnaK does not only possess essential functions in metabolism of stressed cells, but important functions in normal metabolism of unstressed cells as well although we cannot exclude that DnaK is indirectly rather than directly involved in some of these processes.

Analysis of suppressors of cellular defects of Δ*dnaK52* mutants

Isolation of sidB suppressor mutations: We took advantage of the fact that prolonged incubation of Δ*dnaK52* transductant colonies isolated at 30°C led to frequent appearance of papillae. After about 7 days at 30°C only cells from papillae retained their viability and could be restreaked well. They were thus suppressed in the cellular defects of Δ*dnaK52* mutants which cause their low viability at 30°C. Moreover, cells contained within 5 papillae that we chosed for further analysis grew considerably faster than newly isolated Δ*dnaK52* transductants. Four of the five isolates had generation times that were similar to that of wild type cells, while one isolate had a generation time that was between that of wild type cells and newly isolated Δ*dnaK52* transductants. Here we focus on the four isolates BB1553, BB1554, BB1556, BB1557 that showed the greatest improvement of growth as judged by their generation times. Based on their similar growth behaviour at 30°C and their map position (see below) we refer to the suppressor mutations present in these four isolates as *sidB1*, *sidB2*, *sidB3*, and *sidB4*, respectively (suppressor at intermediate temperature of Δ*dnaK52*) (Bukau and Walker, 1991).

*sidB mutations suppress cellular defects of Δ*dnaK52* mutants at 30°C*: presence of *sidB* mutations in Δ*dnaK52* mutants suppressed several major cellular defects at 30°C (Bukau and Walker, 1991). Microscopic analysis revealed that the cell division defects of Δ*dnaK52* mutants were very efficiently suppressed in the *sidB1* Δ*dnaK52* isolate (Fig. 1) and in the other three *sidB* Δ*dnaK52* isolates as well. Furthermore, fluorescent staining of the chromosomes of cells of the *sidB1* Δ*dnaK52* isolate revealed that virtually all cells contained nucleoids, demonstrating that the chromosome segregation defects of Δ*dnaK52* mutants were efficiently suppressed in this isolate. However, methodological limitations do not allow us to completely exclude the possibility that a small fraction of the cell population might be anucleated.

We then tested whether presence of the *sidB1* mutation suppresses defects of Δ*dnaK52* mutants in replication of λ-mini-P1. However, stability tests revealed that λ-mini-P1 is lost in *sidB1* Δ*dnaK52* mutants at a high rate (2.6% plasmid loss/generation) that is only

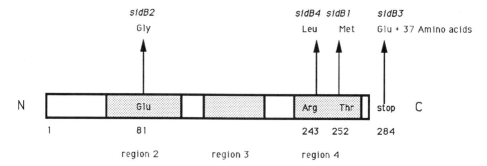

Figure 4. Localization of sidB mutations within σ³². A schematic representation of σ³² protein is shown. Darkened boxes represent regions of homology of σ³² to other sigma factors (Bukau and Walker, 1991).

slightly lower than the rate of loss of this plasmid in ΔdnaK52 mutants lacking any suppressor. These data indicate that presence of the sidB1 mutation fails to efficiently suppress defects in replication of λ-mini-P1 in ΔdnaK52 mutants.

sidB mutations suppress cold sensitivity but not temperature sensitivity of ΔdnaK52 mutants: besides suppression of major cellular defects at intermediate (30°C) temperature, sidB mutations also suppress cold sensitivity of ΔdnaK52 mutants (Bukau and Walker, 1991). However, all four ΔdnaK52 sidB mutants analyzed remained temperature sensitive (at 42°C) for growth. The growth temperature range for sidB1 ΔdnaK52 cells is shown in Figure 3.

Attempts to isolate at 30°C additional suppressors of the sid class that allowed ΔdnaK52 mutants to grow also at high temperatures failed. However, by selection for growth at 42°C of ΔdnaK52 sidB1 mutants we were able to obtain suppressors, named std, that allowed growth of the cells at high (42°C) temperature. Although we cannot exclude the possibility that suppressors exist that suppress growth defects of ΔdnaK52 mutants at intermediate temperature as well as cold sensitivity and temperature sensitivity, our data suggest that multiple mutational events are required to allow ΔdnaK52 mutants to grow within the entire growth temperature range of E. coli. These data provide genetic evidence that at least some cellular requirements for DnaK at high temperature are different from those at intermediate temperature.

sidB mutations are alleles of rpoH: the locations of sidB mutations on the E. coli chromosome were determined using Hfr crosses and P1 transductions. The sidB1, sidB2, sidB3, and sidB4 mutations all were located at around 75-76 min on the genetic map and appeared sufficient to suppress the cellular defects of ΔdnaK52 mutants as described earlier. The *rpoH* gene, which encodes the activator of heat shock gene expression, σ^{32}, has similar linkage to Tn*10* insertions in the 75-76 min region of the chromosome as the *sidB* mutations. This proximity of the map positions of *rpoH* and *sidB*, as well as additional physiological evidence (see below) led us to consider that *sidB* mutations are mutated alleles of *rpoH*. In fact, multiple copies of *rpoH*⁺ complement major *sidB* suppressor phenotypes at 30°C suggesting that *sidB* mutations are recessive mutant alleles of *rpoH*. Furthermore, in marker rescue experiments the *sidB1* mutation could be rescued by plasmids encoding functionally inactive forms of the *rpoH* gene product. These data provide good genetic evidence that at least *sidB1* maps within *rpoH*.

To confirm our hypothesis that *sidB* mutations are alleles of *rpoH* we amplified the *rpoH* genes of these mutants using PCR and then

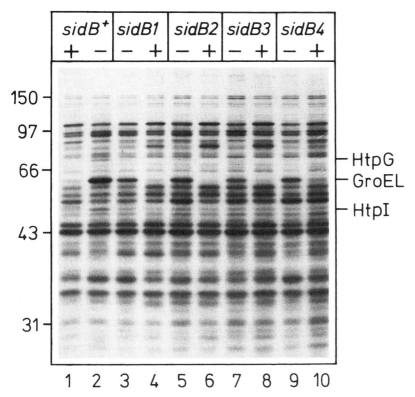

Figure 5. sidB mutations decrease heat shock gene expression at 30°C. Cells of sidB mutants carrying dnaK⁺ (+) or ΔdnaK52 (-) alleles were grown at 30°C, pulse-labeled with ³⁵S-methionine, and the proteins separated on SDS-PAGE and visualized by autoradiography. Migration positions of some major heat shock proteins (right side) and of molecular weight standards (left side) are indicated. Strains used are MC4100 (lane 1), fresh ΔdnaK52 transductants lacking any suppressor mutation (lane 2), BB1553 (lane 3), BB1699 (lane 4), BB1554 (lane 5), BB1700 (lane 6), BB1556 (lane 7), BB1701 (lane 8), BB1557 (lane 9), and BB1702 (lane 10) (Bukau and Walker, 1991).

determined the DNA sequences of the amplified genes (Bukau and Walker, 1991). All four *sidB* mutations analyzed were located within different regions of the coding sequence of the *rpoH* gene (Fig. 4). The *sidB1, sidB2,* and *sidB4* mutations alter conserved amino acids present in three regions of σ^{32} that are highly conserved among sigma factors. The *sidB3* mutation is a T to G transversion in the TAA stop codon of *rpoH* leading to addition of additional 38 amino acids to the C-terminus of σ^{32}.

sidB mutations decrease heat shock gene expression at 30°C as well as after heat shock: sidB mutations affect activity of σ^{32} in transcription of heat shock genes. We determined the rate of synthesis at 30°C of heat shock proteins in both Δ*dnaK52* mutants and *dnaK⁺* cells (Fig. 5). All four *sidB* Δ*dnaK52* mutants were clearly reduced in synthesis of GroEL and other major heat shock proteins as compared to Δ*dnaK52* mutants lacking any suppressor. GroEL synthesis is about 2-4 fold reduced in the *sidB* Δ*dnaK52* mutants as compared to *sidB⁺* Δ*dnaK52* mutants. 2-D-gel electrophoresis of labelled cell extracts of the *sidB1* Δ*dnaK52* mutant revealed that expression of other known heat shock proteins is decreased in *sidB* mutants as well.

Presence of *sidB* mutations also causes defects in the heat shock response of *dnaK⁺* cells. *sidB1 dnaK⁺* mutants and *sidB4 dnaK⁺* mutants exhibited a reduced but still detectable heat shock induction whereas *sidB2 dnaK⁺*mutants and *sidB3 dnaK⁺*mutants showed very little induction of major heat shock proteins upon shift to 43.5°C. The presence of *sidB2* and *sidB3* mutations in *dnaK⁺* cells causes temperature sensitivity of growth of the cells. Taken together, these results indicate general functional defects of the mutated σ^{32} proteins.

Cellular concentration at 30°C and apparent mol. wt of σ^{32} proteins of sidB mutants: to test whether the defects of *sidB* mutants in heat shock gene expression are caused by alterations in the cellular concentration of the mutant σ^{32} proteins we performed Western blots of cell extracts of the mutants using σ^{32}-specific antiserum (Bukau and Walker, 1991). Presence of *sidB1, sidB2* and *sidB4* mutations in Δ*dnaK52* and *dnaK⁺* cells did not significantly alter the cellular concentration of σ^{32} as compared to the respective *sidB⁺* cells indicating that these three *sidB* mutations affect the activity rather than the cellular concentration of σ^{32}. However, we cannot exclude the possibility that small alterations in the cellular concentration of σ^{32} not detected by Western might contribute to the observed defects in heat shock gene expression. In contrast, the *sidB3* mutation not only caused an increase in mol. wt of the σ^{32} protein of this mutant as a consequence of the addition of 38 amino acids to the C terminus but also caused a several fold decrease in cellular concentration of σ^{32} in both Δ*dnaK52* and *dnaK⁺* cells as compared to the respective *sidB⁺* cells. These changes in concentration of σ^{32} might be sufficient to account for the observed defects of *sidB3* mutants in heat hock gene expression.

Conclusions

Our analysis of ΔdnaK52 mutants (Bukau and Walker, 1989a; Bukau and Walker, 1989b) revealed that the DnaK function is required for normal growth within the entire growth temperature range of *E. coli.* While DnaK is essential for growth at high as well as at low temperatures, it is not absolute essential for growth at intermediate temperature although ΔdnaK52 mutants possess multiple severe defects at that temperature. The requirement for DnaK for growth at temperatures from 16°C to over 42°C can be circumvented by suppressor mutations and is therefore not absolute. Suppressor mutations appear to improve growth of ΔdnaK52 mutants in a temperature dependent fashion (Bukau and Walker, 1991) which indicates that at least some of the functions of DnaK change with temperature. Finally, the finding that *sidB* mutations improve growth of ΔdnaK52 mutants at low and intermediate temperatures by reducing expression of the remaining heat shock proteins suggests that the physiologically most significant function of DnaK in metabolism of unstressed cells is its function in heat shock gene regulation.

Acknowledgments

We thank members of the Walker lab for helpful discussions and H. Bujard for generous support which allowed completion of this work. This study was supported by Public Health Service grant GM28988 from the National Institute of General Medical Sciences and by a fellowship of the Deutscher Akademischer Austauschdienst to B.B.

References

Bardwell, JCA and Craig, EA, (1984) Major heat shock gene of *Drosophila* and the *Escherichia coli* heat-inducible *dnaK* gene are homologous. Proc. Natl. Acad. Sci. USA, 81: 848-852.

Bukau, B and Walker, GC, (1989a) Cellular defects caused by deletion of the *Escherichia coli dnaK* gene indicate roles for heat shock protein in normal metabolism. J. Bacteriol., 171: 2337-2346.

Bukau, B and Walker, GC, (1989b) ΔdnaK52 mutants of *Escherichia coli* have defects in chromosome segregation and plasmid maintenance at normal growth temperatures. J. Bacteriol., 171: 6030-6038.

Bukau, B and Walker, GC, Mutations altering heat shock specific subunit of RNA polymerase suppress major cellular defects of *E. coli* mutants lacking the DnaK chaperone. EMBO J., in press.

Bukau, B, Donnelly, CE and Walker, GC, (1989) DnaK and GroE proteins play roles in *E. coli* metabolism at low and intermediate temperatures as well as at high temperatures, p. 27-36. *In* Pardue, ML, Feramisco, JR and Lindquist, S, eds., Stress Induced Proteins. Alan. R. Liss, Inc., New York.

Chiang, H-L, Terlecky, SR, Plant, CP and Dice, JF, (1989) A role for a 70-kilodalton heat shock protein in lysosomal degradation of intracellular proteins. Science, 246: 382-385.

Echols, H, (1986) Multiple DNA-protein interactions governing high-precision DNA transactions. Science, 233: 1050-1056.

66

Funnell, BE, (1988) Mini-P1 plasmid partitioning: excess ParB protein destabilizes plasmids containing the centromere *parS*. J. Bacteriol., 170: 954-960.

Grossman, AD, Straus, DB, Walter, WA and Gross, CA (1987) σ^{32} synthesis can regulate the synthesis of heat shock proteins in *Escherichia coli.* Genes Dev., 1: 179-184.

Hirota, Y, Ryter, A and Jacob, F, (1968) Thermosensitive mutants of *E. coli* affected in the processes of DNA synthesis and cellular division. Cold Spring Harbor Symp. Quant. Biol., 33: 677-694.

Itikawa, H and Ryu, J-I, (1979) Isolation and characterization of a temperature-sensitive *dnaK* mutant of *Escherichia coli* B. J. Bacteriol., 138: 339-344.

Liberek, K, Georgopoulos, C and Zylicz, M, (1988) Role of the *Escherichia coli* DnaK and DnaJ heat shock proteins in the initiation of bacteriophage λ DNA replication. Proc. Natl. Acad. Sci. USA, 85: 6632-6636.

Lindquist, S and Craig, EA, (1988) The heat-shock proteins. Annu. Rev. Genet. 22: 631-677.

Morimoto, R I, Tissieres, A and Georgopoulos, C, eds. (1990) Stress Proteins in Biology and Medicine. Cold Spring Harbor monograph Series Vol. 19, Cold Spring Harbor Press, Cold Spring Harbor, N.Y.

Paek, K-H and Walker, GC, (1987) *Escherichia coli dnaK* null mutants are inviable at high temperature. J. Bacteriol., 169: 283-290.

Sakakibara, Y, (1988) The *dnaK* gene of *Escherichia coli* functions in initiation of chromosome replication. J. Bacteriol., 170: 972-979.

Sell, SM, Eisen, C, Ang, D, Zylicz, M and Georgopoulos, C, (1990) Isolation and characterization of *dnaJ* null mutants of *Escherichia coli.* J. Bacteriol., 172: 4827-4835.

Sternberg, N and Austin, S, (1983) Isolation and characterization of P1 minireplicons, λ-P1:5R and λ-P1:5L. J. Bacteriol., 153: 800-812.

Straus, DB, Walter, WA and Gross, CA (1987) The heat shock response of *E. coli* is regulated by changes in the concentration of σ^{32}. Nature, 329: 348-351.

Straus, DB, Walter, WA and Gross, CA, (1988) *Escherichia coli* heat shock gene mutants are defective in proteolysis. Genes Dev., 2: 1851-1858.

Tilly, K, McKittrick, N, Zylicz, M and Georgopoulos, C, (1983) The *dnaK* protein modulates the heat shock response of *Escherichia coli.* Cell, 34: 641-646.

Tilly, K, Spence, J and Georgopoulos, C, (1989) Modulation of stability of the *Escherichia coli* heat shock regulatory factor σ^{32}. J. Bacteriol., 171: 1585-1589.

Tilly, K and Yarmolinsky, M, (1989) Participation of *Escherichia coli* heat shock proteins DnaK, DnaJ, and GrpE in P1 plasmid replication. J. Bacteriol., 171: 6025-6029.

Zylicz, M, Ang, D, Liberek, K and Georgopoulos, C, (1989) Initiation of lambda DNA replication with purified host- and bacteriophage-encoded proteins: the role of the dnaK, dnaJ and grpE heat shock proteins. EMBO J., 8: 1601-1608.

DnaJ and DnaK Heat Shock Proteins Activate Sequence Specific DNA Binding by RepA

S. Wickner, J. Hoskins[#], K. McKenney[#]
Laboratory of Molecular Biology
National Cancer Institute
National Institutes of Health
Bethesda, MD 20892
USA

Introduction

Two families of heat shock proteins, HSP60s and HSP70s, have been implicated in mediating the folding and unfolding of proteins and the assembly and disassembly of oligomeric protein structures (for review see Rothman, 1989). The mechanisms by which heat shock proteins (HSPs) catalyze these reactions have not been elucidated and are the object of great interest, particularly in view of their ubiquity and high conservation during evolution (for reviews see Neidhardt and Van Bogelen, 1987; Lindquist and Craig, 1988; Munro and Pelham, 1986). DnaK is the *E. coli* HSP70 homolog. It is involved in chromosome segregation at normal temperatures (Bukau and Walker, 1989) and is essential for growth (Itikawa and Ryu, 1979; Sato and Uchida, 1978) and DNA replication (Sakakibara, 1988) at high temperatures. It is also involved in replication of phage l and plasmids mini-F and mini-P1. Two other heat shock proteins, DnaJ and GrpE are also involved in the replication of these three replicons (Georgopoulos, 1977; Sunshine et al., 1977; Saito and Uchida, 1977; Zylicz et al., 1989; Mensa-Wilmot et al., 1989; Tilly and Yarmolinsky, 1989; Bukau and Walker, 1989; Wickner, 1990; Kawasaki et al., 1990; Ezaki et al., 1989).

We have been studying the *in vitro* replication of plasmid mini-P1. The structure of the P1 replicon is diagramed in Figure 1. The origin of replication is flanked on one end by two 9 bp DnaA binding sites and on the other end by five 19 bp RepA binding sites. The P1 *repA* gene, which codes for the initiator protein, is beside the origin with its promoter embedded in the RepA binding sites. A replication control region downstream from the *repA* gene contains nine RepA binding

---

[#] Center for Advanced Research in Biotechnology, National Institute of Standards and Technology, Rockville, MD 20850, USA.

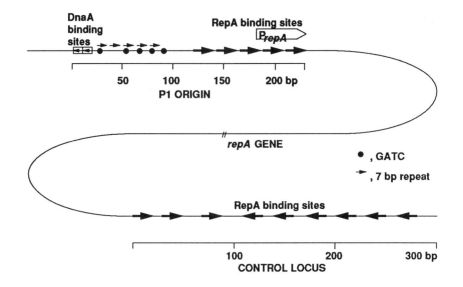

Figure 1. Structure of the P1 plasmid replicon (Abeles et al., 1984).

sites and is required for the plasmid to be maintained at a copy number of 1 or 2. Thus RepA, by binding to DNA, is involved in three functions: initiation of replication, regulation of its own synthesis and regulation of replication.

Our previous studies showed that DNA replication of plasmid DNA carrying the P1 origin is catalyzed by crude protein fractions of *E. coli*. *In vitro* replication requires, in addition to RepA, the *E. coli* DnaA initiator, DnaB helicase, DnaC, DnaG primase, DNA polymerase III holoenzyme, DNA gyrase, and RNA polymerase (Wickner and Chattora, 1987). We also found that three heat shock proteins are required, DnaK, DnaJ and GrpE (Wickner, 1990). We discovered that RepA exists in a stable protein complex with DnaJ, composed of a dimer each of RepA and DnaJ. This complex can be isolated from cell extracts or reconstituted from purified proteins (Wickner, 1990).

DnaK and DnaJ

We have examined DNA binding by RepA. Although RepA binds specifically to P1 origin DNA (Abeles, 1986), about a 100-fold molar excess of RepA dimers to binding sites (or about a 500-fold excess of RepA dimers to DNA fragments) is required for retention of *ori*P1 DNA on nitrocellulose filters and also for retardation of *ori*P1 DNA on gel electrophoresis (Fig. 2). These results suggested that much of the protein in the preparation was inactive or there was a low proportion of bound versus free protein at equilibrium in the reactions. Since we discovered that DnaJ and RepA existed as a complex, we asked

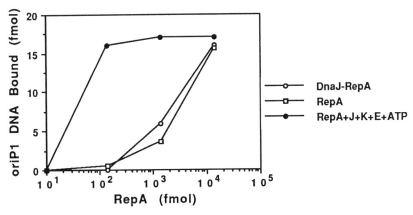

Figure 2. *oriP*1 DNA binding by RepA alone, DnaJ-RepA complex alone and RepA in combination with DnaJ (30 ng), DnaK (100 ng) and ATP. DNA binding was measured by retention on nitrocellulose filters. Reaction mixtures (20 ml) contained 20 mM Tris·HCl, pH 7.5, 40 mM KCl, 100 mM NaCl, 10 mM MgCl$_2$, 0.1 mM EDTA, 1 mM dithiothreitol, 50 µM ATP, 50 µg/ml bovine serum albumin, 50 µg/ml calf thymus DNA, 5 ng of labeled *oriP*1 DNA fragment, DnaK, DnaJ, and RepA. Proteins were prepared as previously described (Wickner, 1990). *oriP*1 DNA was labeled by using a fill in reaction with [α^{32}P]dATP (3000 Ci/mM), unlabeled dGTP, dCTP, dTTP and T7 DNA polymerase (Sequenase). Incubations were for 30 min at room temperature. Reaction mixtures were then filtered through 0.45 mµ nitrocellulose filters. Filters were washed with 2 ml of 20 mM Tris·HCl, pH 7.5, 1 mM EDTA, 1 mM dithiothreitol, 10 mM MgCl$_2$, 40 mM KCl and 100 mM NaCl, dried, and radioactivity was measured by liquid scintillation counting.

whether the complex was more active for DNA binding than RepA alone. Our experiments showed that DnaJ-RepA bound as poorly as RepA alone (Fig. 2).

We then asked whether the other two heat shock proteins required for P1 replication would activate RepA. We found that the combination of DnaJ, DnaK, GrpE, and ATP with RepA increased *oriP*1 DNA binding by about 100-fold (Fig. 2). DnaJ, DnaK and ATP were required; GrpE was dispensable (Fig. 3 and 4).

Unlike many other site specific DNA binding proteins, *oriP*1 DNA binding by RepA in the presence of DnaJ, DnaK, and ATP proceeded linearly at room temperature for 20 min and there was very little DNA binding when the reactions were carried out at 0°C (Fig. 5).

Using temperature shift experiments, we were able to separate the reaction into two steps. The first reaction was for 15 min at room temperature and the second reaction was for 15 min at 0°C. When DnaJ, DnaK, RepA, and ATP were incubated in the first reaction, DNA binding occurred upon the addition of *oriP*1 DNA during the second reaction at 0°C (Table 1). A time course of each step revealed that the first reaction in the absence of DNA proceeded linearly for 20 min and the second reaction at 0°C was complete by 30 sec (Fig. 6). These experiments show that DnaJ and DnaK, in an ATP-dependent reaction,

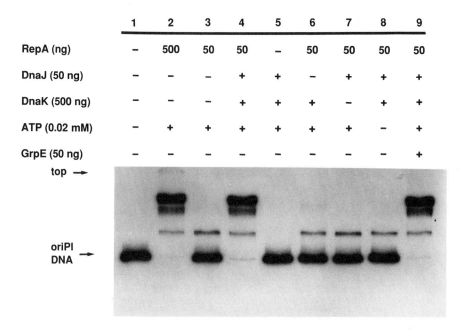

Figure 3. *ori*P1 DNA binding by RepA as measured by gel retardation. DNA binding reactions were carried out as described in Figure 2, but with 25 mM ATP, 5 μg/ml calf thymus DNA, and the proteins in amounts as indicated. Reactions were analyzed by 5% polyacrylamide gel electrophoresis in 89 mM Tris borate and 1 mM EDTA, pH 8.8. The gels were fixed for 15 min in 10% acetic acid containing 10% methanol, transferred to Whatman 3MM paper, dried and autoradiographed.

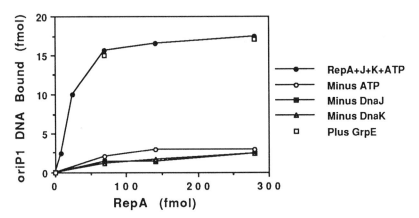

Figure 4. Requirements for activation of RepA for *ori*P1 DNA binding. *ori*P1 DNA binding was measured by retention on nitrocellulose filters as described for Figure 2. Reactions contained RepA, 30 ng of DnaJ, 50 ng of GrpE, and 100 ng of DnaK where indicated.

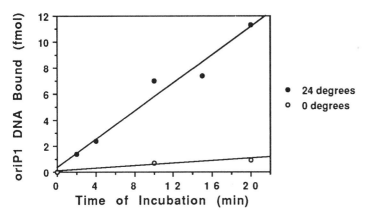

Figure 5. Time course of RepA activation. DNA binding was measured by retention on nitrocellulose filters as described in Figure 2 with 30 ng of RepA, 30 ng of DnaJ, and 200 ng of DnaK. Reactions were for the times indicated at 0°C or at room temperature.

activate RepA in the absence of DNA such that it is capable of rapidly and specifically binding to ori P1 DNA.

Activation of the DnaJ-RepA complex also required ATP and incubation at room temperature for 20 min. We had previously observed that DnaJ and RepA complex formation occurred rapidly at 0°. Thus the rate limiting and temperature dependent step in the activation of RepA

Table 1		
Temperature requirement for activation of RepA		
Incubation at 24°C (15 min)	Incubation at 0°C (15 min)	ori P1 DNA bound (%)
Complete	No additions	26
No additions	Complete	<2
Minus ATP	+ ATP	<2
Minus ori P1 DNA	+ ori P1 DNA	32
Minus DnaJ	+ DnaJ	4
Minus DnaK	+ DnaK	4
Minus RepA	+ RepA	3

DNA binding was measured by retention on nitrocellulose filters. Reaction mixtures for the first incubation at 24° C contained the components described in Figure 2 with the omissions indicated and with 20 ng of RepA, 30 ng of DnaJ, and 100 ng of DnaK. After 15 min, the reactions were chilled on ice, the omitted components were added, and the reactions were continued for 15 min at 0°C.

A.

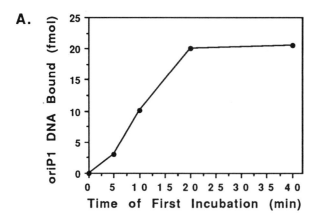

B.

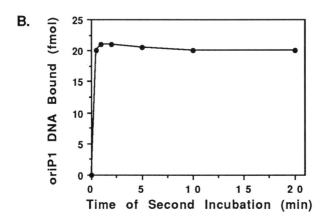

Figure 6. (A) Time course of activation reaction. Reactions as described in Figure 5 were preincubated for the times indicated at 24°C in the absence of DNA. Then at 0°C, *oriP1* DNA was added and bound DNA was measured after 10 min at 0°C. (B) Time course of DNA binding reaction. Reactions as described in Figure 5 were preincubated for 20 min at 24°C in the absence of DNA. Then at 0°C, *oriP1* DNA was added and at the times indicated bound DNA was determined.

is the action of DnaK on the DnaJ-RepA complex.

From titration curves of RepA, DnaJ, and DnaK, with saturating amounts of the other two proteins, we determined that optimal stimulation of RepA requires an equimolar amount of DnaJ and a 20 to 30-fold molar excess of DnaK relative to RepA. This calculation is based on the determinations that RepA and DnaJ are dimers with M_r of 32,000 (Abeles, 1986) and 37,000 (Zilicz et al., 1985), respectively, and DnaK is a monomer with a M_r of 72,000 (Zilicz et al., 1983).

We found that the optimal ATP concentration was about 2 mM. ATP hydrolysis is probably required since the nonhydrolyzable ATP analogue, ATP-γ-S, was a competitive inhibitor. Of the proteins required for RepA activation, only DnaK has an associated ATPase activity (Zilicz et al., 1983). The mixture of DnaJ and RepA had very little effect on DnaK ATPase activity. DnaK also has autophosphorylating activity (Zilicz et al., 1983). There was no effect on phosphorylation of DnaK by DnaJ and RepA. In addition, there was no detectable phosphorylation of DnaJ or RepA.

Conclusions

It has been observed that dnaJ, dnaK, and grpE mutant strains are unable to stably maintain plasmid P1, suggesting that these three proteins are also involved in P1 replication in vivo (Bukau and Walker, 1989; Tilly and Yarmolinsky, 1989). These proteins are also involved in regulation of the repA gene. It has been recently found that repression of the autoregulated repA promoter by RepA is 2 to 10-fold less in dnaJ, dnaK and grpE mutant strains compared to wild-type

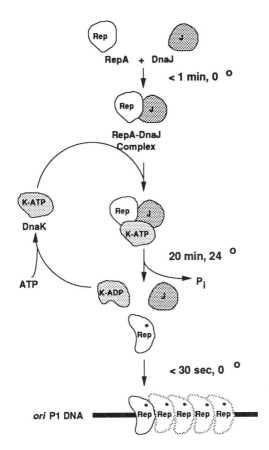

Figure 7. Model for the chaperone role of DnaJ and DnaK in DNA binding by RepA.

strains (Tilly, Sozhamannan and Yarmolinsky, personal communication). Both of these in vivo observations could be explained at the molecular level by our in vitro results that DnaJ and DnaK activate the DNA binding function of RepA. These three proteins are very likely also involved in the regulation of plasmid copy number by the control locus. One remaining puzzle is the biochemical function of GrpE. We have been able to demonstrate a requirement for GrpE in in vitro P1 replication but have not been able to show a requirement for it in the activation of RepA. Possibly GrpE acts indirectly or perhaps it acts at some other stage in P1 replication.

A model for RepA activation is shown in Figure 7. The inactive form of RepA binds tightly and rapidly to DnaJ at 0° to form a 2:2 tetramer (Wickner, 1990). The role of DnaJ may be as a protein cofactor that targets RepA for recognition and activation by DnaK-ATP. ATP hydrolysis is probably used to activate RepA and release DnaJ. The activation step may be a change in conformation of RepA or it may be dissociation of RepA dimers into monomers. RepA alone is found bound to the DNA (our unpublished observation), suggesting that DnaJ and DnaK function as true chaperones in this reaction.

References

Abeles, AL, (1986) P1 plasmid replication. J. Biol. Chem., 261: 3548-3555.

Abeles, AL, Snyder, M and Chattoraj, D, (1984) P1 plasmid replication:replication structure. J. Mol. Biol., 173: 307-324.

Bukau, B and Walker, G, (1989) ΔdnaK52 mutants of Escherichia coli have defects in chromosome segregation and plasmid maintenance at normal growth temperatures. J. Bacteriol., 171: 6030-6038.

Ezaki, B, Ogura, T, Mori, H, Niki, H and Hiraga, S, (1989) Involvement of DnaK protein in mini-F plasmid replication: temperature-sensitive seg mutations are located in the dnaK gene. Mol. Gen. Genet., 218: 183-189.

Georgopoulos, C, (1977) A new bacterial gene (groPC) which affects lambda DNA replication. Mol. Gen. Genet., 151: 35-39.

Itikawa, H and Ryu, J-I, (1979) Isolation and characterization of a temperature-sensitive dnaK mutant of Escherichia coli B. J. Bacteriol., 138: 339-344.

Kawasaki, Y, Wada, C and Yura, T, (1990) Roles of Escherichia coli heat shock proteins DnaK, DnaJ and GrpE in mini-F plasmid replication. Mol. Gen. Genet., 220: 277-282.

Lindquist, S and Craig, EA, (1988) The heat-shock proteins. Ann. Rev. Genet., 22: 631-677.

Mensa-Wilmot, K, Seaby, R, Alfano, C, Wold, MS, Gomes, B and McMacken, R, (1989) Reconstitution of a nine-protein system that initiates bacteriophage lambda DNA replication. J. Biol. Chem., 264: 2853-2861.

Munro, S and Pelham, HRB, (1986) An Hsp70-like protein in the ER: identity with the 78 kd glucose-regulated protein and immunoglobulin heavy chain binding protein. Cell, 46: 291-300.

Neidhardt, F and Van Bogelen, R, (1987) in Escherichia coli and Salmonella typhimurium, Neidhardt, F, Ingraham, J, Low, K, Magasanik, B, Schaechter, M and Umbarger, H eds., American Society for Microbiology, Washington, D.C., pp. 1334-1345.

Rothman, JE, (1989) Polypeptide chain binding proteins: catalysts of protein folding and related processes in cells. Cell, 59: 591-601.

Saito, H and Uchida, H, (1977) Initiation of the DNA replication of bacteriophage lambda in E.coli K12. J. Mol. Biol., 113: 1-25.

Saito, H and Uchida, H, (1978) Organization and expression of the dnaJ and dnaK genes of E.coli K12. Mol. Gen. Genet., 164: 1-8.

Sakakibara, Y, (1988) The *dnaK* gene of *Escherichia coli* functions in initiation of chromosome replication. J. Bacteriol., 170: 972-979.

Sunshine, M, Feiss, M, Stuart, J and Yochem, J, (1977) A new host gene (groPC) necessary for lamda DNA replication. Mol. Gen. Genet., 151: 27-31.

Tilly, K and Yarmolinsky, M, (1989) Participation of *Escherichia coli* heat shock proteins DnaK, DnaJ, and GrpE in P1 plasmid replication. J. Bacteriol., 171: 6025-6029.

Wickner, S and Chattoraj, D, (1987) Replication of mini-P1 plasmid DNA in vitro requires two initiation proteins, encoded by the repA gene of phage P1 and dnaA gene of E.coli. Proc. Natl. Acad. Sci. USA, 84: 3668-3672.

Wickner, S, (1990) Three *Escherichia coli* heat shock proteins are required for P1 plasmid DNA replication: formation of an active complex between E. coli DnaJ protein and the P1 initiator protein. Proc. Natl. Acad. Sci. USA, 87: 2690-2695.

Zylicz, M, Yamamoto, T, McKittrick, N, Sell, S and Georgopoulos, C, (1985) Purification and properties of the dnaJ replicationb protein of E.coli. J. Biol. Sci., 260: 7591-7598.

Zylicz, M, LeBowitz, J, McMacken, R and Georgopoulos, C, (1983) The dnaK protein of E.coli posses an ATPase and autophosphorylating activity and is essential in an in vitro DNA replication system. Proc. Natl. Acad. Sci. USA, 80: 6431-6435.

Zylicz, M, Ang, D, Liberek, K and Georgopoulos, C, (1989) Initiation of lambda DNA replication with purified host- and bacteriophage-encoded proteins: the role of the dnaK, dnaJ and grpE heat shock proteins. EMBO J., 8: 1601-1608.

Note: Certain commercial equipment, instruments and materials are identified in this paper in order to specify the experimental procedure as completely as possible. In no case does such identification imply a recommendation or endorsement by the National Institute of Standards and Technology nor does it imply that the material, instrument or equipment identified is necessarily the best available for the purpose.

HSP70 Family of Proteins of the Yeast *S. cerevisiae*

E. Craig, P.J. Kang, W. Boorstein
Department of Physiological Chemistry
University of Wisconsin-Madison
Madison, Wisconsin 53706
USA

Introduction

The most abundant heat shock protein in many organisms is a 70kDa protein called HSP70. In *E. coli* there is a single HSP70-related protein, the product of the *dnaK* gene. However, in most, if not all, eucaryotes, HSP70 is one of a family of related proteins. HSP70 families are usually composed of proteins that are expressed only after stress and those that are present under optimal growth conditions.

HSP70 proteins have been highly conserved, showing 60-78% identity among eucaryotes and 40-60% identity between *E. coli* HSP70, the *dnaK* gene product, and the eucaryotic HSP70s. While regions of identity extend over large regions of the proteins, in general the amino terminal two thirds of the protein is much more highly conserved than the carboxyl-terminal portion. Within this amino terminal two thirds there are regions of particularly striking identity (for reviews, see Craig, 1985; Lindquist, 1986). All HSP70s examined bind ATP. The ATP-binding domain and the ATPase activity is associated with the 44 kDa amino terminal fragment (Chappell et al., 1987), suggesting, as discussed below, that the less conserved carboxyl portion may have evolved to interact with a particular set of protein substrates while the more conserved region contains the regions associated with biochemical properties common to all HSP70s. Work from numbers of laboratories has suggested that HSP70 proteins act by interacting with proteins whose release is driven by ATP hydrolysis (Rothman, 1989).

The HSP70 multigene family of *S. cerevisiae*

Eight members of the HSP70 multigene family of *S.cerevisiae* have been identified. As shown in Figure 1, the structural relationships amongst the proteins encoded by these genes is complex. Amino acid identities range from about 50 to 97%. Analysis of strains containing different combinations of these mutations have allowed the determination that the obvious structural similarities reflect functional similarities as well. Based on structural and genetic analysis, the members of the HSP70 multigene family have been placed into functional groups. The genes isolated in our laboratory have been called Stress Seventy genes and divided into subgroups A, B, C and D. Subgroup A is the most complex, containing four genes, *SSA1*, *SSA2*, *SSA3* and *SSA4*. These genes encode an essential family, in the sense that proteins encoded by one of these genes must be present in cells at relatively high levels for cell growth (Werner-Washburne et al., 1987). Subgroup B contains two genes, *SSB1* and *SSB2*. Although no phenotype of single mutant strains have been observed, double mutant strains (*ssb1ssb2*) are relatively cold-sensitive for growth, growing nearly as well as wild-type strains at 37°C and progressively worse as the temperature is lowered (Craig and Jacobsen, 1985). *SSC1*, the sole member of the third group, is an essential gene (Craig et al., 1987). *SSD1*, the sole member of the fourth group that we isolated, is the same gene as the previously identified essential HSP70 gene, *KAR2* (Rose et al., 1989; Craig et al., 1990).

HSP70s are localized to different compartments of the cell. The *SSA* proteins are abundant cytoplasmic proteins (Deshaies et al., 1988; Nelson and Craig, unpublished results); Kar2p is in the lumen of the endoplasmic reticulum (ER) (Rose et al., 1989) and Ssc1p is in the matrix of the mitochondria (Craig et al., 1989; Kang et al., 1990). Although less well characterized, the *SSB* protein appear, by biochemical fractionation experiments, to be a part of some complex, suggesting that they are not soluble components of the cytoplasm (Nelson and Craig, unpublished results). Therefore, HSP70 proteins of yeast carry out essential functions in at least three, and perhaps as many as four, compartments of the cell. HSP70s localized in the cytoplasm (Welch and Feramisco, 1984), mitochondria (Engman et al., 1989; Leustek et al., 1989; Mizzen et al., 1989) and endoplasmic reticulum (Munro and Pelham, 1986) have been identified in other eucaryotes as well.

Ssc1p is required for the translocation of proteins into mitochondria

Ssc1p is an abundant protein localized in the matrix of the mitochondria. Fractionation experiments showed that it is a soluble protein of the matrix (Kang et al., 1990). To allow analysis of the essential role of Ssc1p, we isolated temperature-sensitive mutants of *SSC1*. A strain containing an insertion mutation in the chromosomal copy of *SSC1* and a centromeric plasmid containing a wild-type copy of the *SSC1* gene fused to the glucose-repressible *GAL1* promoter was used in a mutant screen. Since *SSC1* is an essential gene, the strain was maintained on galactose-based media to allow *SSC1* expression from the *GAL1* promoter. Transformants were screened for growth on glucose-based media at 23° and 37°C. Those permissive for growth at 23°, but not 37°C were candidates for strains carrying *ssc1* temperature-sensitive mutations (ts). After tests to determine that the mutation conferring temperature-sensitivity was in the *SSC1* gene, one plasmid carrying a mutation in the protein coding region was chosen for further analysis.

We reasoned that an HSP70 might interact with proteins destined for

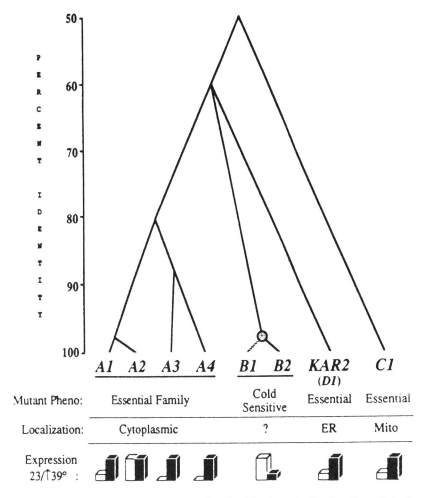

Figure 1. The *S. cerevisiae* HSP70 family. The tree indicates the relatedness of the different HSP70s. Gene names are indicated at the bottom of each branch. Per cent identity at the amino acid level (predicted from the DNA sequence) between two proteins is indicated by the position at which the two branches join on the vertical axis. The dashed line represents data from a partially sequenced gene (Ssb2p, amino acids 12-290, based on comparison with Ssa1p). The grey circle indicates a junction that can not be precisely positioned due to partial sequence data. Properties of individual HSP70s or of closely related subfamilies are shown at the bottom, including phenotypes of disruption mutations in the conjugate genes and subcellular localization of the proteins.

Abbreviations: ER, endoplasmic reticulum; mito, mitochondria. Sources: *SSA1* and SSA2, (Slater and Craig, 1989); *SSA3* (Boorstein and Craig, 1990); *SSA4*, (Boorstein and Craig, 1990); *SSB2* (Stinson and Craig, unpublished results) *SSC1*, (Craig et al., 1989); *KAR2*, (Normington et al., 1989; Rose et al., 1989). Figure adapted from Craig et al., 1990.

the mitochondrial matrix as they cross the membranes. To screen whether a defect in SSC1 results in a defect in the import of proteins into mitochondria, protein extracts were prepared from ssc1 mutant and wild-type cells, and analyzed using immunoblotting techniques with antibodies against nuclear-encoded mitochondrial proteins. After shift to the nonpermissive temperature the precursor form of proteins imported into the mitochondria accumulated, including those for HSP60, β-subunit of the F_1ATPase and Ssc1p. Pulse-labeling experiments indicated that the processing of precursors was inhibited within minutes of temperature shift.

Fractionation of cellular components showed that the accumulated precursor was associated with the mitochondria. However, since the precursor remained sensitive to exogenously added protease, it was concluded that they were localized on the cytoplasmic side of the outer membrane. Together these results indicated that inactivation of Ssc1p leads to a block in mitochondrial import (Kang et al., 1990), therefore is essential for the translocation of proteins across the mitochondrial membrane.

Conclusions

The data discussed above concerning the accumulation of precursor proteins in ssc1 mutants indicates that one or more early steps in import is blocked in the absence of functional Ssc1p. Interestingly, KAR2 (whose gene product is found in the lumen of the ER) temperature-sensitive mutants accumulate precursors of proteins normally translocated across the ER membrane at the nonpermissive temperature (Vogel et al., 1990). These accumulated precursors were found to be on the cytoplasmic side of the ER membrane. These genetic results suggest that HSP70 in the lumen of the ER may be directly involved in the translocation of proteins into the ER, as is HSP70 in the mitochondrial matrix. Perhaps these HSPs have very similar roles in the translocation of precursor proteins across membranes.

The in vivo experiments discussed above do not allow a determination of the role of Ssc1p in translocation. However, experiments have been carried out in vitro with mitochondria from the temperature-sensitive ssc1 mutant. As reported (Kang et al., 1990), mitochondria isolated from ssc1[ts] cells grown at 23°C were defective in import of in vitro-synthesized precursor proteins if they were preincubated at 37°C for 10 min prior to the addition of precursor.

The translocation of precursor across the mitochondrial membranes can be experimentally divided into two sequential reactions. First, the precursor protein is inserted into mitochondrial contact sites such that the amino-terminal presequence reaches the matrix space where it is proteolytically cleaved, while a major portion of the precursor is still located on the cytosolic side. Secondly, the remainder of the precursor unfolds and also is translocated through contact sites, leading to complete movement of the precursor into the matrix. Only the initial steps, but not the completion of translocation, depends on the membrane potential across the inner membrane (reviewed in Hartl and Neupert, 1990). In the mutant mitochondria, the precursor proteins were inserted into and efficiently processed; however, complete translocation to a protease-protected location was impaired. These results indicated that precursor proteins accumulated in contact sites, spanning the inner and outer membranes, suggesting a requirement for Ssc1p in movement across the membranes. *In vivo*, unprocessed precursor accumulated at the nonpermissive temperature. This accumulation is probably due to saturation of import (contact) sites by arrested proteins such that precursors accumulate at an earlier step in the import process.

Since the precursor protein must be unfolded before translocation is completed, the effect of denaturing a precursor protein on translocation in mutant mitochondria was tested. A fusion protein of the complete dihydrofolate reductase (DHFR) and a mitochondrial leader sequence was denatured with urea and then added to mitochondria. While undenatured protein was not translocated efficiently into mutant mitochondria, as measured by susceptibility to exogenously added protease, denatured precursor was translocated as efficiently as into wild-type mitochondria. Protein that was imported into mutant mitochondria was found to be partially unfolded, as evidenced by increased susceptibility to protease as compared to active, folded enzyme. These results suggested that Ssc1p is required for the completion of transport of precursor proteins across contact sites, including the unfolding of a portion of the precursor on the cytosolic side. In addition, refolding of proteins after translocation requires functional Ssc1p.

Genetic and biochemical analysis of another HSP, HSP60 encoded by the MIF4 gene, has provided compelling evidence for its role in the assembly of multimeric complexes and the folding of proteins (Cheng et al., 1989; Ostermann et al., 1989). What is the relationship between Ssc1p and hsp60? Since mutations in SSC1 and *MIF4* both appear to affect at least some of the same proteins, it would be reasonable to propose that these two hsps function in the same pathway. In such a model, a precursor protein, perhaps bound to Ssaps and an NEM-sensitive factor which assist in the unfolding of proteins or in maintaining an unfolded conformation, would be inserted across the outer and inner membranes at contact sites. In the matrix this protein would interact with Ssc1p. Ssc1p would assist translocation, perhaps by actively "pulling" the protein through the membrane. Once completely in the matrix and processing of the precursor had occurred, folding would proceed. Some of this process might occur while the protein is still in contact with Ssc1p, or it might be "passed off" to HSP60, where folding occurs.

In conclusion, both genetic and biochemical data indicate an essential role of HSP70s in protein translocation across membranes. The data obtained *in vivo* and with isolated organelles fits well with *in vitro* data showing that HSP70s interact with a variety of peptides in a cyclical manner, with the release of the peptide being ATP-dependent (Flynn et al., 1989). The challenge facing workers in the field is to understand the basis of these interactions and the specific role HSP70s play in determining the conformation of proteins.

References

Boorstein, W and Craig, EA, Structure and regulation of the *SSA4* gene of *Saccharomyces cerevisiae*. in press.

Chappell, TG, Konforti, BB, Schmid, SL and Rothman, JE, (1987) The ATPase core of a clathrin uncoating protein. J. Biol. Chem., 262: 746-751.

Cheng, M, Hartl, F-U, Martin, J, Pollock, R, Kalousek, F, Neupert, W, Hallberg, E, Hallberg, R and Horwich, A, (1989) Mitochondrial heat-shock protein hsp60 is essential for assembly of proteins imported into yeast mitochondria. Nature, 337: 620-625.

Craig, EA, (1985) The heat shock response. CRC Crit. Revs. in Biochem., 18: 239-280.

Craig, EA and Jacobsen, K, (1985) Mutations in cognate gene of *Saccharomyces cerevisiae* HSP70 result in reduced growth rates at low temperatures. Mol. Cell Biol., 5: 3517-3524.

Craig, EA, Kramer, J and Kosic-Smithers, J, (1987) SSC1, a member of the 70-kDa heat shock protein multigene family of *Saccharomyces cerevisiae*, is essential for growth. Proc. Natl. Acad. Sci. USA, 84: 4156-4160.

Craig, EA, Kramer, J, Shilling, J, Werner-Washburne, M, Holmes, S, Kosic-Smither, J and Nicolet, CM, (1989) SSC1, an essential member of the *S. cerevisiae* HSP70 multigene family, encodes a mitochondrial protein. Mol Cell Biol., 9: 3000-3008.

Craig, EA, Kang, PJ and Boorstein, W, (1990) A review of the role of 70kDa heat shock proteins in protein translocation across membranes Ant van Leeuw Intl. J. Gen. Mol. Microbiol., 58: 137-146.

Deshaies, R, Koch, B, Werner-Washburne, M, Craig, E and Schekman, R (1988) A subfamily of stress proteins facilitates translocation of secretory and mitochondrial precursor polypeptides. Nature, 332: 800-805.

Engman, D, Kirchhoff, LV and Donelson, JE, (1989) Molecular cloning of mtp70, a mitochondrial member of the HSP70 family. Mol. Cell Biol., 9: 5163-5168.

Flynn, GC, Chappell, TG and Rothman, JE, (1989) Peptide binding and release by proteins implicated as catalysts of protein assembly. Science, 245: 385-390.

Hartl, F-U and Neupert, W, (1990) Protein sorting to mitochondria: Evolutionary conservations of folding and assembly. Science, 247: 930-938.

Kang, PJ, Osterman, J, Shilling, J, Neupert, W, Craig, EA and Pfanner, N, (1990) Requirements for HSP70 in the mitochondrial matrix for translocation and folding of precursor proteins. Nature (London), 348: 137-143.

Leustek, T, Dalie, B, Amir-Shapira, D, Brot, N and Weissbach, H, (1989) A member of the hsp70 family is localized in mitochondria and resembles *Escherichia coli* DnaK. Proc. Natl. Acad. Sci. USA, 86: 7805-7808.

Lindquist, S, (1986) The heat-shock response. Ann. Rev. Biochem., 55: 1151-1191.

Mizzen, LA, Chang, C, Garrels, JI and Welch, W, (1989) Identification, characterization, and purification of two mammalian stress proteins present in mitochondria: One related to HSP70, the other to GroEL. 264: 20664-20675.

Munro, S and Pelham, HRB, (1986) An hsp70-like protein in the ER: Identity with the 78kd glucose-regulated protein and immunoglobulin heavy chain binding protein. Cell, 46: 291-300.

Ostermann, J, Horwich, AL, Neupert, W and Hartl, F-U, (1989) Protein folding in mitochondria requires complex formation with HSP60 and ATP hydrolysis. Nature, 341: 125-130.

Rose, MD, Misra, LM and Vogel, JP, (1989) KAR2, a karyogamy gene, is the yeast homolog of the mammalian BiP/GRP78 gene. Cell, 57: 1211-1221

Rothman, J, (1989) Polypeptide chain binding proteins: Catalysts of protein folding and related processes in cells. Cell, 59 (135): 591-601.

Slater, MR and Craig, EA, (1989) The SSA1 and *SSA2* genes of the yeast *Saccharomyces cerevisiae*. Nucl. Acid. Res., 17 (79): 805-806.

Vogel, JP, Misra, LM and Rose, MD (1990) Loss of BiP/grp78 function blocks translocation of secretory proteins in yeast. J. Cell Biol., 110: 1885-1895.

Welch, W and Feramisco, J, (1984) Nuclear and nucleolar localization of the 72,000-dalton heat shock protein in heat-shocked mammalian cells. J Biol. Chem., 259: 4501-4513.

Werner-Washburne, M, Stone, DE and Craig, EA, (1987) Complex interactions among members of an essential subfamily of HSP70 genes in *Saccharomyces cerevisiae*. Mol. Cell Biol., 7: 2568-2577.

Ubiquitin-Conjugating Enzymes Mediate Essential Functions of the Stress Response

S. Jentsch, W. Seufert, J. Jungmann, B. Klingner
Friedrich-Miescher-Laboratorium
der Max-Planck-Gesellschaft
Spemannstr. 37-39
D-7400 Tübingen
FRG

Introduction

Ubiquitin-mediated proteolysis is a major pathway for selective protein degradation in eukaryotic cells. This pathway involves the processive covalent attachment of ubiquitin to proteolytic substrates and their subsequent degradation by a specific ATP-dependent protease complex. We have cloned the genes and characterized the function of seven ubiquitin-conjugating enzymes (UBCs) from the yeast *Saccharomyces cerevisiae*. From this collection, UBC1, UBC4 and UBC5 enzymes were found to mediate degradation of short-lived and abnormal proteins. These enzymes have overlapping functions and constitute a UBC subfamily essential for growth. UBC1 is specifically required at early stages of growth after germination of spores. UBC4 and UBC5 enzymes are heat shock proteins and essential components of the eukaryotic stress response: mutants lacking both UBC4 and UBC5 are unable to grow at elevated temperatures or in the presence of an amino acid analog, constitutively express major heat shock proteins and are constitutively thermotolerant to a severe and acute heat shock.

The ubiquitin system

Modification of proteins by the covalent attachment of ubiquitin is universal to all eukaryotes. Ubiquitin is a small, abundant, very highly conserved intracellular protein. Ubiquitin conjugation has been implicated in a variety of different cellular functions. Biochemical and genetic evidence suggests that ubiquitin-protein conjugates are necessary intermediates in a major ATP-dependent cytosolic proteolytic pathway (reviewed in Finley and Varshavsky, 1985 and in Hershko, 1988). Proteins have strikingly different half-lives from few minutes to several hours. In several cases protein degradation has a regulatory role e.g., in reducing the levels of key enzymes of metabolic pathways and in the down-regulation of cellular regulators. In particular, some transcription factors, certain products of cellular and viral oncogenes and crucial cell cycle regulators such as the cyclins are short-lived *in vivo*. Turnover rates for individual proteins can vary considerably depending on the cell type, nutritional and other influences and the position within the cell cycle. Another function of intracellular protein degradation is the elimination of misfolded, misassembled, mislocalized, damaged, or other abnormal proteins.

Aside from its participation in selective protein degradation, an role for ubiquitin in the direct modification of the structure and function of proteins has been suggested. Support for this idea comes from the discovery of metabolically stable ubiquitin-protein conjugates. In the nucleus of higher eukaryotes, a large fraction of the chromosomal histones H2A and H2B exist in a monoubiquitinated form (Wu et al., 1981). Furthermore, ubiquitin has been found to be attached to certain integral membrane proteins (Leung et al., 1987; Siegelman et al., 1986; Yarden et al., 1986), *Drosophila* actin (Ball et al., 1987), and viral coat proteins (Dunigan et al., 1988). Enzymes have also been described which precisely cleave ubiquitin from conjugates (ubiquitin carboxy-terminal hydrolases; Wilkinson et al., 1989), suggesting that some of these conjugation events are reversible.

Ubiquitin conjugation is a multistep process and requires the activities of a ubiquitin-activating enzyme, E1, and a family of ubiquitin-conjugating enzymes, E2s (reviewed in Jentsch et al., 1990). Some conjugation reactions require additional factors, known as ubiquitin-protein ligases or E3s, to mediate substrate recognition and subsequent ubiquitination. Studies of the mouse cell line ts85, which has a thermolabile ubiquitin-activating enzyme, indicate that ubiquitin conjugation is essential for cell viability (Cjiekanover et al., 1984; Finley et al., 1984). Moreover, deletion of the gene encoding ubiquitin-activating enzyme (UBA1) in yeast results in cell death (McGrath, Jentsch and Varshavsky, manuscript submitted). To further our understanding of the ubiquitin system, we began a systematic genetic analysis aimed at determining the functions of individual ubiquitin-conjugating enzymes. We have purified several of the enzymatic components of the ubiquitin-protein ligase system from the yeast *Saccharomyces cerevisiae*, and isolated the corresponding genes. These studies led to the discovery that the previously characterized yeast DNA repair gene RAD6 and the cell cycle gene *CDC34* both encode ubiquitin-conjugating enzymes (Goebl et al., 1988; Jentsch et al., 1987).

A major pathway for selective protein degradation in eukaryotes is ubiquitin-dependent (Finley and Varschavsky, 1985; Hershko, 1988). This proteolysis pathway involves the processive coupling of a branched, multiubiquitin chain to proteolytic substrates (Chau et al., 1989) and their subsequent degradation by an ATP-dependent protease complex (Matthews et al., 1989). Here, we review our studies (Seufert and Jentsch, 1990; Seufert et al., 1990) on the isolation and functional characterization of three ubiquitin-conjugating enzymes which are key elements in the ubiquitin-dependent proteolysis pathway, UBC1,

UBC4, and UBC5. These enzymes are closely related in sequence, have overlapping cellular functions and constitute a UBC family essential for cell viability.

Cloning of UBC genes

Using ubiquitin-Sepharose affinity chromatography, we have purified several enzymes of the yeast ubiquitin-protein ligase system (Jentsch et al., 1987). To clone the genes encoding these enzymes, two different strategies were utilized. One approach involved the determination of partial amino acid sequences of the purified enzymes and synthesis of the corresponding oligonucleotides for use in library screening. This technique was applied to isolate *UBC4* and *UBC5* genes for the E216K enzymes (Seufert and Jentsch, 1990). *UBC4* and *UBC5* genes encode almost identical proteins (137 identical residues out of total 148, Fig. 1). Both genes contain a single intron located at identical positions in the amino-terminal part of the coding region. The alternative cloning strategy consisted of raising antibodies against the mixture of purified E1 and E2 enzymes and using this antiserum to screen a yeast expression library. Immunoreactive phage clones were identified which carried the gene *UBC1* encoding the E230K ubiquitin-conjugating enzyme (Seufert et al., 1990). The *UBC1* coding region was localized by transposon tagging and its nucleotide sequence was determined. The predicted amino acid sequence (Fig. 1) shows that UBC1 is similar in sequence to all other characterized members of the ubiquitin-conjugating enzyme family. In particular, sequences surrounding the single active site cysteine residue required for ubiquitin-enzyme thiolester formation are highly conserved. Within a conserved domain common to all ubiquitin-conjugating enzymes about 45% of the amino acids are identical between UBC1 and UBC4/UBC5 proteins (Fig. 1). UBC1 protein differs in its structure from UBC4/UBC5 in possessing a highly charged carboxy-terminal extension. UBC4 and UBC5 have been defined as class I ubiquitin-conjugating enzymes (Jentsch et al., 1990) which consist almost exclusively of the conserved "E2 domain". In comparison, UBC1 belongs to the class II E2 enzymes which bear carboxy-terminal extensions. Both previously characterized class II enzymes, UBC2/RAD6 and UBC3/CDC34, have extensions which are known to be necessary for substrate interaction (Goebl et al., 1988 Sung et al., 1988). Following this line, the carboxy-terminal domain of the UBC1 protein may also be involved substrate interaction. The class I enzymes UBC4 and UBC5 most likely require supplementing E3 factors for conjugation activity.

The expression of genes encoding components of the ubiquitin system is highly regulated. We therefore examined the expression of *UBC* genes under different growth conditions and in response to environmental stress (Seufert and Jentsch, 1990; Seufert et al., 1990). *UBC1* transcription was strongly induced when cells enter stationary phase. *UBC4* expression was high in growing cells but hardly detectable

88

```
UBC1  M SRAK R IMK EI QAUKDD PAAH I TLEFU SESD I HHLKGTFLG PPG TP YEG G

UBC4  M SS SK R IAK ELSDLERD PPTSCSAGPUGD DLYHUQASIMG PADSP YAG G

UBC5  M SS SK R IAK ELSDLGRD PPAACSAGPUGD DLYHUQASIMG PASSP YAG G


UBC1  KFUUDI EUP MEY P FKP PKMQF DT KUY H P N ISSUTGAI CLD I LKNAU S PU I

UBC4  UF FLSI HFP TDY P FKP PK I SFTT KIY HPN INA NGNI CLD I LKDQU S PAL

UBC5  UF FLSI HFP TDY P FKP PUNSF TT KIY HPN INS SGNI CLD I LKDQU S PAL


UBC1  T LKSAL I SLQAL LQSPEPND PQDAEUAQHYL RDRESFNKT AALU TRLY AS

UBC4  T LSKULLS I CSL LTDANPDD PLUPE I AH IYKTDRPKYEAT AREU TKKY AU

UBC5  T LSKULLS I CSL LTDANPDD PLUPE I AQ IYKTDKAKYEAT AKEU TKKY AU


UBC1  ETSNGQKGNUEESDLYG I DHDL I DEFESQGFEKDK I UEULRRLGUKSLDP


UBC1  NDNNTANRI I EELLK
```

Figure 1. Deduced amino acid sequence of UBC1, UBC4, and UBC5 proteins (Seufert and Jentsch, 1990; Seufert et al., 1990). Identical amino acid residues are shown in bold face.

during the stationary phase. In contrast, *UBC5* was only weakly expressed during exponential growth but was highly induced upon entry into stationary phase. Interestingly, expression of *UBC4* and *UBC5* (but not of *UBC1*) increased substantially in response to heat shock.

Phenotypes of ubc mutants

To assess the *in vivo* function of the yeast UBC1, UBC4, and UBC5 enzymes the corresponding genes were disrupted and the phenotypes of the resulting mutants were analyzed (Seufert and Jentsch, 1990; Seufert et al., 1990). ubc1 mutants displayed only a moderate slow growth phenotype (Table 1), however, such mutants were markedly impaired in growth after germination of spores. Tetrad analysis following sporulation of ubc1/UBC1 heterozygotes showed that unlike the wild-type spores, the ubc1 mutant spores gave rise to tiny, amorphously shaped colonies indicative of slow growth and poor cell viability. However, cells from these colonies recovered after several divisions and formed colonies of normal appearence. Such colonies were only slightly smaller than wild-type colonies, reflecting the 1.5-fold increase in doubling time of exponentially growing ubc1 cultures. The specific requirement for UBC1 function upon resumption of growth following a resting state also manifested itself after prolonged nutrient deprivation or extended storage on plates. These data suggest that in addition to a role for exponentially growing cells, UBC1-mediated functions are critical for growth and viability at a specific point in the yeast life cycle: during the transition period after a state of quiescence.

Disruption of either gene, UBC4 or UBC5, caused only moderate defects. However, disruption of both genes resulted in more severe phenotypes indicating overlapping or similar functions of both enzymes. Growth of ubc4ubc5 mutants was drastically impaired suggesting that these enzymes mediate important cellular functions during vegetative growth. Given the heat inducible expression of UBC4 and UBC5 genes we investigated the resistance of mutants to chronic heat stress. ubc4ubc5 mutants did not form colonies on plates at 37°C. Thermosensitivity of ubc4ubc5 mutants was not merely due to growth arrest but was caused by cell death. Thus, UBC4/UBC5 proteins are essential for viability at elevated temperatures. Moreover, loss of UBC4/UBC5 function led to constitutive thermotolerant cells which are able to survive an acute heat shock (5 min at 52°C) and which constitutively expressed major heat shock proteins.

Table 1

Phenotypes of ubc1 mutants (Seufert et al., 1990)			
	Doubling time (hours)[a]	Resistance to canavanine (%)b	Degradation of canavanyl-peptides (%)c
wild-type	1.5	82	100
ubc1	2.3	12	76

[a] Cells were grown in YPD liquid medium at 30°C. OD_{600} was followed for determination of doubling times.

[b] Resistance to canavanine was determined with cells grown in SD medium containing required nutrients. Appropriate aliquots were spread on supplemented SD plates with or without canavanine at 1.5 μg/ml. Resistance is given as the fraction of colonies formed in the presence of the amino acid analog.

[c] Cells were pre-treated for 90 min with canavanine at 20 μg/ml and labelled for 5 min with ^{35}S-methionine. Protein degradation was measured as the fraction of total incorporated radioactivity released from cells during the chase period (Seufert and Jentsch, 1990). The wild-type value at 3 hr chase was defined as 100%.

Previous studies have indicated that selective protein degradation required covalent attachment of a multiubiquitin chain to proteolytic substrates (Chau et al., 1989; Hershko, 1988). In a Western analysis using anti-ubiquitin antibodies we showed that UBC4 and UBC5 enzymes generate high molecular weight ubiquitin-protein conjugates in vivo, suggesting a role of these enzymes in protein degradation. To evaluate this possibility the sensitivity of mutants to canavanine was analyzed. Cells incubated with canavanine, an amino acid analog, generate abnormal proteins *in vivo*. Wild-type cells as well as ubc5 mutants efficiently formed colonies on plates containing canavanine. ubc4 mutants were moderately sensitive. ubc4ubc5 double mutants, however, were unable to form colonies under these conditions, consistent with a function of these enzymes in the elimination of abnormal proteins.

We directly analyzed the turnover of naturally short-lived proteins or canavanyl-peptides in wild-type and mutant cells with pulse-chase experiments. Proteins labelled during a 5 min pulse were efficiently degraded in wild-type cells and protein degradation in ubc4 and ubc5 single mutants was barely affected. In ubc4ubc5 double mutants however, turnover of short-lived proteins was drastically reduced. Moreover, UBC4 and UBC5 enzymes mediate proteolysis of abnormal proteins. In ubc4ubc5 double mutants proteolysis of canavanyl-peptides was severely restricted and even single mutants exhibited significant defects.

To evaluate a possible involvement of UBC1 in ubiquitin-mediated protein turnover, we examined the sensitivity of ubc1 mutants to canavanine. When compared to wild-type cells, ubc1 mutants were more sensitive to canavanine (Table 1). The involvement of UBC1 in selective protein degradation was further supported by direct measurements of the turnover of canavanyl-proteins which showed a slight but significant reduction of selective protein degradation in ubc1 mutants (Table 1).

The moderate defects in protein degradation observed in ubc1 mutants might indicate a functional overlap of UBC1 with the UBC4 and UBC5 enzymes which mediate most of the ubiquitin-dependent protein degradation in yeast (Seufert and Jentsch, 1990). To address this possibility, double and triple mutants in these UBC genes were constructed, either by direct disruption of genes in existing haploid ubc mutants, or by crossing different ubc mutant strains followed by meiotic segregation. Disruption of UBC1 in a ubc4 mutant background resulted in double mutants exhibiting more severe phenotypes than that of the single mutants combined. In particular, ubc1ubc4 double mutants showed significantly prolonged doubling times (5.5 hr), suggesting important overlapping functions for these two genes during exponential growth. The critical role of UBC1 after sporulation and germination was again evident when the identical ubc1ubc4 haploid was constructed by mating and meiotic segregation: ubc1ubc4 spores failed to form visible colonies even after extended incubation. Microscopic inspection revealed that most of the ubc1ubc4 spores had germinated and formed microcolonies of a few abnormally shaped cells. Since no viable cells could be recovered from these microcolonies, we conclude that UBC1/UBC4-mediated functions are essential for cell viability after germination of ascospores. In contrast, ubc1ubc5 double mutants showed no apparent phenotypic difference from ubc1 single mutants. This is consistent with the previous observation that UBC5 is dispensible as long as UBC4 is present (Seufert and Jentsch, 1990). Neither by crossing and meiotic segregation, nor by direct gene disruption, could ubc1ubc4ubc5 triple mutants be recovered, indicating that the combined loss-of-function of these three genes was incompatible with cell viability. To confirm this result, UBC1, UBC4 and UBC5 genes were disrupted in a strain carrying a 2μ plasmid with a functional UBC4 gene and a URA3 marker. Spontaneous plasmid loss during growth in non-selective medium should give rise to cells that are ura3 and therefore resistant to the toxic effects of 5-fluoro-orotic acid (5-FOA). Mitotic loss of the UBC4 plasmid was readily observed in wild-type cells and various combinations of ubc mutants, however, growth of the chromosomal triple ubc1ubc4ubc5 mutant was found to be plasmid-dependent. These cells failed to give rise to 5-FOA resistant colonies and no uracil auxotrophs were observed after growth in nonselective medium. Therefore, UBC1, UBC4, and UBC5 constitute a gene family essential for cell growth.

The similarity of UBC1 gene function to UBC4 and UBC5 was substantiated by the observation that overexpression of UBC1 partially complemented ubc4ubc5 mutant phenotypes. Approximately a 10-fold overproduction of UBC1 protein was achieved by placing the gene on a high-copy number 2μ plasmid. In ubc4ubc5 mutants, the overexpression of UBC1 improved growth. In particular, high level expression of UBC1 restored growth of ubc4ubc5 mutants at elevated temperatures and increased resistance of these mutants to canavanine at least 500-fold. Apparently, the defects of ubc4ubc5 mutants in the ubiquitin-mediated proteolysis pathway can be complemented by overexpression of UBC1.

A functional interrelationship of UBC1, UBC4, and UBC5 genes is also reflected at the transcriptional level. Mutations in genes of this subfamily resulted in a compensatory induction of the remaining genes: e.g., deletion of UBC4 led to elevated expression of UBC5 and in ubc4ubc5 mutants, the transcription of UBC1 was strongly induced (Jungmann and Jentsch, unpublished data).

Recent evidences suggest that not only ubiquitin but also the enzymatic components of the ubiquitin-protein ligase system are highly conserved. Antibodies raised against UBC4 cross-react with proteins of similar sizes in extracts of mouse, Drosophila and C.elegans (Klingner and Jentsch, unpublished data), suggesting that similar proteins are functioning in these organisms. This implies that UBC4/UBC5-dependent functions are relevant to all eukayotes.

Conclusions

We have characterized the ubiquitin-conjugating enzymes UBC1, UBC4, and UBC5 from *S.cerevisiae* as key components of the ubiquitin-mediated proteolytic system. UBC1 is specifically required for resuming growth after a resting state. The closely related enzymes UBC4 and UBC5 enzymes are stress-inducible, perform important functions during mitotic growth and are essential components of the eukaryotic stress response. Our genetic analysis has shown that *UBC1*, *UBC4*, and *UBC5* have overlapping functions and constitute an *UBC* subfamily essential for viability.

Acknowledgements

We thank Ute Ehringer and Ute Nußbaumer for technical assistance. Research of this laboratory is supported by the Deutsche Forschungsgemeinschaft (Je 134/2-1).

References

Ball, E, Karlik, CC, Beall, CJ, Saville, DL, Sparrow, JC, Bullard, B and Fyrberg, EA, (1987) Arthrin, a myofibrillar protein of insect flight muscle, is an actin-ubiquitin conjugate. Cell, 51: 221-228.

Chau, V, Tobias, JW, Bachmair, A, Marriott, D, Ecker, DJ, Gonda, DK and Varshavsky, A, (1989) A multiubiquitin chain is confined to specific lysine in a targeted short-lived protein. Science, 243: 1576-1583.

Ciechanover, A, Finley, D and Varshavsky, A, (1984) Ubiquitin dependence of selective protein degradation demonstrated in the mammalian cell cycle mutant ts85. Cell, 37: 57-66.

Dunigan, DD, Dietzgen, RG, Schoelz, JE and Zaitlin, M, (1988) Tobacco mosaic virus particles contain ubiquitinated coat protein subunits. Virology, 165: 310-312.

Finley, D and Varshavsky, A, (1985) The ubiquitin system: functions and mechanisms. Trends Biochem. Sci., 10: 343-346.

Finley, D, Ciechanover, A and Varshavsky, A, (1984) Thermolability of ubiquitin-activating enzyme from the mammalian cell cycle mutant ts85. Cell, 37: 43-55.

Goebl, MG, Yochem, J, Jentsch, S, McGrath, JP, Varshavsky, A and Byers, B, (1988) The yeast cell cycle gene CDC34 encodes a ubiquitin-conjugating enzyme. Science, 241: 1331-1335.

Hershko, A, (1988) Ubiquitin-mediated protein degradation. J. Biol. Chem., 263: 15237-15240.

Jentsch, S, McGrath, JP and Varshavsky, A, (1987) The yeast DNA repair gene RAD6 encodes a ubiquitin-conjugating enzyme. Nature, 329: 131-134.

Jentsch, S, Seufert, W, Sommer, T and Reins, H-A, (1990) Ubiquitin-conjugating enzymes: novel regulators of eukaryotic cells. Trends Biochem. Sci., 15: 195-198.

Leung, DW, Spencer, SA, Cachianes, G, Hammonds, RG, Collins, C, Menzel, WJ, Barnard, R, Waters, WJ and Wood, WI, (1987) Growth hormone receptor and serum binding protein: purification, cloning and expression. Nature, 330: 537-543.

Matthews, W, Tanaka, K, Driscoll, J, Ichihara, A and Goldberg, AL, (1989) Involvement of the proteasome in various degradative processes in mammalian cells. Proc. Natl. Acad. Sci. USA, 86: 2597-2601.

Seufert, W and Jentsch, S, (1990) Ubiquitin-conjugating enzymes UBC4 and UBC5 mediate selective degradation of short-lived and abnormal proteins. EMBO J., 9: 543-550.

Seufert, W, McGrath, JP, and Jentsch, S, (1990) UBC1 encodes a novel member of an essential subfamily of yeast ubiquitin-conjugating enzymes involved in protein degradation. EMBO J., 9: 4535-4541.

Siegelman, M, Bond, MW, Gallatin, WM, St. John, T, Smith, HT, Fried, VA and Weissman, IL, (1986) Cell surface molecule associated with lymphocyte homing is a ubiquitinated branched-chain glycoprotein. Science, 231: 823-829.

Sung, P, Prakash, S and Prakash, L, (1988) The RAD6 protein of *Saccharomyces cerevisiae* polyubiquitinates histones, and its acidic domain mediates this activity. Genes Dev., 2: 1476-1485.

Wilkinson, KD, Lee, K, Deshpande, S, Duerksen-Hughes, P, Boss, JM and Pohl, J, (1989) The neuron-specific protein PGP 9.5 is a ubiquitin carboxyl-terminal hydrolase. Science, 246: 670-673.

Wu, RS, Kohn, KW and Bonner, WM, (1981) Metabolism of ubiquitinated histones. J. Biol. Chem., 256: 5916-5920.

Yarden, Y, Escobedo, JA, Kuang, WJ, Yang-Feng, TL, Daniel, TO, Tremble, PM, Chen, EY, Ando, ME, Harkins, RN, Francke, U, Fried, VA, Ullrich, A and Williams, LT, (1986) Structure of the receptor for platelet- derived growth factor helps define a family of closely related growth factor receptors. Nature, 323: 226-232.

Analysis of Heat Shock Protein Functions

Early Effects of Heat Shock on Enzymes: Heat Denaturation of Reporter Proteins and Activation of a Protein Kinase which Phosphorylates the C-terminal Domain of RNA Polymerase II

O. Bensaude, M.-F. Dubois, V. Legagneux, V.T. Nguyen,
M. Pinto, M. Morange
Biologie Moléculaire du Stress
Institut Pasteur
75724 Paris Cedex 15
France

Introduction

It is a common statement that heat-shock impairs cellular functions because some essential proteins are heat-denaturated. This statement relies mostly on indirect arguments developped by several workers (Hightower, 1980). The fate of such denatured proteins is matter for discussion: are they degraded or renatured? A priming non-lethal heat-shock stimulates transiently the synthesis of the heat-shock proteins (HSP) and increases transiently the cell resistance against a second challenging stress (Subjeck and Sciandra, 1982). Is this thermotolerant state due to a protective effect against thermal denaturation or to a better repair of the damaged proteins? What is the behaviour of the heat-shock proteins?

Early effects of heat shock on enzymes

During the past few years, we have decided to examine these questions and to follow the behaviour of individual proteins. Soluble well-defined enzymes were choosen because they afforded afforded two simple criteria for heat-induced conformational changes: solubility and activity. Two classical reporter exogenous enzymes were easy to monitor: the genes coding for β-galactosidase from *E.coli* or luciferase from *P.pyralis* were introduced into mouse cells. The p68 protein-kinase induced by interferon was an endogenous protein easy to monitor; its dsRNA dependent autokinase activity was predominant in extracts from interferon-treated cells (Galabru and Hovanessian, 1987).

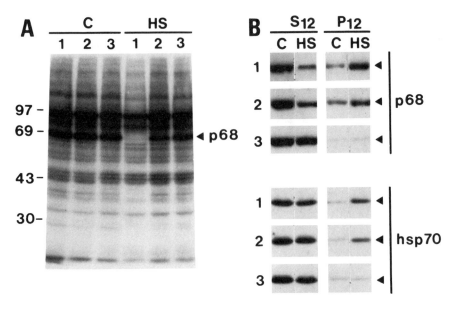

Figure 1: The dsRNA dependent protein-kinase p68 is inactivated and insolubilized by heat-shock in control cells (1). Incubation of the cells with glycerol (2) or D_2O (3) protects p68 against insolubilization and inactivation. The dsRNA-dependent activity was assayed in cytoplasmic extracts (A). Supernantants (S12) and 12,000 g pellets (P12) were analyzed by Western blot using specific monoclonal antibodies against p68 kinase or HSP70.

All three enzymes are essentially found in the postmitochondrial supernatants (12,000 g) of control cell lysates. When cells are lysed immediately after a heat-shock, the enzymatic activities are diminished and the corresponding molecules accumulate with the nuclear pellet (Nguyen et al., 1989). The distribution between supernatant and pellet depends on the intensity of the stress (duration, temperature) and on the reporter: luciferase is the most sensitive, it is inactivated and insolubilized by very mild heat-shock conditions, whereas β-galactosidase insolubilization requires acute heat-shocks and the insolubilized enzyme remains active. When thermoprotectors such as glycerol or D_2O are added to the culture medium, insolubilization and inactivation are attenuated (Fig. 1). The insolubilized enzymes seem to localize on the collapsed intermediate filament network. But disruption of the cytoskeleton with cytochalasin and colchicine does not affect the distributions into supernatants and pellets.

The behaviour of the reporter also depends on the cell in which it is expressed (Nguyen et al., 1989). Luciferase is much more heat-sensitive within a Drosophila cell than within a mouse cell. All reporters are found more heat-resistant in thermotolerant cells and this observation supports the hypothesis that thermotolerant cells are less damaged by stress than control cells.

To study the fate of the heat-denatured proteins, the cells were maintained in cycloheximide during and after stress. In such condition, in the absence of protein synthesis, a complete resolubilization of β-galactosidase and a partial resolubilization of luciferase and p68-kinase is demonstrated (Fig. 1). In the meantime, luciferase and p68 kinase activities are partially recovered. The renaturation process may involve heat-shock proteins but in any case the constitutive heat-shock proteins present in the unstressed cell are capable to do the job. It might be questionned which are the most heat-insolubilized endogenous proteins. The proteins present in nuclear pellets from control and heat-shocked cells were separated by polyacrylamide gel electrophoresis, the major differences concerned the accumulation of 70 kDa and 90 kDa proteins in the heat-shocked cell pellets (Bensaude et al., 1990). These proteins were identified by comigration and monoclonal antibody recognition as the constitutive members of the HSP70 and HSP90 family. The HSP70 is found to aggregate around the nucleus like the p68 kinase. Insolubilization of the various heat-shock proteins had been reported previously to occur during stress, in particular the 20-30 kDa ones and ubiquitin which accumulates covalently bound to high molecular weight complexes (Collier and Schlesinger, 1986). Here we emphasize that heat-shock proteins are the most dramatically insolubilized proteins within the heat-shocked cell.

The heat-induced protein insolubilization can be schematized in a simple model which takes into account the chaperonning role of the heat-shock proteins. Proteins exist as a an equilibrium of various conformations. Increasing the temperature favors the open conformations which unravel "sticky" domains to the solvent. Under non heat-shock conditions, the constitutive heat-shock proteins would chaperon the sticky domains and prevent aggregation. Under heat-shock conditions, too many sticky domains would appear and titrate out the constitutive heat-shock proteins (Bensaude et al., 1990). This titration would have various consequences: a) proteins would start to aggregate; b) heat-shock proteins would be trapped in these aggregates and thus be insolubilized; c) heat-shock protein titration would be the signal for transcriptional activation of the heat-shock genes and would account for the apparent self-regulation of heat-shock gene expression (Di Domenico et al., 1982). Increasing the amount of heat-shock proteins after a priming heat-shock would delay the titration of the heat-shock proteins and delay protein heat-denaturation. Conversely, since in *Drosophila* the heat-shock response is triggered at a lower temperature than in mammals, heat-shock protein titration would also occur at lower temperatures in *Drosophila* cells than in mammalian cells, leading to the heat-denaturation of reporter enzymes also at lower temperatures.

Protein denaturation, enzyme inactivation are negative events which occur during stress. Meanwhile cells react to stress by mounting the heat-shock response. Thus some cellular functions must be activated.

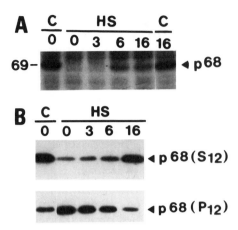

A C   HS   C
    0 0 3 6 16 16
69−   ◄ p68

B C   HS
    0 0 3 6 16
                    ◄ p68(S12)

                    ◄ p68(P12)

Figure 2: Both solubility and activity of the p68 kinase are recovered after heat-shock even if protein synthesis is inhibited. Interferon-treated HeLa cells control (C) or heat-shocked 1 hr at 44°C (HS) were lysed immediately (0) or allowed to recover 3, 6 or 16 hrs in the presence of 20μg/ml of cycloheximide. The lysates were incubated with labelled ATP in the presence of polyIC during 30 minutes at 30°C and analyzed by SDS polyacrylamide gel electrophoresis (part A). Labelled phosphate incorporation into a 68 kDa protein is the result of the autokinase activity. The lysates were also fractionnated by centrifugation at 12.000 g 15 min into a supernatant (S12) and a pellet (P12) which were analyzed by Westerb blot using specific monoclonal antibodies (part B).

In an attempt to understand how stress can activate such functions, we decided to analyze protein-kinase activities within the heat-shocked cell. Protein-kinase activities are rather easy to assay in crude cell extracts and protein-kinases are good candidates to contribute rapidly to the modulation of cellular functions. Various kinases substrates were added to cell extracts and assayed for *in vitro* ^{32}P incorporation (Legagneux et al., 1990). Two substrates showed an increased incorporation in extracts from heat-shocked cells: histone H1 and the peptide hepta-4. Similar increases were also obtained after treating the cells with sodium arsenite.

Histone H1 is a classical substrate of the mitotic promoting factor related to the yeast *cdc2* gene product. We demonstrated that the histone H1 kinase activated by stress in HeLa cells has the characteristics of a *cdc2* related kinase : it binds to the *suc1* yeast gene product and to antibodies directed against the *cdc2* C-terminal peptide. The *cdc2* related kinase had also been shown to phosphorylate *in vitro* the C-terminal domain of RNA polymerase II large subunit (Cisek and Corden, 1989). The C-terminal domain of eucaryotic RNA polymerase II consists in multiple repetitions of the heptapeptide motif: Ser-Pro-Thr-Ser-Pro-Ser-Tyr. Phosphorylation of the C-terminal

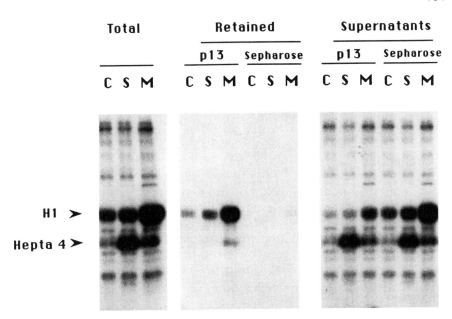

Total **Retained** **Supernatants**

p13 Sepharose   p13 Sepharose

C S M   C S M C S M   C S M C S M

H1 >

Hepta 4 >

Figure 3: Histone H1 and hepta-4 kinase activities are higher in lysates from arsenite treated cells (S) and nocodazole treated cells (M) than from control cells (C). HeLa cells were incubated 1 hr with 800μM sodium arsenite or 18 hr with 10μM nocodazole prior to lysis. The lysates were incubated with p13-sepharose or control sepharose beads and fractionnated by centrifugation into retained material or supernatants. Total lysates, and fractionnated material were incubated at 30°C during 15 min with labelled ATP in the presence of histone H1 and hepta-4 and analyzed by SDS polyacrylamide gel electrophoresis.

domain of RNA polymerase II has been suggested to be required for elongation of transcription, while the formation of the transcription initiation complex would require the unphosphorylated polymerase (Payne et al., 1989). Interestingly, transcription of the *Drosophila hsp70* gene is blocked at the elongation step (Rougvie and Lis, 1988). To examine the kinase activities able to phosphorylate the RNA polymerase C-terminal domain, we used the peptide hepta-4 which is a four-mer repeat of the above mentionned motive.

The hepta-4 kinase activity is strongly enhanced in lysates from stressed cells. Hepta-4 peptide had previously found to be a substrate of *cdc2* related kinases (Cisek and Corden, 1989). Indeed, histone H1 and hepta-4 kinase activities are enhanced in nocodazole treated cells (Fig. 2). The nocodazole treated cells accumulate active *cdc2*-like kinase because they are blocked in mitosis. Most of the histone H1 activities induced by stress or nocodazole are retained by either the suc 1 gene product or the anti-*cdc2* antibodies. In contrast, the stress-induced hepta-4 kinase activity is not retained. Thus, the heat-induced hepta-4 kinase is distinct from *cdc2*-like. It is likely to be an RNA polymerase C-terminal domain kinase (CTD kinase) since it is also able to phosphorylate a fusion protein consisting of *E.coli* ß-galactosidase fused to the C-terminal domain of yeast RNA polymerase II (Lee and Greenleaf, 1989).

Conclusions

As a conclusion heat-shock and related stress promote the heat-denaturation of several proteins. However, a few proteins are activated : HSF binding to HSEs, the *cdc2* related kinase and a CTD kinase. It remains a open question how these proteins are activated : titration of heat-shock proteins is a possible mean to activate HSF through a yet unknown mecanism. What would be the consequence of the *cdc2* kinase and the CTD kinase activation by heat-shock remains a challenge.

References

Bensaude, O, Pinto, M, Dubois, MF, Nguyen, VT and Morange, M, Protein denaturation during heat-shock and related stress. In Heat shock proteins. Schlesinger, MJ and Santoro, G, eds., Springer Verlag, Berlin, in press.

Cisek, LJ and Corden, JL, (1989) Phosphorylation of RNA polymerase by the murine homologue of the cell-cycle control protein cdc2. Nature, 339: 679-684.

Collier, NC and Schlesinger, MJ, (1986) The dynamic state of heat shock proteins in chicken embryo fibroblasts. J. Cell Biol., 103: 1495-1507.

Di Domenico, BJ, Bugaisky, GE and Lindquist, S, (1982) The heat shock response is self-regulated at both the transcriptional and post-transcriptional levels. Cell, 31: 593-603.

Galabru, J and Hovanessian, A, (1987) Autophosphorylation of the Protein Kinase Dependent on Double-stranded RNA. J. Biol. Chem., 262: 15538-15544.

Hightower, LE, (1980) Cultured animal cells exposed to amino acid analogues or puromycin rapidly synthesize several polypeptides. J. Cell. Physiol., 102: 407-427.

Lee, JM and Greenleaf, AL, (1989) A protein kinase that phosphorylates the C-terminal repeat domain of the largest subunit of RNA polymerase II. Proc. Natl. Acad. Sci. USA, 86: 3624-3628.

Legagneux, V, Morange, M and Bensaude, O, (1990) Heat-shock and related stress enhance RNA polymerase II C-terminal-domain kinase activity in HeLa cell extracts. Eur. J. Biochem., 193: 121-126.

Nguyen, VT, Morange, M and Bensaude, O, (1989) Protein denaturation during heat shock and related stress. J. Biol. Chem., 264: 10487-10492.

Payne, JM, Laybourn, PJ and Dahmus, ME, (1989) The transition of RNA polymerase II from initiation to elongation is associated with phosphorylation of the carboxyl-terminal domain of subunit IIa. J. Biol. Chem., 264: 19621-19629.

Rougvie, AE and Lis, JT, (1988) The RNA polymerase II molecule at the 5' end of the uninduced *hsp70* gene of *D. melanogaster* is transcriptionally engaged. Cell, 54: 795-804.

Subjeck, JR and Sciandra, JJ, (1982) Coexpression of thermotolerance and heat-shock proteins in mammalian cells. *In* Heat Shock From Bacteria to Man. Schlesinger, MJ, Ashburner, M and Tissières, A, eds. Cold Spring Harbor Laboratory, Cold Spring Harbor, pp 405-411.

Interaction of HSP47 with Newly Synthesized Procollagen, and Regulation of HSP Expression

K. Nagata, A. Nakai, N. Hosokawa, M. Kudo, H. Takechi,
M. Sato, K. Hirayoshi
Department of Cell Biology, Chest Disease Research Institute,
Kyoto University
Sakyo-ku, Kyoto 606
Japan

Introduction

We have found a novel stress (heat shock) protein whose molecular size is 47kDa. As hsp47 binds to collagen, it can be easily purified using gelatin-Sepharose. Indirect immunofluorescence and immunoelectron microscopic studies using polyclonal and monoclonal antibodies against HSP47 indicated the localization of HSP47 in the endoplasmic reticulum (ER) of the fibroblast. Immunoprecipitation studies using anti-HSP47 antibody revealed that two chains (α and β) of type I procollagen were co-precipitated with HSP47.

Expression of HSP47

We previously found a novel 47kDa heat shock (stress) protein, HSP47, in chick embryo fibroblasts (Nagata et al., 1986). Its isoelectric point is so high (approximately 9.0) that it cannot be analysed by two dimensional gel electrophoresis using conventional isoelectric focusing as the first dimension (Fig. 1).

In addition to the induction after heat shock, the expression of HSP47 was also sensitive to malignant transformation of the cells. When chick embryo fibroblasts were transformed with Rous sarcoma virus, the synthesis of HSP47 by the transformed cells was about 1/3 - 1/10 of that in the normal cells (Nagata and Yamada, 1986). Similar decrease after transformation was also observed in BALB/3T3 cells transformed with simian virus 40 and in NIH3T3 cells trans-formed with activated c-Ha-ras oncogene (Nakai et al., 1990, and unpublished observation). Both the induction of HSP47 after heat shock and the suppression of synthesis of HSP47 following transformation were regulated at the level of mRNA (Nagata et al., 1988). Recently, we succeeded in cloning a full-length cDNA of HSP47 from the cDNA library of chick embryo fibroblasts, and we confirmed that regulation

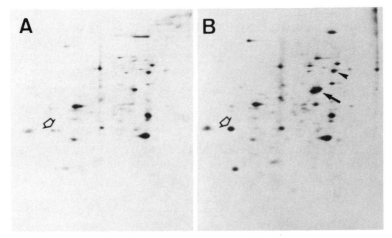

Figure 1. Two-dimensional analysis by NEPHGE/SDS-PAGE of extracts from BALB/3T3 cells treated at high temperature. Cells were preincubated at 42°C for 80 min and then labeled at 37°C for 1 h with 0.1 mCi/ml of [35S]-methionine. A, control and B, heat shocked cells. Thick arrow and arrow head indicate the HSP70s and the HSP90, respectively.

by heat shock and transformation is at the transcriptional level by performing a nuclear run on assay using this cDNA clone as a probe, as well as by northern blot analysis.

In addition to transformation-sensitivity, we found that the synthesis of HSP47 was also regulated during the process of cell differentiation. F9 cells, a murine teratocarcinoma cell line, were induced to differentiate into visceral endoderm or parietal endoderm cells when respectively treated with retinoic acid alone or combined with dibutyryl cyclic AMP. Although the synthesis of HSP47 in undifferentiated F9 cells was barely detectable, its synthesis in differentiated F9 cells was increased 20 - 30 fold. The synthesis of HSP47 was higher in parietal endoderm cells than in visceral endoderm cells.

In all the cell lines we have investigated, constitutive expression of HSP47 has been closely correlated with that of collagen (Table 1). When fibroblasts were transformed, the synthesis of both HSP47 and type I collagen was decreased. During the differentiation of F9 cells, the synthesis of HSP47 was concomitantly increased to the same extent as that of type IV collagen. In the cell lines which do not synthesize detectable level of collagen such as mouse myeloid leukemic cells (M1), and mouse pheochromocytoma cells (PC12), HSP47 synthesis was not observed, either.

Interaction of HSP47 with newly synthesized procollagen within endoplasmic reticulum

HSP47 was originally found to be a collagen-binding protein (Nagata

Table 1

HSP47 synthesis always correlates with collagen synthesis				
		HSP47	Collagen type I	type IV
Fibroblast	CEF	+++	+++	
	RSV-transformed CEF	+	+	
	BALB/3T3	+++	+++	
	SV40-transformed 3T3	+	+	
	NIH3T3	++	++	
	c-Ha-ras-transformed 3T3	+	+	
Adipocyte	3T3-L1	+++	+++	+
	Differentiated 3T3-L1	++	++	+
Myelocyte	M1	-	-	-
	Differentiated M1	-	-	-
Neurocyte	PC12	-	-	-
	NGF-treated PC12	-	-	-
Embryonic carcinoma	F9	+		+
	F9 treated with RA	++		++
	F9 treated with RA + dBcAMP	+++		+++

and Yamada, 1986), and was easily purified using gelatin- or collagen-coupled affinity chromatography. The affinity is so high that HSP47 cannot be dissociated from collagen-coupled Sepharose even in a buffer containing 2 M NaCl. However, HSP47 can be eluted from the collagen- or gelatin-Sepharose when the pH of the elution buffer is lowered to pH 6.3 (Saga et al., 1987). These observations suggest that the binding of HSP47 to collagen is regulated by pH.

Indirect immunostaining using polyclonal or monoclonal antibodies against HSP47 revealed HSP47 in the endoplasmic reticulum (ER) of chick embryo fibroblasts. Immuno-electronmicroscopy also demonstrated HSP47 in ER (Saga et al., 1987). These observations were consistent with recent results of cDNA sequencing data (manuscript submitted). HSP47 has an RDEL (Arg-Glu-Asp-Leu) sequence at the C-terminus of the polypeptide. Pelham and his colleagues showed that a specific mechanism exists for the retention of proteins in ER (Munro and Pelham, 1987). All the luminal proteins in ER so far examined, such as GRP78, GRP94 and PDI (protein

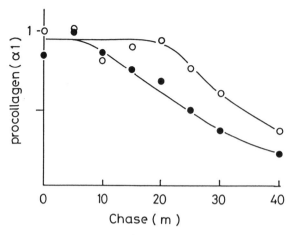

Figure 2. Kinetics of disappearance of total cellular procollagen and co-precipitated procollagen with HSP47 in CEF. Cells were pulse labeled for 10 min with [³⁵S]methionine, and chased for various periods in the presence of excess cold methionine. After trypsinized and treated with collagenase to remove extracellular collagen, cells were cross-linked in vivo for 30 min at 4°C with DSP. DSP was blocked with glycine, and cells were extracted with 1% NP40 containing excess gelatin which inhibits the interaction of procollagen with HSP47 after cells were lysed. Cell extract were immunoprecipitated with anti-HSP47 or anti-collagen. Open circle shows the kinetics of disappearance of total procollagen α1 band which was immunoprecipitated with anti-collagen, and thus it indicates the secretion of procollagen from the cell. Closed circle shows the procollagen bound to HSP47, which was co-precipitated with HSP47 using anti-HSP47.

disulfide isomerase), have KDEL (Lys-Glu-Asp-Leu) or similar sequences at the C-terminus; this signal functions as the retention signal of the proteins to be retained within the ER. The RDEL sequence in HSP47 might be an alternative version of KDEL.

Immunoprecipitation studies using anti-HSP47 antibodies revealed that two chains (a and b) of type I procollagen were co-precipitated with HSP47 (Nakai et al., 1990). To clarify the interaction of HSP47 and procollagen within the cells, we performed in vivo cross-linking experiments using permeable, thiol-cleavable cross-linker [DSP, dithiobis(succinimidylpropionate)] after pulse labelling of the cells with [³⁵S]-methionine. In vivo cross-linking and immunoprecipitation analysis using anti-HSP47 and anti-type I collagen further indicated that HSP47 binds to procollagen within the cells, presumably in the ER. Pulse label and chase experiments showed that the binding of procollagen to HSP47 in ER occurred immediately after its synthesis, and that procollagen was dissociated from the HSP47-procollagen complex after 10 min of pulse labelling, and then gradually disappeared from the cells due to its secretion (Fig. 2). When cells were treated with α,α'-dipyridyl, an iron-chelating reagent that inhibits the formation of triple helix collagen, abnormally-folded procollagen remained bound to HSP47 up to 2 hr. These results suggest that HSP47 binds to newly-

basic   acidic

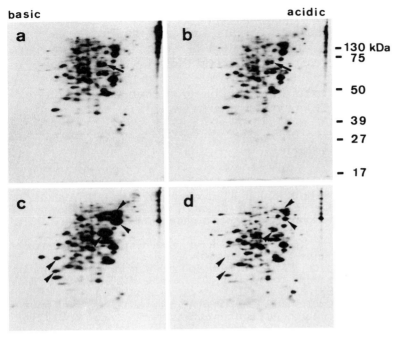

Figure 3. Inhibition in the induction of heat shock proteins in COLO 320DM cells by the treatment with quercetin. Cells were pretreated with 100 μM quercetin or 0.25% DMSO for 6 h, followed by heat shock at 43°C for 1.5 hr. After incubation at 37°C for 2 h, cells were labeled with [³⁵S]methionine for 1 h. Quercetin or DMSO as a vehicle was present during heat shock and recovery as well as the pretreatment period. A, Vehicle (0.25% DMSO) without heat treatment; B, 100 μM quercetin without heat treatment; C, vehicle with heat shock; D, 100 μM quercetin with heat shock. Arrows indicate the HSC70 (p72), and arrowhead show inducible-type hsps (HSP90, HSP70, HSP47 and HSP40).

synthesized procollagen in ER until the procollagen is secreted from ER. This function of HSP47 has been referred to as "molecular chaperon".

Inhibition of the expression of HSPs by the treatment with quercetin

The induction of HSP47 by heat shock in human cell lines was inhibited by the presence of quercetin, a bioflavonoid. Quercetin also inhibited the induction of the other stress proteins, namely, HSP100, HSP90, HSP70s, HSP40 (Ohtsuka et al., 1990), and HSP28 in HeLa cells and human colon carcinoma cell line, COLO 320DM after heat shock (Fig. 3). The inhibiton in the synthesis of HSP90, HSP70s, HSP47 and HSP28 after heat shock was confirmed by immunoprecipitation of the cell lysate from HeLa and COLO 320DM cells. The inhibition of the induction of HSP70 in quercetin-treated cells after heat shock was regulated at the level of transcription, which was examined by northern blot and CAT analysis (Hosokawa et al., 1990). In addition, the acquisition of thermotolerance was also inhibited by the presence of quercetin (Koishi et al., manuscript in preparation).

We are now trying to clarify the inhibitory mechanism of inhibition of the expression of HSPs by quercetin-treatment.

References

Hosokawa, N, Hirayoshi, K, Nakai, A, Hosokawa, Y, Marui, N, Yoshida, M, Sakai, T, Nishino, H, Aoike, A, Kawai, K and Nagata, K, Flavonoides inhibit the expression of heat shock proteins. Cell Struct. Funct., 15: in press.

Munro, S and Pelham, HRB, (1987) A C-terminal signal prevents secretion of luminal ER proteins. Cell, 48: 899-907.

Nagata, K and Yamada, KM, (1986) Phosphorylation and transformation sensitivity of a major collagen-binding protein of fibroblasts. J. Biol. Chem., 261: 7531-7536.

Nagata, K, Saga, S and Yamada, KM, (1986) A major collagen-binding protein of chick embryo fibroblasts is a novel heat shock protein. J. Cell Biol., 103: 223-229.

Nagata, K, Hirayoshi, K, Obara, M, Saga, S and Yamada, KM, (1988) Biosynthesis of a novel transformation-sensitive heat-shock protein that binds to collagen. Regulation by mRNA levels and in vitro synthesis of a functional precursor. J. Biol. Chem., 263: 8344-8349.

Nakai, A, Hirayoshi, K and Nagata, K, (1990) Transformation of BALB/3T3 cells by simian virus 40 causes a decreased synthesis of a collagen-binding heat-shock protein (hsp47). J. Biol. Chem., 265: 992-999.

Ohtsuka, K, Masuda, A, Nakai, A and Nagata, K, (1990) A novel 40-kDa protein induced by heat shock and other stresses in mammalian and avian cells. Biochem. Biophys. Res. Commun., 166: 642-647.

Saga, S, Nagata, K, Chen, W-T and Yamada, KM, (1987) pH-dependent function, purification, and intracellular location of a collagen-binding glycoprotein. J. Cell Biol., 105: 517-527.

In vitro Inhibition of Nascent Polypeptide Formation by HSP70 Proteins

M.J. Schlesinger, C. Ryan, S. Sadis*, L.E. Hightower*
Department of Molecular Microbiology
Box 8230, Washington University School of Medicine
660 South Euclid
St. Louis, MO 63110
USA

Introduction

Highly purified preparations of three kinds of HSP70 can inhibit formation of polypeptides made *in vitro* utilizing either extracts of rabbit reticulocytes or wheat germ and several different mRNAs. Inhibition was dose-dependent over a range of 0.1-0.4 nmoles of HSP70 and more pronounced at low temperatures. When bromo mosaic virus mRNAs were tested, inhibition was greater for the larger polypeptides indicating that HSP70 blocks nascent polypeptide elongation. With Sindbis virus 26S mRNA, the addition of HSP70 affected the formation of the capsid autoprotease activity leading to larger translation products. All of these data suggest that high concentrations of HSP70 can perturb the normal folding of nascent polypeptides and could explain why cells carefully autoregulate the level of HSP70.

Effect of *HSP70* on *in vitro* formation of nascent polypeptides

There is now abundant evidence that temperature stress rapidly leads to increased levels of unfolded or partially folded proteins in the cell. Two possible fates are postulated for such proteins: one of these results in ubiquitin conjugation and degradation of the polypeptide but the second allows for rescue and renaturation of the protein by virtue of the intervention of HSP70, a highly conserved heat shock protein found in almost all cells. Several isoforms of HSP70 have been described and they function in different compartments of the cell to form complexes with different kinds of proteins (Table 1).
In most of the examples described thus far, members of the HSP70

---

* Department of Molecular and Cell Biology, University of Connecticut, Storrs, CT, USA.

Table 1
Proteins that Form Complexes with HSP70
Clathrin-coated vesicles
Nuclear-encoded proteins targeted for insertion into mitochondria
Secreted proteins that do not enter the endoplasmic reticulum via a signal sequence
Prokaryotic enzymes forming the initiation complex for DNA replication
Newly imported proteins in the lumen of the endoplasmic reticulum and compartments of the mitochondria
Forms of tumor-suppressor proteins
Proteins present in the nucleolus and nucleus after a heat shock (possibly preribosomal complexes, splicosome complexes, chromatin proteins)
Steriod hormone receptors

family of proteins act as chaperones (Ellis, 1990) whereby they transiently bind to a cellular protein and then are released from the complex by hydrolysis of ATP.

The recognition of a polypeptide by HSP70 is believed to occur when hydrophobic domains which are normally buried inside the matured fully-folded protein are exposed to solvent in the partially folded structure. These same kinds of abnormal polypeptide structures are believed to be sites recognized by an isoform of the E3 ubiquitin conjugation enzyme (Hershko, 1988). Based on these hypotheses, we predicted that an HSP70 protein and an E3 ubiquitin conjugation enzyme might compete for unfolded proteins. A test of this model required an experimental system in which misfolded polypeptides were formed in the presence of both an active ubiquitin conjugation system and homogenous preparations of HSP70 that could be added to the system.

A suitable *in vitro* ubiquitination system is currently being studied in one of our laboratories. We had discovered that *in vitro* translation of truncated mRNAs in a rabbit reticulocyte lysate produced abnormal proteins that were multiubiquitinated (Agell et al., 1988). Thus, we could determine if the addition of a highly purified preparation of HSP70 to this *in vitro* system would affect ubiquitination of the newly

Table 2	
Inhibition of *in vitro* Ubiquitination and Translation of Polypeptides Encoded by a Ubiquitin mRNA	
Protein Band Analyzed	Radioactivity (cpm)
Polyubiquitinated 30 kDa (+BSA)	2,300
" " (+HSC70)	1,100
30 kDa + 14 kDa + 6 kDa (+BSA)	6,500
" " " " " (+HSC70)	4,600

For *in vitro* translation, the reaction mixture contained a commercial preparation (Promega) of wheat germ or rabbit reticulocyte lysate. BMV mRNAs were supplied with the lysate kits and other mRNAs were prepared by *in vitro* transcription of plasmid cDNAs containing an SP6 promoter immediately upstream of the translational start site (Agell et al., 1988; Wen and Schlesinger, 1986). RNA was isolated by phenol extraction and ethanol precipitation in the presence of wheat germ tRNA (10 µg/ml). The HSC70 was purified from bovine brain (Sadis et al., 1990) to yield a preparation of 7.1 mg/ml that contained about 60 pmoles ATP/µl. A typical wheat germ translation mixture contained 12.5 µl of lysate, 1.7 µl 1 mM amino acid mixture (-met), 0.75 µl RNAsin (40 units µl), 0.85 µl 35S-met (1000 Ci/mm; 15 µCi/µl), 2.95 µl M potassium acetate, 5.25 µl water and 1 µl mRNA (0.5 µg/µl). A typical reticulocyte mixture contained 17.5 µl lysate, 0.5 µl 1 mM amino acid mixture (-met), 1 µl RNA-sin, 1.35 µl 35S met, 3.65 µl water and 1 µl mRNA. A 5 µl sample of the above was used and supplemented with either 2 µl buffer, BSA or HSP70 at concentrations to give 0.1 to 0.4 nmoles protein. Incubation times ranged from 15 to 60 min at temperatures from 23° to 37°C. Reactions were stopped by addition of Laemmli (Laemmli, 1970) gel loading buffer and samples boiled 3 min before separation by SDS/PAGE. Gels were fixed with dimethyl sulfoxide, impregnated with PPO, dried and autoradiographed. Radioactive bands were excised and amounts of label measured by scintillation counting.

made polypeptide translated from the mRNA. The initial results of this kind of experiment showed that the HSP70 not only blocked ubiquitination but also inhibited overall nascent polypeptide chain formation (Table 2). To determine if the *in vitro* inhibition by HSC70 was a more general effect we tested several different mRNAs, two different in vitro protein synthesizing systems and three different preparations of HSP70. In all these experiments control samples containing similar amounts of bovine serum albumin (BSA) and amounts of ATP that were present in the HSP70 preparations were included. The latter was necessary because extra amounts of ATP (0.5 to 1 nmole) added to the in vitro translation lysates were inhibitory. Purified preparations of HSP70s often contain ATP because the final stage in purification involves affinity chromatography with an ATP-sepharose column and elution with ATP. We measured the amounts of ATP in all HSP70 samples by a luciferase assay. Based on these levels which ranged from 10 to 60 pmoles, the control sample were supplemented with an equivalent level of ATP.

The results with different preparations of HSP70 added to either reticulocyte or wheat germ lysates supplemented with bromo mosaic virus (BMV) mRNAs showed that formation of an *in vitro* translation product of 100 kDa was inhibited from 20 to 80 per cent (Table 3).

Both lysates showed inhibition with all three preparations of HSP 70. With a reticulocyte lysate, the inhibition of a 58 kDa BMV mRNA product was temperature dependent with a 65% block at 23°C and less than 10% at 37°C. At the higher temperature and an equivalent amount of time of *in vitro* translation, larger molecular weight proteins were formed and these were preferentially inhibited (data not shown). With a preparation of bovine brain HSC70, the extent of inhibition of a BMV mRNA 35 kDa polypeptide was linearly dependent on the amounts of HSP70 added from 0.1 to 0.4 nmoles.

Based on this value and an estimate of the amount of nascent chains formed *in vitro* (a value calculated from the levels of ^{35}S methionine incorporated and the specific activity of the label) we calculated that

Table 3	
Inhibition of *in vitro* translation products from BMV mRNA by preparations of HSP70 and different *in vitro* translation systems	
Preparation	**Inhibition (%)**

(1)Reticulocyte Lysate: HSC70
                               BiP
                               Uncoating ATPase
	65
	69
	84

(2)Wheat Germ Lysate: HSC70
                               BiP
                               Uncoating ATPase
	28
	46
	20

(3)Reticulocyte Lysate: HSC70 (0.2 nm)
                               58 kDa, 23°C
                               58 kDa, 30°C
                               58 kDa, 37°C
	
	65
	24
	8

In (1) and (2), the amounts of HSC70, BiP and uncoating ATPase were 0.4 nmoles and *in vitro* translations were carried out for 60 min at 23°C with BMV mRNA. BiP, purified from bovine liver microsomes, and uncoating ATPase from bovine brain (Flynn et al., 1989) were dialyzed and lyophilized to obtain solutions of 7.9 and 3.4 mg/ml, respectively. The ATP levels were about 10 pmoles/µl. A 100 kDa band was analyzed; the levels of cpms ranged from 30,000 to 144,000 in samples with BSA. In (3), BMV mRNA was incubated with lysate for 15 min. The level of cpms in the 58 kDa bands were 800, 5000 and 5100 at 23°, 30° and 37°C, respectively.

the ratio of added HSP70 molecules to nascent chains was about 1000. These data indicate that large amounts of HSP70 molecules can affect nascent polypeptide chain formation in a detrimental manner. In a recent report, Beckman et al., (Beckman et al., 1990) noted that HSP70 molecules could be isolated from polyribosomes obtained from Hela cells given short pulses of ^{35}S-methionine *in vivo*. The data were consistent with their conclusions that HSP70 forms complexes with nascent polypeptides and they proposed that this complex was part of a folding mechanism for the nascent chains. The *in vitro* data presented in our work also suggests that nascent polypeptide chains can form a complex with HSP70; however, some of these complexes appear to interfere with normal completion of the polypeptide synthesis.

One additional example from our work argues strongly that an HSP 70 is capable of perturbing nascent polypeptide chain folding. In this particular case, we tested an mRNA that encodes a polyprotein containing the structural proteins of Sindbis virus, a small RNA enveloped virus. This mRNA has a single site for initiation of translation and the first protein (noted as the capsid of the virus) synthesized from this mRNA folds to form a serine-protease catalytic site that autocleaves

Table 4	
Inhibition of Sindbis virus Autoprotease Formation by HSP70	
Protein Band Analyzed	Radioactivity (cpm)
30 kDa Virus Capsid (+BSA)*	236,000
40 kDa (+BSA)	3,700
30 kDa Virus Capsid (+HSC70)	102,000
40 kDa (+HSC70)	51,000
30 kDa Virus Capsid (+ inact. HSC70)	141,000
40 kDa (+ inact. HSC70)	3,800

* Indicates addition of either BSA or HSC70 to reaction mixture. HSC70 was heated at 60°C for 5 min to inactivate it. The identification of the 40 kDa band as a larger product of translation was based on V-8 protease mapping. Samples from a wheat germ translation that had been incubated with virus mRNA and 0.4 nm of HSP70 for 60 min at 30°C were separated by SDS/PAGE. The gel was washed twice with water for 1 hr, dried and autoradiographed. The position of the 40 kDa and capsid bands were excised, cut and incubated with .125 mM Tris, pH 6.8, 0.1% SDS, 1 mM EDTA for 1h. Gel pieces and the eluate were placed into wells of a 5-20% acrylamide gel containing a 5% acrylamide stacking gel and supplemented with 0, 0.5 or 5 µg S.aureus V-8 protease (Sigma). After electrophoresis the gel was fixed and fluorographed. Five separable bands of 5 to 14 kDa were found with the capsid and four of these were also detected in the digest of the 40 kDa band.

the protein co-translationally (Cancedda and Schlesinger, 1974). Temperature sensitive mutants with single amino acid changes in this protein lead to a misfolded protein at the nonpermissive temperature and in vitro translation of mRNAs from such mutants produce larger molecular weight proteins. When we added HSP70 to in vitro translation of the virus mRNA, we found both an inhibition of the capsid and appearance of a larger protein (Table 4).

Based on a V-8 protease digestion of the 40 kDa band and the capsid, it was clear that the larger protein was a readthrough translation product (see legend to Table 4). Thus, the HSP70 inhibited the autoprotease activity of the newly formed capsid. The most likely mechanism to account for these observations is the transient formation of a complex between HSP70 and a domain of the capsid protein that has not been able to completely fold.

It is known that cells autoregulate levels of HSP70 (Di Domenico et al., 1982; Tilly et al., 1983) and excessive amounts of the protein artificially introduced into a cell can block cell growth and viability. One explanation for this is offered by the data presented here which show a deleterious effect on protein synthesis by the promiscuous behaviour of the HSP70-heat shock protein chaperone.

Aknowledgments

We thank G. C. Flynn and J. E. Rothman for kindly supplying preparations of BiP and uncoating ATPase.

Supported by a grant from the National Science Foundation to MJS.

References

Agell, N, Bond, U and Schlesinger, MJ, (1988) In vitro proteolytic processing of a diubiquitin and a truncated diubiquitin formed from in vitro generated mRNAs. Proc. Natl. Acad. Sci. USA, 85: 3693-3697.

Beckman, RP, Mizzen, LA and Welch, WJ, (1990) Interaction of hsp70 with newly synthesized proteins: implications for protein folding and assembly. Science, 248: 850-854.

Cancedda, R and Schlesinger, MJ, (1974) Formation of Sindbis virus capsid protein in mammalian cell-free extracts programmed with viral messenger RNA. Proc. Natl. Acad. Sci. USA, 71: 1843-1847.

Di Domenico, BJ, Bugaisky, GE and Lindquist, S, (1982) The heat shock response is self-regulated at both the transcriptional and posttranscriptional levels. Cell, 31: 593-603.

Ellis, RJ, ed, (1990) Molecular chaperones. Semin. Cell Biol., 1: 1-72.

Flynn, GC, Chappell, TG and Rothman, JE, (1989) Peptide binding and release by proteins implicated as catalysts of protein assembly. Science, 245: 385-390.

Hershko, A, (1988) Ubiquitin-mediated protein degradation. J. Biol. Chem., 263: 15237-15240.

Laemmli, UK, (1970) Cleavage of structural proteins during assembly of the head of bacteriophage T4. Nature, 227: 680-685.

Sadis, S, Raghavendra, K and Hightower, LE, (1990) Secondary structure of the mammalian 70-kilodalton heat shock cognate protein analyzed by

circular dichroism spectroscopy and secondary structure prediction. Biochemistry, 29: 8199-8206.

Tilly, K, McKittrick, N, Zyulicz, M and Georgopoulos, C, (1983) The dnaK protein modulates the heat shock response of *Escherichia coli*. Cell, 34: 641-646.

Wen, D and Schlesinger, MJ, (1986) Regulated expression of Sindbis and vesicular stomatitis virus glycoproteins in *Saccharomyces cerevisiae*. Proc. Natl. Acad. Sci. USA, 83: 3639-3643.

HSP90, a Carrier of Key Proteins that Regulates Cell Function

I. Yahara, Y. Miyata, Y. Minami, Y. Rimura, S. Matsumoto,
S. Koyasu, N. Yonezawa*, E. Nishida*, H. Sakai*
The Tokyo Metropolitan Institute of Medical Science
Tokyo 113
Japan

Introduction

We have previously made two key observations which raised the possibility that HSP90 might be an actin-binding protein. First, a heat shock-resistant variant of CHO cell line that expressed HSP90 at relatively high level had more elongated morphology and higher ability to migrate than the parental strain (Yahara et al., 1986). This variant normally expresses other heat shock proteins such as HSP70. Second, immunofluorescence staining of human KB cells with anti-HSP90 has revealed that HSP90 is uniformly distributed throughout the cytoplasm and, in addition, is enriched in ruffling membranes (Koyasu et al., 1986). Ruffling membranes are brightly stained either with anti-actin antibody or with phalloidin, suggesting that ruffling membranes contain actin filaments to a large extent.

HSP90 is a Ca^{2+} -calmodulin-regulated actin-binding protein

We have examined the above possibility and found that purified HSP90 co-precipitated with polymerized actin (Fig. 1). The binding appeared to be specific (see below). A dissociation constant of the binding was calculated by reciprocal plot analysis of the binding data to be 2 to 3 x 10^{-6} M (Nishida et al., 1986), indicating that the interaction is relatively weak as compared to interactions of actin and other actin-binding proteins such as α-actinin. Hydrodynamic properties associated with HSP90 revealed that HSP90 exists as a homo dimer in the native state. It was also indicated that each dimeric form of HSP90 binds to at most 10 actin molecules existing in polymerized form under actin-excess conditions (Nishida et al., 1986). The low shear viscosity of polymerized actin solution was increased by HSP90 in a dose-dependent manner, suggesting that HSP90 cross-

---

*Faculty of Science, University of Tokyo, Tokyo 113

120

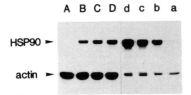

A B C D d c b a

HSP90 ►

actin ►

Figure 1: HSP90 was co-precipitated with polymerized actin.

linked actin filaments (Koyasu et al., 1986). Binding of tropomyosin to actin filaments inhibits further binding of HSP90 to the actin filaments. This result is consistent with the observation that HSP90 is not distributed on stress fibers to which tropomyosin binds. Calmodulin inhibits the binding of HSP90 to actin filaments in a Ca^{2+} ion dependent manner. The equilibrium gel filtration method using a Sephadex G-100 column equilibrated with 40 μg/ml calmodulin revealed that calmodulin binds HSP90 in the presence of Ca^{2+}. This result suggests that Ca^{2+}-calmodulin inhibits the binding by interacting with HSP90. Another member of the HSP90 family, HSP100 has been also shown to similarly bind actin filaments (Koyasu et al., 1986; 1990).

8S-glucocorticoid receptor binds to actin filaments through the HSP90 moiety of the receptor

The total cell lysates prepared from mouse Hepa 1 cells in a relatively low ionic strength containing molybdate were labeled with [^{3}H]triamcinolone acetonide and subjected to sucrose gradient centrifugation. The glucocorticoid receptor (GCR) as revealed by the radio-labeled ligand sedimented giving a peak with the sedimentation coefficient of 8S. The sedimentation peak of GCR was found to shift toward the bottom when anti-HSP90 antibody was added to the mixture before the centrifugation, indicating that the GCR complex contained HSP90 as a component (Fig. 2). The molybdate-stabilized 8S-GCR in the crude cell extracts was incubated with actin filaments. 8S-GCR was found to be co-precipitated with actin filaments by centrifugation (Miyata et al., submitted). When purified rabbit skeletal muscle actin was exogenously added to the mixture, GCR co-precipitated with actin filaments increased. These results strongly suggest that 8S-GCR in the crude cell lysates binds to actin filaments. Using partially purified 8S-GCR which was free from actin, we have shown that the binding of 8S-GCR to actin filaments was inhibited by either one of (1) HSP90, (2) tropomyosin, and (3) Ca^{2+}-calmodulin. These results strongly suggest that 8S-GCR binds actin filaments through its HSP90 moiety. Furthermore, 4S-GCR prepared in the absence of molybdate under relatively high ionic conditions was found not to bind actin filaments. This result also support the above conclusion.

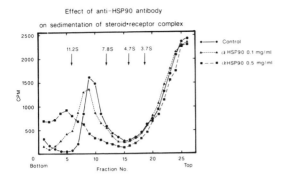

Figure 2: Sucrose density gradient centrifugation of 8S-GCR.

An interaction of 8S-GCR with cytoskeleton provides a novel anchoring mechanism in the citoplasm for proteins which possess nuclear location signals

Our preliminary results have revealed that 8S form of the dioxin receptor consisting of the ligand-binding polypeptide and HSP90 also binds to actin filaments. Other members of the steroid hormone receptor superfamily or the nuclear receptor superfamily including the thyroid hormone receptors and the retinoid receptors exist as low molecular weight complexes in the nucleus rather than in the cytoplasm, do not form complexes with HSP90, and do not interact with actin filaments.

Taken these results altogether, the localization of members of the steroid hormone receptor (or the nuclear receptor) superfamily may be determined as follows. (A) All members of the steroid hormone receptor (or the nuclear receptor) superfamily possess nuclear location signals and are, therefore, present in the nucleus unless they form complexes with HSP90 in the cytoplasm. The receptors forming complexes with HSP90 anchor on the cytoplasm using an interaction between the HSP90 moiety and cytoskeleton. (B) Alternatively, the nuclear location signals of GCR (and also other steroid hormone receptors) and the dioxin receptor may be hidden by HSP90 in the complexes. In either case, when ligands bind the receptor complexes, they are dissociated and the ligands-bound receptors are translocated into the nucleus.

References

Koyasu, S, Nishida, E, Kadowaki, T, Matsuzaki, F, Iida, K, Harada, F, Kasuga, M, Sakai, H and Yahara, I, (1986) Two mammalian heat shock proteins, HSP90 and HSP100, are actin-binding proteins. Proc. Natl. Acad. Sci. USA, 83: 8054-8058.

122

Koyasu, S, Nishida, E, Miyata, Y, Sakai, H and Yahara, I, (1990) HSP100, a 100-kDa heat shock protein, is a Ca^{2+}-calmodulin-regulated actin-binding protein. J. Biol. Chem., 264: 15083-15087.

Nishida, E, Koyasu, S, Sakai, H and Yahara, I, (1986) Calmodulin-regulated binding of the 90 kDa heat shock protein to actin filaments. J. Biol. Chem., 261: 16033-16036.

Yahara, I, Iida, H and Koyasu, S, (1986) A heat shock-resistant variant of Chinese hamster cell line constitutively expressing heat shock protein of Mr 90,000 at high level. Cell Struct. Funct., 11: 65-73.

Genetic Analysis of Heat Shock Protein Functions in Yeast

S. Lindquist
Howard Hughes Medical Institute
Department of Molecular Genetics and Cell Biology
The University of Chicago
Chicago, Illinois 60637
USA

Introduction

My laboratory has investigated the function of heat-shock proteins in *Saccharomyces cerevisiae* by cloning the genes encoding these proteins, creating mutations in the cloned genes by *in vitro* disruption mutagenesis, and replacing wild-type genes with the mutated versions by gene conversion *in vivo* (Rothstein, 1983). Here I will briefly review the results obtained with three genes, those encoding HSP26, HSP82, and HSP104 (Fig. 1).

HSP26

As determined by the heat-induced incorporation of radio-labelled amino acids, yeast cells produce only one major protein in the small hsp class, designated HSP26. Two different types of null mutations were created in the *hsp26* gene (Petko and Lindquist, 1986). In one case, a selectable auxotrophic marker was inserted between the *hsp26* promoter and the start-site of translation. In the other, every amino acid of the coding sequence was removed and replaced with a selectable marker. Cells carrying either mutation were phenotypically indistinguishable from each other and from wild-type cells (Petko and Lindquist, 1986; Petko, Ph.D. dissertation). They grew as well as wild-type cells at both high and low temperatures, in media that supported fermentative metabolism and in media that forced them into respiratory metabolism. They also survived short-term exposure to high temperatures or to high concentrations of ethanol as well as wild-type cells did. Moreover, they sporulated at the same rate as wild-type cells, germinated at the same rate, and withstood equally well heat treatments during sporulation and during germination in either rich media or under conditions of amino acid starvation. Their ability to survive long term storage in stationary phase or as spores was also unaffected. To ensure that the strain we had used for these experiments was representative of the species, the same mutations were introduced

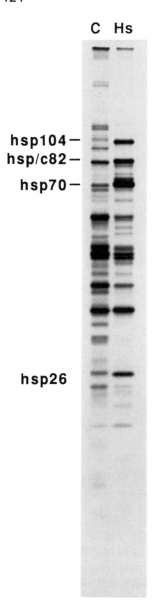

C Hs

hsp104 —
hsp/c82 —

hsp70 —

hsp26

Figure 1. The heat shock proteins of *Saccharomyces cerevisiae*. Log phase cells were pulse labeled with ³H-leucine at 25°C (C) or 30 min after a shift to 39°C (HS).

into two other laboratory strains of *S. cerevisiae* with very different genetic backgrounds. These strains grew at different rates in different media, reached different stationary phase densities, survived exposure to high temperatures at very different rates, and showed marked differences in their ability to sporulate. However, in every case in which cells carrying the *hsp26* mutation were compared to their isogenic wild-type parents, the two were indistinguishable.

This result was unexpected. All eukaryotic cells, and at least some bacteria, produce small proteins of this class in response to heat. These proteins are less conserved in amino acid sequence than the other heat-shock proteins, but they show clear homology with each other. The proteins are also remarkably conserved in their patterns of expression. In all organisms examined they are induced not only in response to heat shock, but in response to specific developmental and metabolic cues. In other words, these proteins, and their patterns of expression, appear to have been conserved over more than a billion years of evolution, strongly suggesting that they have an important function.

We have considered five explanations for the failure to detect a phenotype with the *hsp26* mutations. First, the yeast *hsp26* gene might not be a true member of the conserved, small-hsp gene family. Accordingly, we sequenced the gene and characterized the protein biochemically. The predicted amino acid sequence of the yeast *hsp26* gene reveals that the yeast protein is as closely related to the small hsps of plants, vertebrates, and insects as the proteins of these different species are to each other (Susek and Lindquist, 1989). The protein is also phosphorylated, as has been described for the mammalian species. More compellingly, the small hsp proteins of all organisms examined to date, including those of plants, *Drosophila*, and mammals, form large particles of a distinct size and shape. By size-exclusion chromatography, the yeast proteins form similar structures (Rossi and Lindquist, 1989). Thus, we believe the HSP26 protein of yeast is a *bona fide* member of the small hsp gene family.

Second, a closely related gene in the yeast genome might mask the effect of the *hsp26* mutation. This is quite common in *S. cerevisiae*. For example, haploid cells carry two genes for β-tubulin and three for glyceraldehyde-3-phosphate dehydrogenase. However, our attempts to find another gene that is closely related to *hsp26* were futile. No cross-hybridizing species were detected by low stringency nucleic-acid hybridization; no cross-reacting polypeptides were detected with a polyclonal antibody. Moreover, if yeast cells do contain such a gene it can not be very strongly induced by heat, as no other heat-induced proteins are readily detected in this region of the gel after heat shock. Third, *hsp26* may be functionally redundant with another gene that is not closely related to it. Precedence for this exists in both yeast and in other organisms. One method for uncovering the function of such a gene is to express it ectopically. We placed *hsp26* coding sequences under the control of the *gal 1* promoter so that the protein could be expressed at high levels at normal temperatures. No obvious differences were detected between these cells and wild-type cells. They grew at the same rate at high and low temperatures and were killed with the same kinetics during exposure to extreme temperatures. We have also combined the *hsp26* mutations with mutations in other heat-shock genes. To date, we have uncovered no new phenotypes, but we have not tested all combinations. Although there is no evidence to support this hypothesis, it remains viable. Redundancy is common in biological systems.

Fourth, HSP26 might have a unique cellular function that makes a significant contribution to fitness on an evolutionary time scale but is too subtle for the phenotype of the mutation to be detected by our methods. This is certainly a viable hypothesis, but the strong differences we observe between different strains of yeast in growth and survival at high temperatures are puzzling in this context. When other genetic polymorphisms make such major differences to temperature-related phenotypes, how could such an extremely subtle contribution from HSP26 be selected upon?

Fifth, the HSP26 may not have a specific, conserved cellular function, but might represent a primitive viral or selfish DNA element (Susek and Lindquist, 1989). We are prompted to make this suggestion by the fact that the yeast *HSP26* shows some limited homology to nucleases and to maturases that are implicated in the movement of introns within genomes. Moreover, the amino acid composition of HSP26 is similar to that reported for the proteinaceous component of a cytoplasmically inherited, RNA-containing particle in yeast cells (Wejksnora and Haber, 1978). In many other organisms, HSP26 has been reported to associate with RNAs. If such is the origin of the small hsps, it would not preclude them from having acquired important cellular functions in the course of evolution, as appears to be the case for certain crystallins (Ingolia and Craig, 1982).

HSP82

HSP82 is one of the most highly conserved and abundantly synthesized heat-shock proteins (Lindquist and Craig, 1988). Sequence analysis of homologous genes cloned from several, evolutionarily diverse organisms reveals that the proteins of *S. cerevisiae* have 62, 62, 63, and 43% identity with the proteins of *Homo sapiens*, *Drosophila melanogaster*, *Trypanosoma cruzi*, and *Escherichia coli.* Some organisms, including *D. melanogaster* and *E. coli*, have only one gene in this family. Vertebrate cells contain at least three genes in this family, including one whose protein product is transported into the endoplasmic reticulum. In all organisms investigated the proteins are abundant at normal temperatures and further induced by heat. In many they are developmentally regulated as well.

Biochemical analysis of HSP90, the homolog of HSP82 in vertebrate cells, suggests that it binds to a variety of other cellular proteins, modulating their activity. For example, HSP82 has been found to associate with several retroviral transforming proteins, in particular the oncogenic tyrosine kinases such as pp60[src] (Brugge et al., 1983; Courtneidge et al., 1982). In pulse chase experiments, immediately after synthesis pp60[src] is cytosolic, stoichiometrically complexed with HSP90, and phosphorylated at serine residues, but not at tyrosines. Thereafter, pp60[src] simultaneously loses its association with HSP90,

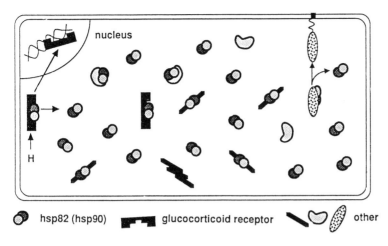

Figure 2. A simple early model for the functions of HSP82. The interaction of HSP82 (also known as HSP90) with steroid hormone receptors (dissociated by hormone), tyrosine kinases (prior to insertion in the membrane), and other cellular proteins (including actin and tubulin) are depicted.

is activated as a kinase, phosphorylated at tyrosine residues, and localized to membranes. HSP90 has also been found complexed with several different steroid hormone receptors, when these receptors are in their inactive state (Sanchez et al., 1985; Catelli et al., 1986). Transformation of these receptors to the active, DNA-binding state coincides with dissociation of HSP90 from the receptor. A major problem in the interpretation of such biochemical studies stems from the vast excess of HSP90 that is found in most tissues. Even at normal temperatures it is one of the 50 most abundant proteins in the cell. Thus, these complexes might be the result of artifactual association during cell lysis. If these interactions with other proteins are biologically meaningful, the vast excess of free HSP90 in the cell begs the question: does this reservoir of free protein serve another function?

To examine these issues, we have created mutations in the two closely related members of this gene family that are found in S. cerevisiae. HSC82 is expressed constitutively at a very high level and its expression is maintained at high temperatures. HSP82, which shares 96% identity with HSC82 at the amino acid level, exhibits low expression constitutively and is strongly induced by heat and by developmental signals (i.e., during the transition to stationary phase growth and during sporulation). Both proteins are cytosolic. Diploid cells have a total of four genes and, by site-directed mutagenesis, we have created a variety of mutant combinations (Borkovich et al., 1989). Cells homozygous for mutations in either gene are viable but cells homozygous for mutations in both genes do not grow at any temperature. Thus, these genes constitute an essential gene pair in yeast cells. Although the two proteins have different patterns of expression, they have equivalent functions. Remarkably, a gene encoding the mammalian

128

Postulated Influence of Temperature on hsp82 (⚭) Functions

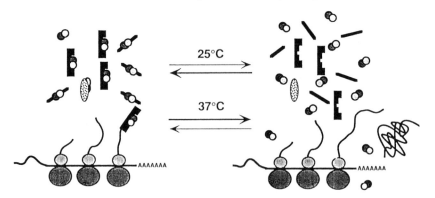

Figure 3. At high temperatures, HSP82-complexes are presumed to be unstable. At the bottom of the figure, HSP82 helps the glucocorticoid receptor to assume an active configuration.

cytosolic protein can fully compensate for the double mutant, producing cells that grow well over a wide range of temperatures (Kursheed, Picard and Lindquist, unpublished). Thus, the essential functional features of the protein have been conserved for over a billion years of evolution.

Cells harboring other combinations of mutations in the *hsp82* and *hsc82* temperatures decreases with the copy number of the genes. Recently, using promoter mutations, we have engineered cells to produce much lower concentrations of the HSP/HSC82 protein than is normally produced by a single gene (Picard et al., 1990; and Taulein and Lindquist, unpublished). These cells, grow well at 20°-25°C but are very temperature sensitive. Thus, HSP82 is normally produced in manyfold excess of the requirements for growth at 20°-25°C. However, growth at higher temperatures requires much higher concentrations of protein.

A simple first model for the function of HSP82 is compatible with both the biochemical experiments of other laboratories and the genetic data in our own lab (Fig. 2). It postulates that HSP82 binds to a variety of important cellular proteins, modulating their activity, until these complexes encounter an over-riding, biological signal. (An example would be the interaction the HSP82:steroid hormone receptor complex with hormone) It further postulates that these complexes are unstable at high temperatures (Fig. 3). Under these conditions, higher concentrations of HSP82 would be required to maintain the proper level of complex formation. The excess protein produced at 25°C, then, presumably acts as a buffer against any change in conditions that might affect these protein-protein interactions.

Experiments we have conducted in collaboration with Didier Picard and Keith Yamamoto have demonstrated, in at least one case, that the interactions of HSP90 with other cellular proteins are, in fact, biologically meaningful. They have also expanded our interpretation of the role of HSP82 in such complexes (Picard et al., 1990). The Yamamoto laboratory has created steriod-responsive yeast cells by transforming them with plasmids encoding the estrogen, mineral corticoid, and glucocorticoid receptors, together with reporter genes consisting of β-galactosidase coding sequences under the control of appropriate receptor binding elements. Wild-type yeast cells carrying these plasmids respond to the addition of hormone by inducing the synthesis of β-galactosidase. The response is equivalent in cells that produce the mammalian protein instead of the yeast protein. And in both cases, immunoprecipitation with anti-receptor antibodies indicates the receptor complexes with HSP82 in a 1:2 ratio, consistent with the work in vertebrate systems. In cells that produce very low concentrations of HSP82 and relatively high concentrations of receptor, most of the receptor is not complexed with HSP82. In this case, the ability of the cells to respond to steroids is severely reduced. These results argue that the interaction of HSP82 with receptor does not simply maintain the receptor in an inactive state. Rather, interaction with HSP82 is required to form a fully active receptor in the first place. We propose that HSP82 helps the receptor to assume a configuration, and to maintain a configuration, from which it can be readily activated by ligand (Fig. 3, lower left). Whether HSP82 is the sole agent in this process is a question for further experimentation.

HSP104

Most cells produce proteins of 100 to 110-kDa in response to heat (see Lindquist and Craig, 1988, and citations therein). (*Drosophila* would appear to be one of the rare exceptions, as no prominent protein in this size range is detected after heat shock.) Immunological characterization of the 100 kDa protein in mammalian cells indicates concentration in the Golgi. Antibodies against the 110 kDa protein reveal localization in or around nucleoli. Treatment with DNase disrupts this staining, while treatment with RNase eliminates it altogether. Both the 100- and 110-kDa proteins are constituents of normal cells and are also glucose regulated. Their induction patterns are complex: under various conditions they and the other glucose-regulated proteins are induced together, independently, or reciprocally.

S. cerevisiae has only one protein in this size class that is strongly heat-inducible. Mutations in this gene, designated HSP104, have a very specific phenotype (Sanchez and Lindquist, 1990). They do not effect growth at 25°C or at 37°C. They also have no effect on basal thermotolerance. That is, *hsp104* cells die at the same rate as HSP104 cells when shifted directly from 25°C to 50°C. However, *hsp104* cells are markedly defective in induced thermotolerance. When given a mild pre-heat treatment at 37°C before being exposed to 50°C, mutant cells die much more rapidly than wild-type cells (Fig. 4). The shape of the

130

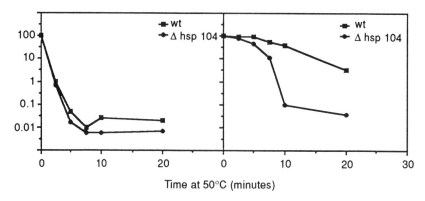

Figure 4. Thermotolerance is reduced in *hsp104* cells. Log phase cells were exposed to 50°C for various times with (right) and without (left) a conditioning pre-treatment at 37°C for 30 min. Viability was assessed by plating. (Reported from *Science,* 248: 1114, 1990; Copyright 1990 by the AAAS)

killing curve indicates that other cellular factors make a contribution to thermotolerance as well. That is, in the first few minutes of exposure to high temperature, both the mutant and the wild-type display equivalent levels of tolerance. After 20 min of exposure, the survival of mutant cells decreases by more than a thousand-fold relative to the wild-type. The *hsp104* mutation confirms a long standing assumption in the heat-shock field, that HSPs play a vital role in induced thermotolerance. The mutants are also compromised in their ability to acquire tolerance to ethanol, demonstrating that at least one hsp has broadly protective functions.

With the exception of yeast, genes encoding the highest molecular weight heat shock proteins of other organisms have not been cloned and the relationships between the proteins of different species are still unclear. Using the yeast HSP104 gene to probe Northern blots of RNAs from control cells and heat shocked cells, we have observed hybridization with a heat-inducible RNA of the correct size in cells as distantly related as those of humans. Thus, it appears that this protein, too, will prove to be very highly conserved in evolution.

References

Borkovich, KA, Farrelly, FW, Finkelstein, DB, Taulien, J and Lindquist, S, (1989) hsp82 is an essential protein that is required in higher concentrations for growth of cells at higher temperatures. Mol. Cell Biol., 9: 3919-3930.
Brugge, J, Yonemoto, W and Darrow, D, (1983) Interaction between the Rous sarcoma virus transferring protein and two cellular phosphoproteins: Analysis of the time-over and distribution of this complex. Mol. Cell. Biol., 3: 9-19.

Courtneidge, SA and Bishop, JM, (1982) Transit of pp60^{v-src} to the plasma membrane. Proc. Natl. Acad. Sci. USA, 79: 7117-7121.

Catelli, MG, Binart, N, Jung-Testas, I, Renoir, J-M, Baulieu, E-E, et al., (1985) The common 90Kd protein component of non-transformed '85' steroid receptors is a heat shock protein. EMBO J., 4: 3131-3135.

Ingolia, TD and Craig, EA, (1982) Four small Drosophila heat shock proteins are related to each other and to mammalian a-cristallin. Proc. Natl. Acad. Sci. USA, 79: 2360-2364.

Lindquist, S, (1986) The heat-shock response. Ann. Rev. Biochem., 55: 1151-1191.

Lindquist, S and Craig, EA, (1988) The heat shock proteins. Annu. Rev. Genet., 22: 631-677.

Petko, L and Lindquist, S, (1986) Hsp26 is not required for growth at high temperatures, nor the thermotolerance, spore development, or germination. Cell, 45: 885-894.

Picard, D, Khursheed, B, Garabedian, MJ, Fortin, MG, Lindquist, S and Yamamoto, KR, (1990) Reduced levels of hsp90 compromise steroid receptor action in vivo. Nature, 348: 166-168.

Rossi, JM and Lindquist, SL, (1989) The intracellular location of yeast HSP26 varies with metabolism. J. Cell Biol., 108: 425-439.

Rothstein, RJ, (1983) One-step gene disruption in yeast. Methods Enzymol., 101: 201-211.

Sanchez, ER, Toft, DO, Schlessinger, MJ and Pratt, WB, (1985) Evidence that the 90-kda phosphoprotein associated with the untrasformed c-cell glucocorticoid is a murine heat shock protein. J. Biol. Chem., 260: 12398-12401.

Sanchez, Y and Lindquist, S, (1990) HSP104 required for induced thermotolerance. Science, 248: 1112-1115.

Susek, RE and Lindquist, SL, (1980) hsp26 of Saccharomyces cerevisiae is related to the superfamily of small heat shock proteins but is without a demonstrable function. Mol. Cell Biol., 9: 5265-5271.

Wejksnora, PJ and Haber, JE, (1978) Ribonucleoprotein particle appearing during sporulation in yeast. J. Bacteriol., 134: 246-260.

The Stress Response in the Freshwater Polyp *Hydra*

T.C.G. Bosch, K. Gellner, G. Praetzel
Zoological Institute
University of Munich
Luisenstraße 14
8000 Munich 2
FRG

Introduction

Although not conspicuous, the freshwater coelenterate *Hydra* is widespread in ponds, lakes and rivers where polyps usually are attached to stones or vegetation. Hydra are phylogenetically old metazoa with a relatively simple tissue structure consisting of only two cell layers, a few cell types and a simple nervous net. Tissue growth in hydra occurs continuously and polyps have an extensive capacity for regeneration. These aspects made hydra to a favorite laboratory model for investigations in developmental biology. In our laboratory hydra serve as model to analyze the mechanisms regulating stem cell differentiation. For manipulating stem cells *in vivo* we began to look for hydra specific, inducible promoters to direct transcription of genes possibly involved in stem cell decisions. This effort was necessitated by the failure of commonly used promoters (TK, SV40, *Drosophila* HSP70, RSV) to activate transcription of the bacterial CAT gene in hydra cells (Bosch, Steele and Bode, unpublished data). Since heat shock genes are an obvious place to look for strong inducible promoters, we started to study the heat shock response in hydra. These studies yielded several surprising findings which we will summarize here. In particular, we will discuss recent work on the characterization of the major heat shock protein of hydra. We will also focus attention on the observation of species specific differences in the stress response in hydra and will argue that differences in the environment might be a critical element in the development of the stress response.

Protein synthesis in heat treated *hydra*

Shifting *Hydra vulgaris* (previously called *H. attenuata*, Holstein et al., 1990) from 18° to 22°C or higher temperatures induces synthesis of a major HSP which appears in 10% SDS/poly-acrylamide gels of molecular mass 60 kDa (Fig. 1; Bosch et al., 1988). This HSP, which

134

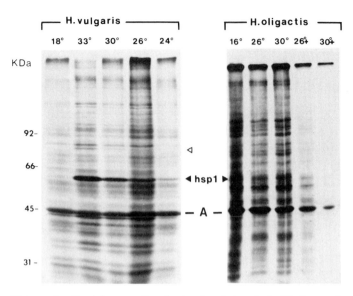

Figure 1. Protein synthesis in heat treated *Hydra vulgaris* and *Hydra oligactis*. 26+, *H. oligactis* at 26°C in medium containing 0.5 mM azide. 30+ *H. oligactis* at 30°C in medium containing 0.5 mM azide. A, actin. (Modified from Bosch et al., 1988).

previously was called HSP60, is now referred to as *hydra* hsp1. Synthesis of this protein occurs rapidly and the level of synthesis depends on the stress temperature. Induction of this protein is detectable not only after exposure to heat but also after exposure to cadmium ions (25µM) and sodium azide (0.5 mM) (Bosch et al., 1988). In addition to the 60 kDa hsp1, several other stress proteins in the 80-90 kDa and 20-30 kDa size range are synthesized in response to stress although at levels much lower than the 60 kDa protein. The finding of the major *hydra* heat shock protein to be of apparent molecular mass of 60 kDa was quite unexpected because in all organisms studied so far the major HSP is of 70 kDa (HSP70, Lindquist and Craig, 1988). There are, in principle, two ways by which a 60 kDa major hsp can be explained: (1) *Hydra* hsp1 is a "truncated version" of the virtually ubiquitous HSP70; or (2) *Hydra* hsp1 is evolutionary unrelated to HSP70. In a first attempt to clarify the relationship between *hydra* 60 kDa stress protein and the evolutionary conserved HSP70 family, we tested several antibodies known to crossreact with this protein family.

Immunological relationship between *hydra* hsp1 and the conserved stress protein families in the 60 to 70 kDa size range

All members of the *hsp70* protein family, including the *E.coli* DnaK protein and the HSP70s from yeast, nematodes, sea urchins,

Drosophila and mouse, are immunologically related (Craig, 1985). Therefore, we used antibodies against various HSP70 proteins and analyzed whether they could cross-react with *hydra* hsp1. The antisera used included polyclonal antiserum against the purified *E.coli* DnaK gene product (provided by Angela Mehlert and Douglas Young, London) as well as a monoclonal antibody specific for *Drosophila* HSP70 (tested by Susan Lindquist, U. of Chicago). Both antisera were previously shown to detect HSP70 related proteins in a wide variety of different organisms (Young et al., 1988; Lindquist, pers. communication). Using extracts from *H.vulgaris* polyps which had been cultured at 18°C or had been subjected to heat shock, we found in Western blots anti-DnaK antiserum not reacting with a protein in the 60 kDa size range but detecting three proteins of apparent molecular weight of 68-75 kDa (Bosch and Praetzel, 1990). There was no detectable increase in the mass of these proteins after heat shock treatment of polyps. Analogous results were obtained when testing a *Drosophila* HSP70 mono-clonal antibody for an evolutionary conserved epitope of HSP70. Again, this antibody detected proteins in the 68-75 kDa size range but did not bind to *hydra* hsp1 (unpublished data). In conclusion, *hydra* hsp1 appears not to be detected by antisera known to cross-react with HSP70 proteins in several different organisms.

Recently a new class of ubiquitous stress proteins in the 60 kDa size range was identified and termed "chaperonins" (Ellis, 1987; Hemmingsen et al., 1988). In light of the molecular mass of 60 kDa and the negative results with the HSP70 antisera, *hydra* hsp1 could be related to the chaperonin class of heat-inducible proteins. If so, it should cross-react with anti-HSP60 antisera. Figure 2 shows that polyclonal antiserum directed against *Tetrahymena* HSP60 (kindly provided by Richard Hallberg, Iowa State U.) indeed cross-reacts with the major heat inducible stress protein in *H.vulgaris*. Identical results were obtained using yeast HSP60 antiserum (data not shown). Thus, *hydra* hsp1 appears to be immunologically related to the 60 kDa chaperonin class of stress proteins.

Cloning and characterization of the gene for the major *hydra* HSP

Based on these immunological observations, to isolate *hydra* cDNA encoding HSP1 a yeast *HSP60* gene (gift from Richard Hallberg, Iowa State U.; for details see Reading et al., 1989) was used to screen a *hydra* cDNA library (provided by Eva Kurz) under low stringency conditions. So far a near full length cDNA clone (*chsp1*) was isolated and characterized (Gellner and Bosch, unpublished results). Northern-blot analysis indicated that in *H.vulgaris* the *hsp1* mRNA level is drastically increased upon heat shock (Gellner and Bosch, unpublished results). This contrasts with the two- to fourfold induction of *hsp60* transcription seen in *Tetrahymena* (McMullin and Hallberg,

136

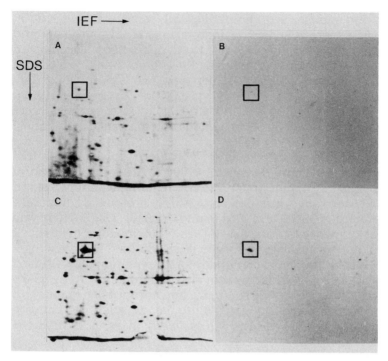

Figure 2. Antigen-crossreactivity of *Hydra vulgaris* hsp1 (square) with antiserum directed against *Tetrahymena* HSP60. A and C, fluorogram of normal (A) and heat shocked (B) polyps. B and D, Western blot analysis from normal (B) and heat shocked (D) polyps. The antiserum was provided by Richard Hallberg, Iowa State U.

1987) and yeast (McMullin and Hallberg, 1988). Sequence analysis revealed striking homologies between the predicted hydra *hsp1* gene product and members of the HSP70 family. *Hydra* hsp1 was found, for example, 73% homologous to *Xenopus* HSP70, 78% homologous to human HSC70 and 69% homologous to *Drosophila* HSP70. No overall homology, however, could be detected to yeast HSP60 which was used as screening probe. This finding was completely unexpected, and also very puzzling. How can a *hsp60* specific probe hybridize to an *hsp70* homologous gene?

Sequence analysis of *hydra* chsp1 reveals that 70 and 60 kDa heat shock proteins share a common carboxy terminal structural motive

Detailed sequence analysis revealed a region of homology between *hydra* hsp1 and yeast HSP60 at the carboxy terminus where both proteins share a tetrapeptide repeat sequence Gly-Gly-Met-Pro. In yeast HSP60 the region from amino acid 558 to 573 is 76% homologous

Human	hsp70	613	A..GGPGPGGFGAQGPKGGSG...............SGPTIEEVD*
Xenopus	hsp70	615	GVPGGV.PGGM.PGSSC.GAQARQGGN.........SGPTIEEVD*
Yeast	SSA2	608	A..GGA.PEGAAPGG.FPGGAPPAPEAE.........GPTVEEVD*
C.elegans	hsp70A	615	A..GGA.PPGAAPGGAA.GGAG...............GPTIEEVD*
Petunia	hsp70	621	A..GGA..TMDEDGPSVGGSAGSQTGA..........GPKIEEVD*
Drosophila	hsc70	614	A...GFPPGGM.PGG..GGGMPGAAGAAGAAGAGGASGPTIEEVD*
Human	hsc70	614	A..GGM.PGGM.PGG.FPGGGAPPSGGAS.......SGPTIEEVD*
Mouse	hsc72	614	A..GGM.PGGM.PGG.FPGGGAPPSGGAS.......SGPTIEEVD*
Trypanosoma	hsp70	629	GMPGGM.PGGM.PGG.M.GGGMGGAAAS........SGPKVEEVD*
Hydra	hsp1	n.d	A.GGGM.PGGM.PGG.M.PGGMPGSGSKA......SGGPTIEEVD*
Yeast	hsp60	558	A..GGM.PGGM.PG..M.PGMM*
E.coli	groEL	535	A..GGM..GGM.GG..M.GGMM*
Human	hsp60	560	AM.GGM.GGGM.GGG.MF*

Figure 3. Alignment of deduced C terminal amino-acid sequences of 60 and 70 kDa stress proteins. Helix-breaking amino acids with bend potential > 1.4 (Chou and Fasman, 1978) are shaded. Alignment was done in the laboratory of Dr. Gojobori, Natl. Inst. Genetics, Mishima, using the algorithm of Hein (Hein, 1990).

to the corresponding region in *hydra* hsp1; homology on the nucleotide level is 81%. When C-terminal sequences of other members of the HSP60 and HSP70 families were compared, we found (Fig. 3) that all 60 and 70 kDa stress proteins share this sequence in more or less conserved form.

Two features of this highly conserved sequence are of interest. First, the abundance of proline and glycine residues at the C-terminus of 60 and 70 kDa stress proteins could provide structural flexibility since these residues are known as disrupters of α-helix and β-strand secondary structures (Chou and Fasman, 1976). It is interesting in this regard that in some members of the stress protein family (e.g., the *E.coli* HSP70 homologue DnaK) proline and glycine are replaced by residues with a similar high potential of polypeptide chain reversal such as asparagine, aspartic acid and serine.

The secondary structure of this C-terminal glycine- and proline-rich sequence recently was analyzed in rat HSC70 and found to contain mostly aperiodic structures and β-turns (Sadis et al., 1990).

Second, strongly hydrophobic residues (Nozaki and Tanford, 1971) such as methionine, phenylanaline, valine and isoleucine occur in a repeated pattern at the C-terminus of both HSP60 and HSP70 proteins. Recently, a methionine-rich domain was suggested to be involved in signal sequence recognition via hydrophobic interactions (Bernstein et al., 1990). By analogy, methionine residues could play a similar role at the C-terminus of stress proteins. The evolutionary conservation of a repeat sequence favoring structural flexibilty and hydrophobic interactions suggests that the C-terminus of both 60 and 70 kDa stress proteins may be an important site for the "folding/ unfolding" function common to both classes of heat shock proteins

138

(Beckmann et al., 1990; Hartl and Neupert, 1990; Phillips et al., 1990).

With respect to the immunological observation in Figure 2 it remains to be shown whether the c*hsp1* gene product can be detected by the yeast HSP60 antiserum. Preliminary data from deletion mutants (Hallberg, pers. communication) seem to indicate that one major epitope of the HSP60 antiserum is at the C-terminus suggesting that in addition to HSP60 other proteins containing this conserved sequence (such as the c*hsp1* gene product) might be detected by the antiserum.

Differences in the stress response in *hydra* species from different habitats

When examining the stress response of various species of hydra we observed species-specific differences in the ability to synthesize hsp1 in response to stress (Bosch et al., 1988): While a strong induction of hsp1 synthesis was observed in *H.vulgaris* and *H. magnipapillata*, no induction could be seen, e.g., in *H.oligactis* (see Fig. 1). We have preliminary evidence indicating that in *H.oligactis* the level of activation of transcription of c*hsp1* in response to stress is much lower than in *H.vulgaris* (Gellner and Bosch, unpublished results). The apparent absence of a heat shock response is correlated in all cases with a low level of resistance to environmental stress. Species which fail to synthesize hsp1 in response to stress are also unable to acquire

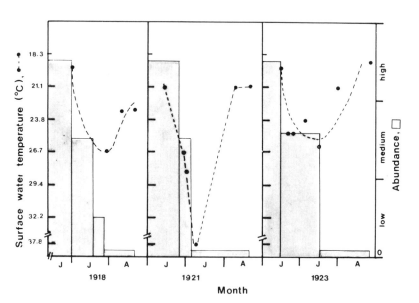

Figure 4. Seasonal decline of *Hydra oligactis* population from surface water is tied to water temperature. Drawn from data in Welch and Loomis (Welch and Loomis, 1924).

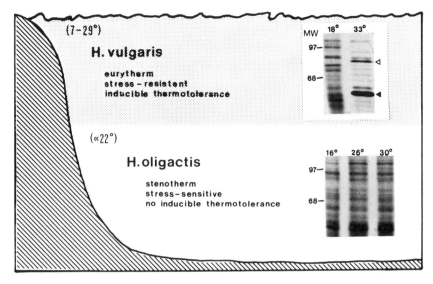

Figure 5. Species-specific differences in the stress response in *Hydra vulgaris* and *Hydra oligactis* are correlated with different habitats.

thermotolerance (Bosch et al., 1988). In light of the reduced stress-resistence of species such as *H. oligactis* one would expect to find these species in thermally more stable habitats than species with a strong stress response. Few ecological field studies have been undertaken in hydra. However, as summarized below, the data available indeed support the assumption that in their natural environment *H. oligactis* and *H. vulgaris* have different temperature preferences. *H. oligactis* is strictly stenotherm (Holstein, 1990) with a low upper lethal temperature of 26°C (Schroeder and Callaghan, 1981) and common in lakes and fast flowing rivers but absent in shallow, warm water habitats (Schroeder and Callaghan, 1981; P. Tardent, personal communication). Daily maximum temperatures above 22°C are accompanied by disappearance of the *H. oligactis* population from surface waters resulting in conspicuous seasonal fluctuations in abundance (Fig. 4; Welch and Loomis, 1924; Ribi et al., 1985).

Whether disappearance is due simply to migration to deeper waters or to death of the polyps remains to be shown. This trend in the vertical distribution of *H. oligactis* to deep and cold water is also evident from Paetkau's (Paetkau, 1967) and Schroeder and Callaghan's (Schroeder and Callaghan, 1981) field studies showing restriction of *H. oligactis* to the metalimnion (12°-15°C) during the summer. In light of this vertical distribution Paetkau (Paetkau, 1967) referred to *H. oligactis* as "deep-water" form. In addition to low temperature, a stable pH between 7.2 and 7.5 was found as second environmental factor characteristic for *H. oligactis* habitat (Paetkau, 1964). The water temperature appears to limit the geographical distribution of *H. oligactis* to habitats in temperate regions of North America, Europe and Japan and to preclude its survival in, e.g., the southeastern states of America (Hyman, 1930). In contrast, *H.vulgaris* and *H. magnipapillata* are eurytherm (7°-29°C; Holstein, 1990) and appear to inhabit habitats with a much wider range of temperatures than *H. oligactis* including shallow warm water ponds. Seasonal fluctuations in abundance are not observed in these species (Holstein, 1990). In conclusion, *H.vulgaris* and *H.oligactis* appear to inhabit habitats different in temperature and, possibly, stability of pH. Therefore, since our comparative analysis of the stress response in hydra revealed strong species specific differences, we propose (Fig. 5) that differences in the environment can be a critical element in the development of the heat shock response. Species such as *H.oligactis*, which live in a low and constant temperature environ-ment with little change in pH might have lost the ability to develop a "classical" heat shock response while eurytherm species such as *H.vulgaris*, which are adapted to a variable habitat with a wide range of temperatures, respond to changes in temperature (or salinity) with the immediate synthesis of stress proteins. Analogous to animals that live in caves and have lost the ability to see, it appears conceivably that animals that live in constant temperature environments have lost the ability to induce the synthesis of heat shock proteins.

Conclusions

The heat shock response is universal from bacteria to man (Lindquist and Craig, 1988). However, experiments in heat shock, for obvious reasons, concentrated on standard laboratory animals or cells. There is little information, for example, on the heat shock response in organisms adapted to freshwater or marine habitats with stable temperatures such as constant-temperature springs or the deep sea. Field studies demonstrating a higher upper lethal temperature for a pupfish from a stream with variable temper-atures than from a constant-temperature spring (Hirshfield et al., 1980) are intriguingly similar to the findings in hydra. It would be particularly interesting to know if the reduced temperature tolerance in the spring pupfish correlates with the failure to develop a stress response. We are convinced that extending the studies we have started in hydra to other taxa and comparing species from constant and variable environments will lead to a better understanding of the biological relevance of the stress response under natural conditions.

What molecular mechanisms cause those species-specific differences? Why are thermosensitive species of hydra not able to increase synthesis of hsp1 in response to stress in a way thermo-resistant species do it? One way of thinking about the molecular "defect" of *H.oligactis* may be in terms of changes in the hsp1 regulatory region, i.e., altered promoter sequences or transacting factors. Alternatively, action of factors controlling translation of hsp1 may be modulated in *H. oligactis*. These possibilities are now being examined in our laboratory.

Acknowledgements

We thank Charles David for discussion and support; Richard Hallberg for discussion, providing molecular probes and sharing unpublished results; Richard Campbell for discussion about ecological work on hydra; T. Gojobori for help with the multiple sequence alignment; and Toshi Fujisawa for critically reading the manuscript. This work was supported by the German Science Foundation (SFB 190).

References

Beckman, RP, Mizzen, LA and Welch, WJ, (1990) Interaction of Hsp70 with newly synthesized proteins: Implications for protein folding and assembly. Science, 248: 850-854.

Bernstein, HD, Poriz, MA, Strub, K, Hoben, PJ, Brenner, S and Walter, P, (1989) Model for signal sequence recognition from amino-acid sequence of 54K subunit of signal recognition particle. Nature, 340: 482-486.

Bosch, TCG, Krylow, SM, Bode, HR and Steele, RE, (1988) Thermotolerance and synthesis of heat shock proteins: These responses are present in *Hydra attenuata* but absent in *Hydra oligactis*. Proc. Natl. Acad. Sci. USA, 85: 7927-7931.

Bosch, TCG and Praetzel, G, The heat shock response in hydra: immunological relationship of hsp60, the major heat shock protein of *Hydra vulgaris* Pallas, to the ubiquitous hsp70 family. Hydrobiologia, in press.

Chou, PY and Fasman, GD, (1978) Empirical predictions of protein conformation. Ann. Rev. Biochem., 47: 251-276.

Craig, EA, (1985) The heat shock response. CRC Crit. Rev. Biochem., 18: 239-280.

Hartl, F-U and Neupert, W, (1990) Protein sorting to mitochondria: Evolutionary conservations of folding and assembly. Science, 247: 930-938.

Hein, J, (1990) Unified approach to alignment and phylogenies. Meth. Enzym., 183: 626-645.

Hemmingsen, SM, Woolford, C, van der Vies, SM, Tilly, K, Dennis, DT, Georgopoulos, CP, Hendrix, V and Ellis, RJ, (1988) Homologous plant and bacterial proteins chaperone oligomeric protein assembly. Nature, 333: 330-334.

Hirshfield, MF, Feldmeth, CR and Soltz, DL, (1980) Genetic differences in physiological tolerances of Amargosa pupfish (*Cyprinidon nevad-ensis*). Science, 207: 999-1001.

Holstein, TW, Cnidaria-Hydrozoa. In Süsswasserfauna Mittel-europas, Schwoerbel, J, ed., Fischer Verlag, Konstanz., in press.

Holstein, TW, Campbell, RD and Tardent, P, (1990) Identity crisis. Nature, 346: 21-22.

Hyman, L, (1930) Taxonomic studies of the hydras of North America. 2. The characters of *Pelmatohydra oligactis*. Pallas Trans. Amer. Micros. Soc., 49: 322-329.

Lindquist, S and Craig, E, (1988) The heat shock proteins. Annu. Rev. Genet., 22: 631-677.

McMullin, TW and Hallberg, RL, (1987) A normal mitochondrial protein is selectively synthesized and accumulated during heat shock in

Tetrahymena thermophila. Mol. Cell. Biol., 7: 4414-4423.

McMullin TW and Hallberg, RL, (1988) A highly evolutionary conserved mitochondrial protein is structurally related to the protein encoded by the *Escherichia coli* groEL gene. Mol. Cell. Biol., 8: 371-380.

Nozaki, Y and Tanford, C, (1971) The solubility of amino acids and two glycine peptides in aqueous ethanol and dioxane solution. J. Biol. Chem., 246: 2211-2217.

Paetkau, P, (1964) The taxonomy and ecology of the hydridae (Hydrozoa) of Alberta and the Northwest territories. Thesis, Dept. Zoology, Univ. Edmonton, Alberta.

Phillips, GJ and Silhavy, TJ, (1990) Heat-shock proteins DnaK and GroEL facilitate export of LacZ hybrid proteins in *E. coli.* Nature, 344: 882-884.

Reading, DS, Hallberg, RL and Myers, AM, (1989) Characterization of the yeast *HSP60* gene coding for a mitochondrial assembly factor. Nature, 337: 655-659.

Ribi, G, Tardent, R, Tardent, P and Scascighini, C, (1985) Dynamics of hydra populations in Lake Zürich, Switzerland, and Lake Maggiore, Italy. Schweiz. Z. Hydrol., 47: 45-56.

Sadis, S, Raghavendra, K and Hightower, LE, (1990) Secondary structure of the mammalian 70-kilodalton heat shock cognate protein analyzed by circular dichroism spectroscopy and secondary structure prediction. Biochem., 29: 8199-8206.

Schroeder, LA and Callaghan, WM, (1981) Thermal tolerance and acclimation of two species of *Hydra.* Limnol. Oceanogr., 26: 690-696.

Young, D, Lathigra, R, Hendrix, R, Sweetser, D and Young, RA, (1988) Stress proteins are immune targets in leprosy and tuberculosis. Proc. Natl. Acad. Sci. USA, 85: 4267-4270.

Welch, PS and Loomis, HL, (1924) A limnological study of *Hydra oligactis* in Douglas Lake, Michigan. Trans. Am. Microscop. Soc., 43: 203-235.

Heat Shock Response during Morphogenesis in the Dimorphic Pathogenic Fungus *Histoplasma capsulatum*

L. Carratù[#], B. Maresca [#*]
[#]International Institute of Genetics and Biophysics
Via Marconi, 12
80125 Naples
Italy

Introduction

Temperature changes (and a variety of other stimuli) coordinately induce in all organisms active transcription of specific genes (heat shock genes). In dimorphic organisms the increase in the temperature of incubation has important consequences since temperature works as a signal for adaptation (induction of heat shock phenomenon) and triggers the phase transition. In fact, parasites and dimorphic organisms in general may exist in multiple hosts that have different body temperatures, and exhibit distinct stages during morphogenesis. Therefore, these organisms are particular cases of developmentally regulated heat shock response. An important characteristic of morphogenesis in some dimorphic fungi is that induction of one phase leads to the form found in infected tissue. Dimorphic pathogenic fungi have in fact the unique ability to colonize host tissue that is parallel to and may be intimately involved with the developmentally regulated morphological transition. In laboratory conditions, dimorphism in the human pathogenic fungus *H.capsulatum* is directly and reversibly controlled by temperature changes: mycelium (25°C) ⟺ yeast (37°) (Schwarz, 1981; Maresca and Kobayashi, 1989). Therefore, in this fungus, temperature is a primary factor that controls phase transition. Since a sudden temperature change to 37°C induces the mycelial to yeast-phase transition, Lambowitz et al. suggested that the early event of the mycelium to yeast transition in *H.capsulatum* was a "heat shock response" which was followed by cell adaptation to the higher temperature (Lambowitz et al., 1983). Furthermore, it has been suggested that also in other diphasic organisms morphogenesis might be a by-product of the heat shock response (Maresca and Kobayashi,

---

[*] Washington University, School of Medicine, Dep. of Internal Medicine, St. Louis, MO, 63110, USA

1989). However, in these systems it is difficult to distinguish the effect of temperature on morphogenesis from the induction of the heat shock response. In fact, in *H.capsulatum* at temperatures at which heat shock proteins are not induced (heat shock mRNAs are not measurable) phase transition from mycelium to yeast does not occur (Maresca, unpublished).

Contrary to the heat shock response in other systems (in which temperature must be lowered to ensure survival), in dimorphic organisms the "heat shock" must be maintained constant and it is an essential part of the life cycle of the organism. Further, heat shock genes are among those which are induced very early during phase transition and in these organisms may play a critical role for adaptation to the new environment. It seems most likely that the developmental control elements are located further upstream to the TATA box than the sites for heat shock gene expression. Further, the regulatory proteins which confer developmental control are distinct from the heat shock transcription factor (HSF; Bienz and Pelham, 1987; Glaser and Lis, 1990). However, recently it has been shown by deletion mutations analysis that in *S.cerevisiae* the *cis*-regulatory elements responsible for *hsp26* heat induction are not separated from those involved during stationary phase-growth (Susek and Lindquist, 1990).

hsp70 and *hsp82* transcription

We have investigated, in several fungal isolates that exhibit different temperature sensitivity and virulence, the regulation and the possible roles that *hsp70* and *hsp82* genes have during morphogenesis by studying their expressions at various temperatures (Caruso et al., 1987; Medoff et al., 1986; Minchiotti et al., submitted). We have defined the degree of thermoresistance as the capacity of maintaining coupled oxidative phosphorylation at a specific temperature during the heat-induced morphogenesis. Further, we and other authors have shown that a strong correlation exists between temperature sensitivity and the degree of virulence in a mouse model (Medoff et al., 1986). The higher the temperature necessary to uncouple respiration (39° to 43°C), the greater is the level of virulence (Maresca and Kobayashi, 1989). We examined the transcription of a member of *hsp70* gene family in a low virulent and temperature sensitive Downs (ATP synthesis is uncoupled from electron transport when mycelia are incubated at 37°C) and in G222B strain, a virulent and temperature resistant strain (in mycelial cells respiration uncouples only when cells are incubated at >40°C; Caruso et al., 1987; Patriarca et al., submitted). We cloned an *hsp70* gene from a cosmid library of the Downs strain and determined its nucleotide sequence. We found that the cloned *hsp70* gene contains two overlapping HSE elements and has a high degree of homology to other cloned *hsp70* sequences (Caruso et al., 1987; Lindquist, 1986; Maresca et al., unpublished). We studied the optimum of heat shock mRNA transcription in strains with different temperature sensitivity. We have found by Northern blot that a 2.3 kb mRNA is transiently induced in the first 6 hr in heat shocked cells during the mycelium to yeast transition. Maximal transcription occurred at 34°C in the less pathogenic and more temperature-sensitive Downs strain. However, a temperature of 37°C was necessary to induce maximal transcription in the more temperature-resistant G222B strain.

We have also cloned a nucleotide sequence from *H.capsulatum* G222B strain corresponding to an *hsp82* gene, as judged by sequence analysis and transcriptional studies. We determined the entire nucleotide sequence and its flanking regions (Minchiotti et al., submitted). The corresponding protein contains 680 amino acids with a predicted molecular weight of ca. 82 kD. The coding sequence closely resembles that of *S.cerevisiae* (>70% homology in protein sequence). The *H.capsulatum* gene contains a single HSE located between -377 and -363 nt with respect to the major transcriptional start site. Further, *hsp82* gene contains two introns: the first is 122 nt long (IVS I, located at 261 downstream to the ATG) and the second 86 nt (IVS II, located 988 nt downstream to the ATG). We have analyzed *hsp82* mRNA expression in two strains of *H.capsulatum* and found the predicted mRNA transcript of 2.8 kb. We found that in the first 3 hr after temperature shift, maximal transcription of *hsp82* mRNA occurred at 34°C in the Downs strain, while a temperature of 37°C was necessary to induce maximal transcription in the G222B strain (Minchiotti et al., submitted). Thus, these results are comparable to those we previously obtained with *hsp70* gene in the same strains of this fungus. Protein data corroborated these results: Shearer et al. have in fact demonstrated that synthesis of heat shock proteins follows a similar profile (Shearer et al., 1987). The pattern of protein synthesis is quite different from strain to strain and depends on the temperature used to induce phase transition. Heat shock protein synthesis peaks at 34°C in Downs, whereas in G184B and G222B it is higher at 37°C. Thus, it seems that the Downs strain reflects a general defect which results in decreased heat shock response both at transcriptional and translational level than in the more virulent strains at the physiological temperature of 37°C.

Similar behavior in heat shock response in virulent and avirulent strains has been recently demonstrated in bacterial systems. In fact, it has been shown that virulent and avirulent strains of *Listeria monocytogenes* (Morange et al., 1990), *Mycobacterium leprae* and *M.tubercolosis* (Young et al., 1990) exhibit a different pattern of heat shock protein synthesis.

Heat shock effect on mitochondrial ATPase

We have compared during heat shock the rate of oxygen consumption and the coupling status of oxidative phosphorylation in the Downs strain of *H.capsulatum*. We have shown that immediately after a shift up to 37°C cellular respiration and concentrations of cytochrome components decrease drastically. Furthermore, a rapid decline in ATP content occurs in the first 30 min and it is no longer detectable within 1 hr (stage 1). After 24 hr, no oxygen consumption is measurable. During the following 3-4 days, RNA and protein synthesis are not detectable (stage 2). In stage 3, RNA and protein syntheses resume and morphological changes take place (Maresca et al., 1981; Lambowitz et al., 1983). We have demonstrated that temperatures among 39° and 43°C are required to mimic similar effects in the more temperature resistant G222B and G184B strains during mycelial to yeast phase transition (Medoff et al., 1986). In fact, in the more virulent strains the induction of phase transition by a temperature shift to 37°C results in physiological and biochemical changes that are similar to those occurring in Downs, but less extreme. Conversely, at 34°C, only partial uncoupling of respiration occurs in the Downs strain. Further, the efficiency of electron transport diminishes to less than 50% of the initial values when compared to the level measured at 37°C. Therefore, it appeared that the Downs strain is more temperature sensitive and does not maintain functional RNA, protein and ATP syntheses during the early stages of the mycelial- to yeast- phase transition. In addition, Medoff et al. showed that the early biochemical events occurring during mycelial- to yeast-phase transition of other dimorphic fungi as *Paracoccidioides brasiliensis* and *Blastomyces dermatitidis*, are fundamentally similar to those found in *H.capsulatum* (Medoff et al., 1987). These data show that a correlation exists between temperature sensitivity (coupling capacity), degree of virulence and level of transcription of *hsp70* and *hsp82* genes.

Thermotolerance

In all organisms tested, the induction of heat shock proteins confers protection against cellular damage upon subsequent exposure to higher temperature (Lindquist, 1986). For example, it has been demonstrated that in *S.cerevisiae* a mild heat treatment induces thermotolerance. In fact, a rapid shift in the incubation temperature from 23° to 36°C, prior to a normally lethal heat shock to 52°C, results in an increase in cell survival (McAlister et al., 1980).

We have studied in the Downs strain of *H.capsulatum* the effect of a short pre-incubation at the intermediate temperature of 34°C on coupling capacity of mycelial cells and have measured respiration after a subsequent shift to 37°C. We also analyzed the effect of incubation at 37°C before exposure to 40°C in the thermoresistant G217B strain. We demonstrated that in *H.capsulatum* induction of thermotolerance prevents the uncoupling of oxidative phosphorylation from subsequent severe increases in temperature (at 37°C in Downs and at 41°C for G217B; Patriarca and Maresca, 1990). Further, the effectiveness of the rescue was correlated to the optimal temperature of transcription of heat shock genes.

We have examined under similar experimental conditions the induction of *hsp70* and *hsp82* mRNA. These two genes, which are induced at very low level at 37° or 40°C respectively in Downs and G217B, are transcribed at high level if thermotolerance is elicited (Patriarca et al., submitted; Minchiotti et al., submitted). Furthermore, mycelial cells incubated for two days in a diluted medium, induced heat shock proteins at 25°C (Patriarca et al., 1987). This condition caused protection of coupling of oxidative phosphorylation as well as an increase in *hsp70* and *hsp82* gene transcription at 37° or 41°C. However, while induction of heat shock genes was transient when the temperature of incubation was raised (heat shock), it was sustained in cells that had been placed in starvation medium (stress). In fact, *hsp70* mRNA was present even after 2 days of incubation at 37°C. Addition of actinomycin D prior to temperature shift inhibited the acquisition of the protected state (Patriarca and Maresca, 1990).

We have measured the effect of protection of coupling of oxidative phosphorylation also in *S.cerevisiae* (strain S288C) and shown that the pre-incubation at the intermediate temperature of 37°C for 60 min prevents the impairment of coupling capacity at high temperatures (44°C). Temperature shifts from 25°C to temperatures ranging among 36° and 39°C induce maximal heat shock gene transcription in *S.cerevisiae*. Inhibition of cytoplasmic RNA or protein synthesis immediately before the induction of thermotolerance prevented protection of the coupled state. These results suggest that heat shock proteins (or for proteins induced during the same period of time) protects ATPase activity during heat shock.

We also showed that the block in the coupling capacity in *S.cerevisiae* was rapidly reversed even in cells that had been returned to the normal growth temperature of 25°C after an exposure to 44°C for 90 min. However, in contrast to the protective phenomenon, the recovery in coupling capacity was practically instantaneous (within 1 min). Differently from what Yost and Lindquist have demonstrated in *Drosophila* cells, in which the recovery from the block in RNA splicing after heat shock requires several hours (Yost and Lindquist, 1986), in *S.cerevisiae* the recovery of coupling of oxidative phosphorylation does not need the synthesis of new proteins. Therefore, we believe that in *S.cerevisiae* the impairment in the synthesis of ATP during heat shock is probably due to a reversible structural (conformational) modification determined by heat on mitochondrial membrane components (e.g., one of the subunits of ATPase).

mRNA splicing

It is well known that heat shock affects RNA metabolism, including RNA processing and mRNA degradation (Sadis et al., 1988; Yost and Lindquist, 1986). It has been demonstrated that the intron-containing *Drosophila hsp83* gene, as well as the intron containing *Adh* gene, are strongly induced during moderate heat shock conditions (29°-33°C) but only poorly under severe conditions (37°-39°C). At these temperatures, mature mRNA is not produced, and intron-containing pre-mRNAs accumulate in the cytoplasm. Such a block in splicing persists for several hours after the cells are returned to permissive temperature (25°C). Similarly, it has been demonstrated that inhibition of mRNA maturation occurs in *Caenorhabditis elegans* (Kay et al., 1987), in *Trypanosomes* (*trans*-splicing of *hsp70* mRNA, while the *trans*-splicing pathway is sensitive to disruption by severe heat shock; Muhich et al., 1989), and in HeLa cells (Bond, 1988). It has also been shown that when *Drosophila* cells were first exposed to mild heat pre-treatments (to temperatures that induce heat shock proteins) prior to a shift to higher temperatures, mRNA precursor splicing continued under non-restrictive conditions (Yost and Lindquist, 1986). More recently, Yost and Lindquist have demonstrated that the maturation of yeast actin mRNA precursor is disrupted at high temperatures and that mild heat pre-treatment protects splicing from disruption (Yost and Lindquist, 1991). However, in yeast, new synthesis of heat shock proteins is not required for the protection of RNA maturation because cycloheximide treatment does not prevent protection of splicing, and strains carrying mutations in four major classes of heat shock genes (*hsp26*, *hsp83/hsc83*, *hsp104* and a member of *hsp70*) are not effected either (Yost and Lindquist, 1991).

On the contrary, we have demonstrated by Northern blots and RNase analysis, that splicing of *hsp82* and β-tubulin mRNA precursors is not blocked in either strains of *H.capsulatum* at the upper temperature range of the heat shock response specific for each strain of the organism (up to 42°C; Minchiotti et al., submitted). We explain this difference by considering that dimorphic organisms such as *H.capsulatum* must adapt to dramatic shifts in temperatures and environmental stresses during host invasion in order to survive in mammalian tissues. In fact, in contrast to other eucaryotic cells, which are exposed to gradual temperature increments, parasites and dimorphic fungi as *H.capsulatum* are organisms that experience a sudden and drastic environmental temperature change, e.g., from one that is poikilothermic to one that is homeothermic. Therefore, a block in splicing during host invasion would severely hamper their capacity to adapt and survive.

Conclusions

We have focused our interest on the relationship between heat shock and mitochondrial activities in *H.capsulatum*. Lambowitz et al. showed that a brief exposure of mycelial cells of the avirulent Downs strain to 37°C causes a rapid decline in intracellular ATP levels that parallels uncoupling of oxidative phosphorylation (Lambowitz et al., 1983). These observations led to the novel idea that in *H.capsulatum* and other dimorphic fungi the process of morphogenesis or adaptation to a temperature of 37°C result in the induction of the heat shock response. We showed that temperatures among 39° and 43°C are required to mimic a similar drastic effects in more pathogenic strains during mycelial to yeast phase transition (Medoff et al., 1986). We have demonstrated that the induction of phase transition by shifting the temperature to 34°C in Downs or to 37°C in the more virulent strains results in physiological and biochemical changes that are similar but less extreme. A sudden and drastic increase in the temperature is associated with respiratory stress, affecting mitochondrial activity. Findly et al. showed that in two *Tetrahymena* species, in which the heat shock response is induced at different temperatures, a 50% decrease in cellular ATP level occurs immediately (within 3 min) when either species was shifted to temperatures that strongly induce the synthesis of heat shock proteins (Findly et al., 1983). These authors also proved that when one of the two species was shifted to 37°C rather than 40°C, only a 25% decline in ATP level was observed (Findly et al., 1983).

We reasoned that if heat shock causes a block or a decrease in ATP synthesis, specific mitochondrial functions would be one of the targets induced by the rescue mechanisms. We found that induction of thermotolerance in *H.capsulatum* prevents the uncoupling of oxidative phosphorylation when exposed to additional severe increase in temperature. Furthermore, the effectiveness of the rescue is correlated to the optimal temperature of transcription of heat shock genes (which in turn is specific for any given strain of *H.capsulatum*) and to new protein synthesis. We have also shown that induction of higher levels of heat shock proteins (obtained either by starvation or pre-incubation at intermediate temperatures) protects the cells from entering stage 2 during the mycelial to yeast phase transition. Our data suggest that, among other functional roles, the gradual expression of the heat shock phenomenon plays a key role in protection of mitochondrial ATPase activity. Protection of the coupling capacity appears to be specific for ATPase since it does not cause an increase in efficiency of electron transport. In fact, we have shown that oligomycin, whose effect has been described as slightly inhibitory on electron transport, under heat shock conditions stimulates oxygen consumption and that the extent of stimulation is dependent on the temperature insult experienced by the cells. In addition, we have found that in *S.cerevisiae* uncoupling of oxidative phosphorylation is an early event that follows incubation above the upper temperature range of the heat shock response (44°C). We also showed that a pre-treatment at the intermediate temperature of 37°C prevents the impairment of coupling capacity at high temperatures.

Pelham has proposed the involvement of protein factors in folding and assembly of polypeptides (Pelham, 1986). He suggested that heat shock proteins may function to refold heat-denatured proteins and to influence the folding or the aggregation state under non-stress conditions (Pelham, 1986). Nuclear-encoded precursors are synthesized in the cytosol and subsequently imported into mitochondria. Precursors would then assume an "unfolded" conformation and be translocated across the mitochondrial membrane. Finally, all imported proteins have to refold and in most cases be assembled into supramolecular complexes to become functionally active (Cheng et al., 1989). HSP70 like-proteins have been shown to provide an unfolding function necessary for translocation into both mitochondria (Deshaies et al., 1988; Kang et al., 1990) and the lumen of endoplasmic reticulum (Chirico et al., 1988). Furthermore, it has also been demonstrated that HSP60 (Ssc1p) is a protein involved in the folding and assembly of polypeptides into yeast mitochondria (Kang et al., 1990; Craig et al., 1990). It is important to note that *hsp60* and *hsp70* yeast mutants accumulate β-subunit of the mitochondrial protein F_1-ATPase at 37°C (Cheng et al., 1989; Deshaies et al., 1988). Recently, Prasad et al. have shown that HSP60 protein forms a stable association with the α-subunit of the multi-component complex F_0-F_1-ATPase, showing that HSP60 can function in the folding and assembly of mitochondrial proteins (Prasad et al., 1989). With this evidence it has been hypothesized that the heat shock response may reflect a cellular mechanism for maintaining functional conformation of mitochondrial proteins during and following heat stress.

Our results give biochemical evidence of an involvement of heat shock protein synthesis in the protection of mitochondrial ATPase during temperature shift up. Data from our laboratory on yeast heat shock mutants suggest that both HSP60 (*mif4*; Cheng et al., 1989) and HSP70 (*Ssc1-2*; Kang et al., 1990) proteins are independently involved in the protection mechanism of oxidative phosphorylation in S.cerevisiae (Carratù et al., manuscript in preparation).

Temperature-induced perturbations in membrane architecture cause severe problems to organisms that live over a wide range of temperatures. In fact, the state of membrane lipids alters several important cellular processes, as membrane associated enzymes, membrane receptors, transport processes etc. (Hazel, 1988). Perturbations of the lipid status caused by increase in temperature has a destabilizing effect on the lipid environment on protein function as shown for Na^+/K^+-ATPase (Hazel, 1988) and in *Bacillus subtilis* uncoupler-resistant mutants (Guffanti et al., 1987). Our data suggest the hypothesis that in eucaryotic cells, which experience temporary fluctuations in the environmental temperature, heat shock effects membranes functionality by altering their fluidity (uncoupling of oxidative phosphorylation) (Guffanti et al., 1987). We have postulated that induction of heat shock proteins "corrects" membrane macromolecular structures rescuing membrane-associated functions like ATPase (Patriarca and Maresca, 1990; Carratù et al., manuscript in preparation).

Therefore, in dimorphic organisms like *H.capsulatum* in which the heat shock response is not an artificial phenomenon but part of the life cycle, thermotolerance probably plays a very little physiological role during morphogenesis. On the contrary, thermotolerance has an important role during daily fluctuations in temperature that do not lead to morphogenesis. Thus, dimorphic pathogens such as

H.capsulatum, that must adapt to sudden temperature changes during phase transition, are valuable model systems for the understanding the particular role that heat shock proteins play in temperature adaptation.

Acknowledgments

This work was supported by a Contract from the Commission of the European Community, TSD2-0132-I, a Contract from Ministero della Sanità - Istituto Superiore di Sanità, Progetto AIDS # 5203 026, Progetto Speciale CNR Proteine da Stress, NIH grants # AI 07015, AI29609 and a NATO travel award grant # 5-205/RG890437

References

Bienz, M and Pelham, HRB, (1987) Mechanism of heat shock gene activation in higher eukaryotes. Adv. Genet., 24: 31-72.

Bond, U, (1988) Heat shock but not other stress inducers leads to the disruption of a sub-set of snRNPs and inhibition of *in vitro* splicing in HeLa cells. EMBO J., 7: 3509-3518.

Caruso, M, Sacco, M, Medoff, G and Maresca, B, (1987) Heat shock 70 gene is differentially expressed in *Histoplasma capsulatum* strains with different level of thermotolerance and patogenicity. Mol. Microbiol., 1: 151-158.

Cheng, MY, Hartl, UF, Martin, J, Pollock, RA, Kalousek, F, Neupert, W, Hallberg, EM, Hallberg, RL and Horwich, AL, (1989) Mitochondrial heat-shock protein hsp60 is essential for assembly of proteins imported into yeast mitochondrial. Nature (London), 337: 620-625.

Chirico, WJ, Waters MG and Blobel, G, (1988) 70K heat shock related proteins stimulate protein translocation into microsomes. Nature (London), 332: 805-810.

Craig, EA, Kang, PJ and Boorstein, W, (1990) A review of the role of 70 kDa heat shock proteins in protein translocation accross membranes Ant. van Leew. International. J. Gen. and Mol. Microbiol., 58: 137-146.

Deshaies, RJ, Koch, BD, Werner-Washburne, M, Craig, EA and Schekman, R, (1988) A subfamily of stress proteins facilitates translocation of secretory and mitochondrial precursor polypeptides. Nature (London), 332: 800-805.

Findly, RC, Gillies, RJ and Shulman, RG, (1983) In vivo Phosphorus-31 nuclear magnetic resonance reveals lowered ATP during heat shock of *Tetrahymena.* Science, 219: 1223-1225.

Glaser, RL and Lis, JT, (1990) Multiple, compensatory regulatory elements specify spermatocyte-specific expression of the *Drosophila melanogaster* hsp26 gene. Mol. Cell. Biol., 10: 131-137.

Guffanti, AA, Clejan, S, Falk, LH, Hicks, DB and Krulwich, TA, (1987) Isolation and characterization of uncoupler-resistant mutants of *Bacillus subtilis.* J. Bacteriol., 169: 4469-4478.

Hazel, JR, (1988) Homeoviscous adaptation in animal cell membranes in Physiological regulation of membrane fluidity. Aloia, RC, Curtain, CC and Gordon, LM, eds., A.R. Liss, Inc., N.Y., pg 149-188.

Kay, RJ, Russnak, RH, Jones, D, Mathias, C and Candido, EP, (1987) Expression of intron-containing *C.elegans* heat shock genes in mouse cells demonstrates divergence of 3' splice site recognition sequences between nematodes and vertebrates, and an inhibitory effect of heat shock on the mammalian splicing apparatus. Nucl. Acid Res., 15: 3723-3741.

152

Kang, PJ, Ostermann, J, Shilling, J, Neupert, J, Craig, EA and Pfanner, N, (1990) Requirement for hsp70 in the mitochondrial matrix for translocation and folding of precursor proteins. Nature (London), 348: 137-143.

Lambowitz, AM, Kobayashi, GS, Painter, A and Medoff, G, (1983) Possible relationship of morphogenesis in pathogenic fungus, *Histoplasma capsulatum*, to heat shock response. Nature (London), 303: 806-808.

Lindquist, S, (1986) The heat shock response. Ann. Rev. Biochem., 55: 1151-1191.

Maresca, B, Lambowitz, AM, Kumar, BV, Grant, GA, Kobayashi, GS and Medoff, G, (1981) Role of cysteine oxidase in regulating morphogenesis and mitochondrial activity in the dimorphic fungus *H.capsulatum*. Proc. Natl. Acad. Sci. USA, 78: 4596-4600.

Maresca, B, Kobayashi, GS, (1989) Dimorphism in *Histoplasma capsulatum*: a model for the study of cell differentiation in pathogenic fungi. Microbiol. Rev., 53: 186-209.

McAlister, L and Finkelstein, DB, (1980) Heat shock proteins and thermal resistance in yeast. Biochem. Biophys. Res. Commun., 93: 819-824.

Medoff, G, Maresca, B, Lambowitz, AM, Kobayashi, GS, Painter, A, Sacco, M and Carratù, L, (1986) Correlation between pathogenicity and temperature sensitivity in different strains of *Histoplasma capsulatum*. J. of Clinical Investigation, 78: 1638-1647.

Medoff, G, Painter, A and Kobayashi, GS, (1987) Mycelial to yeast phase transitions of the dimorphic fungi *Blastomyces dermatitidis* and *Paracoccidioides brasiliensis*. J. Bacteriol., 169: 4055-4060.

Minchiotti, G, Gargano, S and Maresca, B, The intron-containing *hsp82* gene of the dimorphic pathogenic fungus *Histoplasma capsulatum* is properly spliced in severe heat shock conditions, submitted to Mol. Cell. Biol.

Morange, M, Hévin, B and Fauve, R, (1990) The heat shock response in the interaction between macrophages and virulent and avirulent Listeria monocytogenes. Abstract of the 3rd Europ. Cong. Cell Biol., Cell Biol. Intn'l Reports, Abstract Suppl., 14: 8.

Muhich, ML and Boothroyd, JC, (1989) Synthesis of *Trypanosome hsp70* mRNA is resistant to disruption of *trans*-splicing by heat shock. J. Biol. Chem., 264: 7107-7110.

Patriarca, EJ, Sacco, M, Carratù, L, Minchiotti, G and Maresca, B, (1987) Gene expression during the differentiation process in *H.capsulatum* Pasteur Institute, 100th Anniversary Celebration Molecular Biology and Infectious Diseases, pg.145, Paris, France.

Patriarca, EJ and Maresca, B, (1990) Acquired thermotolerance following heat shock protein synthesis prevents impairment of mitochondrial ATPase activity at elevated temperatures in *Saccharomyces cerevisiae*. Exp. Cell Res., 190: 57-64.

Patriarca, EJ, Kobayashi, GS and Maresca, B, Mitochondrial activity and heat shock response during morphogenesis in the dimorphic pathogenic fungus *Histoplasma capsulatum*, submitted to J. Bacteriol.

Pelham, HRB, (1986) Speculations on the functions of the major heat shock and glucose-related proteins. Cell, 46: 959-961.

Prasad, TK, Hack, E and Hallberg, RL, (1990) Function of the maize mitochondrial chaperonin hsp60: specific association between hsp60 and newly synthesized F1-ATPase alpha subunits. Mol. Cell. Biol., 10: 3979-86.

Sadis, S, Hickey, E and Weber, WL, (1988) Effect of heat shock on RNA metabolism in HeLa cells. J. Cell. Physiol., 135: 377-386.

Schwarz, L, (1981) Histoplasmosis. Praeger Science Press. New York.

Shearer, G, Birge, C, Yuckenberg, PD, Kobayashi, G and Medoff, G, (1987) Heat-shock proteins induced during the mycelial-to-yeast transitions of strains of *Histoplasma capsulatum*. J. Gen. Microbiol., 133: 3375-3382.

Susek, RE and Lindquist, S, (1990) Transcritpional derepression of the *Saccharomyces cerevisiae hsp26* gene during heat shock. Mol. Cell. Biol., 10: 6362-6373.

Yost, HJ and Lindquist, S, (1986) RNA splicing is interrupted by heat shock and is rescued by heat shock protein synthesis. Cell, 45: 185-193.

Yost, HJ and Lindquist, S, (1991) Heat shock proteins affect RNA processing during heat shock response of *Saccharomyces cerevisiae*. Mol. Cell. Biol., 11: 1062-1068.

Young, DB, Mehlert, A and Smith, DF, (1990) Stress proteins and infectious diseases. In Morimoto, RI, Tissieres, A, Georgopoulos, C, eds, Stress Proteins in Biology and Medicine. Cold Spring Harbor Press, N.Y., pg 131-165.

Effects of Heat Shock on Development and Actin mRNA Stability in *Drosophila*

N.S. Petersen and P. Young
Department of Molecular Biology
University of Wyoming
Laramie, WY
USA

Introduction

Heat shock during development induces morphological defects in vertebrates and invertebrates. In mammals, heating prior to neural tube closure induces cleft palate, anencephaly, and encepholoceles. These defects can be prevented if embryos are first made thermotolerant (Walsh et al.,1987). In *Drosophila*, a variety of different defects can be induced many of which resemble mutant defects. The heat induced defects are called phenocopies because of their resemblance to mutant phenotypes. Phenocopies in *Drosophila* can also be prevented by thermotolerance inducing treatments (Petersen and Mitchell, 1985). There are similarities in the conditions for induction of developmental defects in *Drosophila* and mammals which lead us to suggest that the molecular mechanism may also be similar (Petersen, 1990). These are: 1) Most treatments which induce developmental defects also induce synthesis of heat shock proteins; 2) The type of defect induced depends on the time of treatment rather than on the environmental stress (chemical or heat); 3) The induction of a defect depends on the genetic background as well as the environmental treatment; 4) Heat induced defects can be prevented by a thermotolerance inducing treatment before the environmental insult. We are using *Drosophila* as a model system to study the molecular basis for thermotolerance and for the induction of defects by heat.

Because heat induced developmental defects resemble mutant phenotypes, and because the sensitive periods for induction of the defects precede the the event which is being affected by several hours, it has been suggested that the defects are due to effects of heat on gene expression (Mitchell, 1966). Recently we have obtained more direct evidence that this is the case. In *Drosophila*, phenocopies of three recessive mutations, *forked*, singed, and *multiple wing hair* can be induced in each mutant heterozygote by heating; however, these phenocopies are not induced in wild type (Mitchell and Petersen, 1985; Petersen and Mitchell, 1987). Since the mutant heterozygote would be expected to produce half the wild type level of a gene product, this implies that the level of expression of specific gene products is critical for phenocopy induction. Furthermore, since heterozygote phenocopies are prevented in thermotolerant pupae (Petersen and Mitchell, 1989), the prevention of defects in the thermotolerant state must also be due to effects on gene expression.

Heat shock and development

Heating *Drosophila* pupae to temperatures which induce phenocopies inhibits both RNA and protein synthesis during the heat treatment and for several hours during the recovery at lower temperatures (Mitchell and Lipps, 1978; Petersen and Mitchell, 1982). There is a developmental delay of 15-20 hr when phenocopies are induced, and a delay of 8-12 hr in thermotolerant pupae. In thermotolerant cells, mRNA synthesis, mRNA processing, and protein synthesis (as well as many other cellular functions) recover much more rapidly (Lindquist, 1986; Craig, 1985). If mRNA were unstable during this recovery period, the loss of specific mRNAs could be responsible for phenocopy induction. It has been suggested by several investigators that mRNA decays during or following a severe heat shock, and that the effect of the thermotolerance inducing treatment is to prevent this loss of mRNA (Mitchell et al., 1979; Velazquez et al., 1980; Haass et al., 1989). We have tested this hypothesis by looking at the effects of heat on actin mRNA stability.

During normal development, actin mRNA levels in pupal wings decrease at least 10 fold between 48 and 52 hr of pupal development (Petersen et al., 1985). This normal drop in actin mRNA concentration is completed in less than four hours as shown in Figure 1. When 48 hr pupae (time 0) are heated at 40.5°C for 30 min and then returned to 25°C, protein synthesis is completely inhibited following the 40°C heat shock for a period of about eight hr. As recovery proceeds at 25°C, heat shock proteins are synthesized first, followed by the recovery of the normal 48 hr pattern of protein synthesis including the synthesis of actin. The normal developmental decrease in actin synthesis is delayed and now occurs between 9 and 15 hr during recovery as shown in Figure 1. It has been shown previously, under milder heat shock conditions which do not induce phenocopies, that even though mRNA is not translated during and following a heat treatment, it is still present in cells and can be translated in vitro (Storti et al., 1980; Kruger and Beneke, 1981; Lindquist, 1981; Petersen and Mitchell, 1981).

In order to determine whether actin mRNA is present during recovery from these more drastic heat shock conditions, and in a developing system, we have measured actin mRNA levels in two ways. RNA was isolated from wings at different stages during recovery from heat shock and either translated in vitro, or blotted onto modified nylon paper and detected hybridization to an actin 5C gene coding region probe (a gift of B. Bond). As shown in Figure 2, both kinds of experiments give

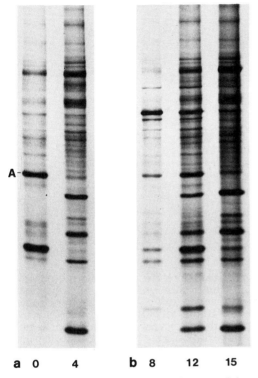

A-

a 0   4   **b** 8   12   15

Figure 1. The drop in actin synthesis is delayed in pupae which have been heat shocked. This figure is an autoradiogram of an SDS gel (Petersen and Mitchell, 1981) showing ^{35}S-methionine labeled proteins synthesized in wings from pupae during normal development (a), or following a 30 min heat shock at 40.5°C (b). The time is indicated in hours after the start of the heat shock, or after 48 hr in controls. Pupae were staged precisely at the start of these experiments by eye color. At 48 hr just before the drop in actin synthesis, the eyes start to turn yellow. (Animals were raised in a 25°C incubator.)

similar results. After a small initial drop, actin mRNA levels remained fairly constant, and actin mRNA decays only after normal protein synthesis has recovered. Since new RNA synthesis is inhibited by 40°C heat shock (Spradling et al., 1977; Mitchell and Lipps, 1978; Findley and Pedersen, 1981), this indicates that actin mRNA is stabilized by heat shock and that the programmed decay in actin mRNA does not take place until after recovery of the normal pattern of actin synthesis. By contrast, in thermotolerant wings, actin mRNA decays almost as fast as it would without a heat shock. This clearly shows that the actin mRNA is not protected from decay in thermotolerant cells.

Is the effect of heat shock on actin mRNA typical of the effects of heat shock on other messages? Experiments in cell lines indicate that even 37°C heat shocks stabilize histone mRNAs (Farrell-Towt and Sanders, 1984). We have previously shown by in vitro translation that at least two unidentified mRNAs normally synthesized at 38 hr of development are stabilized by a high temperature heat shock in a similar way to

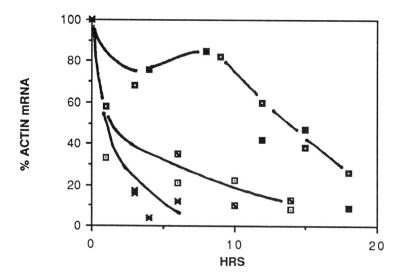

Figure 2. Actin mRNA levels following heat shock compared with actin mRNA levels during normal development. ✳ no treatment; ▫ ■ 48 hr pupae were heated 30 min at 40.5°C and then returned to 25°C; ▣ ◙ pupae were heated 30 min at 35°C before being heated at 40.5°C for 30 min. ▣ ◘ mRNA levels measured by blot hybridization; ◙■ mRNA levels measured by *in vitro* translation. Zero time is 48 hr after pupariation. RNA was isolated from pupal wings at different times following heat shock. For hybridization experiments, each RNA sample was loaded into slots on zeta-probe paper at three concentrations: 1 mg, 2 mg, and 5 mg. The RNA was hybridized to an actin coding region probe (Petersen et al., 1985) (a 1.8 kb Hind III fragment of the coding region of the actin 5C gene in pBR322, a gift of B. Bond) and then stripped and hybridized to a ribosomal RNA probe (PY22C from M. Pelligrini). The hybridization to ribosomal RNA was used to correct for loading efficiency. Actin mRNA hybridization was measured using a densitometer to quantitate data from autoradiograms and the areas under the peaks were plotted as a function of total RNA loaded. The slope of this line is proportional to actin mRNA concentration. All of the slopes were normalized to a value of 100% for the actin concentration at time zero and plotted as seen above. For in vitro translation mRNAs were translated in a rabbit reticulocyte system as described previously (Petersen and Mitchell, 1981).

actin mRNA (Petersen and Mitchell, 1982). Figure 3 shows *in vitro* translation of mRNA immediately before heat shock and at 10 hr following either a 40.5°C heat shock or a 35° to 40.5°C shock. It is clear that there is no dramatic general loss of translatable message, and that the amount of mRNA per O.D.260 unit of ribosomal RNA is similar under both conditions. What is different are the translation products encoded by the messages isolated under the two heat shock conditions. The RNA from wings which received a single heat shock is making heat shock proteins and proteins characteristic of 48 hr wings, while RNA from thermotolerant wings is synthesizing proteins characteristic of a later stage in development.

Thermotolerant cells are known to recover many processes more rapidly than non-tolerant cells (Lindquist, 1986). One of these is now

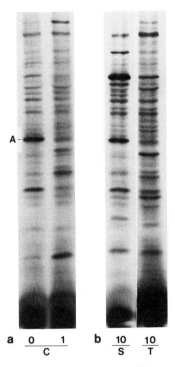

a $\underline{\quad 0 \qquad 1 \quad}$ b $\underline{\quad 10 \quad}$ $\underline{\quad 10 \quad}$
   $\quad$ C $\qquad$ S $\qquad$ T

Figure 3. In vitro translation of wing RNA showing that the change in actin mRNA concentration occurs sooner following heat shock in thermotolerant pupae. RNA was isolated from pupal wings during normal development (a) or following heat shock (b) and translated in a rabbit reticulocyte lysate. The translation products were separated on SDS poly-acrylamide gels and detected by autoradiography. The times indicated are the time in hours after heat shock or after 48 hr in the unheated control. In control wings, actin mRNA concentration is reduced significantly in 4 hr. The amount of actin mRNA present 10 hours after heat shock was dependent on whether the animals were first made thermotolerant by heating 30 min at 35°C. Lane S shows the translation products from wings of pupae which were heated for 30 min at 40.5°C only, and lane T shows translation products from wings of pupae that were heated 30 min at 35°C immediately prior to the 30 min, 40.5°C heat treatment. There is less actin mRNA in the wings from the thermotolerant pupae and also less developmental delay.

the normal developmental decay of actin mRNA. This observation is in direct conflict with the hypothesis that cells are thermotolerant in part because the mRNA has been stabilized. In fact, actin mRNA is less stable in thermotolerant wing cells. Since mRNA in general appears to be stabilized by heat shock, then we can further conclude that mRNA concentrations are not likely to be the limiting factor in gene expression under conditions where phenocopies are induced. The critical step in recovery of gene expression in thermotolerant cells appears to be at a later step, possibly the recovery of protein synthesis or protein processing. HSP70 appears to be required for acquisition of thermotolerance (Johnston and Kucey, 1988; Riabowol et al., 1988). Since members of the HSP70 family play a role in protein transport and processing (Chirico et al., 1988; Deshaises et al.,1988), this suggests that heat shock proteins themselves may be important in protein processing and transport and that phenocopies may be due to failure of gene expression at the level of protein synthesis or processing. Further analysis of this question requires that the limiting gene product be identified and the effects of heat on its normal expression pattern be determined.

The *forked* phenocopy is an example of a phenocopy where the gene is identified. Failure to express the *forked* gene appears to be directly responsible for the induction of the *forked* phenocopy. We are currently studying the effects of heat on the gene products responsible for this phenocopy. In the *forked* mutant, the large bristles on the fly are twisted and misshapen. The *forked* gene codes for a 2.4 kb mRNA which is first synthesized in whole pupae at 34 hr after white prepupa formation. Thirty four hours is also the sensitive period for the induction of the *forked* phenocopy. The *forked* mRNA codes for a protein of about 65 kD which accumulates in pupae from 40-54 hr. Antibody to the protein stains fibers which are present in developing bristles from 40 to 54 hr. These fibers are missing during the same time period in homozygous *forked* pupae. It appears that the *forked* protein is a component of these fibers and that the fibers themselves play a role in making bristles straight. We are currently determining the time and tissue specific expression of the *forked* mRNA and protein in both the mutant and the wild type as a background for looking at the effects of heat on the expression the *forked* gene. We expect heat to affect the amount or timing of synthesis of the bristle fibers leading to the observed bristle defects.

Acknowledgements

This work was supported by an NSF-EPSCoR grant # RII-8610680.

References

Chirico, WJ, Waters, MG and Blobel, G, (1988) 70K heat shock related proteins stimulate protein translocation into microsomes. Nature, 332: 805-810.

Craig, EA, (1985) The heat shock response. In "CRC Critical Reviews in Biochemistry". 18 Issue 3, pg 239-280.

Deshaies, RJ, Koch, BD, Werner-Washburne, M, Craig, EA and Schekman, RA, (1988) A subfamily of stress proteins facilitates translocation of secretory and mitochondrial precursor peptides. Nature, 332: 800-805.

Farrell-Towt, J and Sanders, M, (1984) Noncoordinate histone synthesis in heat-shocked *Drosophila* cells is regulated at multiple levels. Mol. Cell. Biol., 4: 2676-2685.

Findley, CR and Pedersen, T, (1981) Regulated Transcription of the Genes for Actin and Heat-shock proteins in Cultured Drosophila Cells. J. Cell Biol.,

88: 323-328.

Haass, C, Falkenburg, PE and Kloetzel, PM, (1989) In "Stress Induced Proteins". Pardue, ML, Feramisco,JR and Lindquist, S, eds., Alan R. Liss Inc. New York, N.Y., pg 175-185.

Johnston, RN and Kucey, BL, (1988) Competitive inhibition of hsp70 gene expression causes thermosensitivity. Science, 242: 1551-1554.

Kruger, C and Beneke, B, (1981) In vitro translation of Drosophila heat shock and non-heat shock mRNAs in heterologous and homologous cell free systems. Cell, 23: 595-604.

Lindquist, S, (1981) Regulation of protein synthesis during heat shock. Nature, 293: 311-314.

Lindquist, S, (1986) The heat-shock response. Ann. Rev. Biochem., 55: 1151-1191.

Mitchell, HK, (1966) Phenol oxidases and Drosophila development. Insect Physiol., 12: 755-765.

Mitchell, HK and Lipps, LS, (1978) Heat shock and phenocopy induction in Drosophila. Cell, 15: 907-919.

Mitchell, HK, Moller, G, Petersen, NS and Lipps-Sarmiento, L, (1979) Specific protection from phenocopy induction by heat shock. Dev. Genet., 1: 181-192.

Mitchell, HK and Petersen, NS, (1985) The recessive phenotype of forked can be uncovered by heat shock in Drosophila. Dev. Genet., 6: 93-100.

Petersen, NS, (1990) Effects of Heat and Chemical Stress on Development. In "Advances in Genetics vol. 28: Genomic Responses to Environmental Stress". Scandalios, JG, ed., Acad. Press, San Diego, CA.

Petersen, NS, Bond, BJ, Mitchell, HK and Davidson, N, (1985) Stage-specific regulation of actin genes in Drosophila wing cells. Dev. Genet., 5: 219-225.

Petersen, NS and Mitchell, HK, (1981) Recovery of protein synthesis following heat shock; preheat treatment affects mRNA translation. Proc. Natl. Acad. Sci. USA, 78: 1708-1711.

Petersen, NS and Mitchell, HK, (1982) Induction and prevention of the multihair phenocopy in Drosophila. In "Heat shock from bacteria to man." Schlessinger, M, Ashburner, M and Tissieres, A, eds., N.Y., Cold Spring Harbor Press, pp 345-352.

Petersen, NS and Mitchell, HK, (1985) Heat shock proteins. In "Comprehensive Insect Physiology, Biochemistry, and Pharmacology." Kerkut, GA and Gilbert, LI, eds., Vol.X. Biochemistry Oxford: Pergamon Press, pp 347-365.

Petersen, NS and Mitchell, HK, (1987) The induction of a multiple wing hair phenocopy by heat shock in mutant heterozygotes. Dev. Biol., 121: 335-341.

Petersen, NS and Mitchell, HK, (1989) The forked phenocopy is prevented in thermotolerant pupae. In: "Stress Induced Proteins". UCLA Symposia on Molecular and Cellular Biology, New Series, Vol.96, Pardue, ML, Feramisco, J and Lindquist, S, eds., Alan R. Liss Inc. New York, N.Y., pp. 235-244.

Spradling, A, Pardue, ML and Penman, S, (1977) Messenger RNA in heat-shocked Drosophila cells. J. Mol. Biol., 109: 559-587.

Storti, RV, Scott, MP, Rich, A and Pardue, ML, (1980) Translational control of protein synthesis in response to heat shock in D. melanogaster cells. Cell, 22: 825-834.

Riabowol, KT, Mizzen, LA and Welch, WJ, (1988) Heat shock is lethal to fibroblasts injected with antibodies to hsp70. Science, 242: 433-436.

Velazquez, JM, Di Dominico, BJ and Lindquist, S, (1980) Intracellular localization of heat shock proteins in Drosophila. Cell, 20: 679-689.

Walsh, DA, Klein, NW, Hightower, LE and Edwards, MJ, (1987) Heat shock and thermotolerance during early rat embryo development. Teratology, 36: 181-191.

Heat Shock Proteins and Translocations

Role of HSP60 in Folding/Assembly of Mitochondrial Proteins

A.L. Horwich, F.-U. Hartl*, M.Y. Cheng
Howard Hughes Medical Institute and Department of Human
Genetics, Yale School of Medicine
333 Cedar St.
New Haven, CT 06510
USA

Introduction

It has long been believed that polypeptides in the intact cell assume their biologically active conformations through steps of spontaneous folding (Anfinsen, 1973). A host of *in vitro* experiments support such behaviour, demonstrating refolding of proteins diluted from urea or guanidine into their biologically active forms (e.g., Sela et al., 1957; Garel et al., 1876). Yet the time required for such reactions is often hours or days, in contrast with only minutes required to produce biologically active polypeptides in intact cells. Further, the concentrations of polypeptide at which one can observe refolding *in vitro* are usually less than those present in the intact cell. Additionally, only certain proteins can undergo refolding *in vitro*: in general, proteins with complex domain structure and multiple subunits fail to fold *in vitro*.

Recent experiments indicate that in intact cells the folding of polypeptides is a process catalyzed by other proteins, notably heat shock proteins (reviewed in Ellis, 1987; Rothman, 1989). The first intimation of such a process derived from experiments of Georgopoulos and coworkers who observed that mutants affecting the groE operon of *E. coli* impaired the assembly of bacteriophage heads (Georgopoulos et al., 1983). More recently, this operon has been shown to be essential for viability of *E. coli* at all temperatures, indicating that it encodes a vital cellular function (Fayet et al., 1989). Additional studies have supported a role for groE proteins in the process of protein assembly - overproduction of a cyanobacterial groE operon in *E. coli* facilitated the assembly of a cyanobacterial dimeric RUBISCO enzyme (Gouloubinoff et al., 1989a); when urea-denatured subunits of the same dimeric RUBISCO were incubated *in vitro* with purified groE gene

* Institut fur Physiologische Chemie der Universität München, Goethestrasse
33, 8000 München 2, FRG

products, groEL and groES, assembly into active RUBISCO enzyme was observed (Gouloubinoff et al., 1989b).

In organelles endosymbiotically-related to bacteria, chloroplasts and mitochondria, components homologous to groEL have been identified: in the stroma of plant chloroplasts the RUBISCO subunit binding protein is found associated with the chloroplast-encoded large subunits of RUBISCO, but not with the mature L8S8 RUBISCO enzyme, implying a role for the binding protein in assembly (Barraclough and Ellis, 1980). In mitochondria, Richard Hallberg and coworkers identified a heat-inducible protein of molecular size 60K, the same size as groEL (McMullen and Hallberg, 1987); antiserum against the 60K protein derived from *T. thermophila* could recognize groEL (McMullen and Hallberg, 1988).

Folding/assembly of imported mitochondrial proteins requires the nuclear-encoded matrix component HSP60

Our own studies began with a search for a specific conditional yeast strain in which proteins could be translocated from the cytosol into mitochondria but once inside would fail to fold or assemble into active forms. Such a strain was identified, and came to be called *mif4* ("mif" for mitochondrial import function defect) (Cheng et al., 1989). *mif4* yeast are temperature-sensitive for growth on all media. At 37°C, matrix proteins are imported, and processed to mature size, but they fail to assemble into active forms. In particular, the programmed human matrix enzyme ornithine transcarbamylase (OTC) failed to assemble into its normal homotrimeric form and thus failed to exhibit enzymatic activity. The endogenous yeast matrix protein Fl-ATPase β subunit was imported and processed but failed to assemble into the FlATPase complex. Two additional endogenous proteins that normally enter the matrix and are then reexported through the inner membrane in a process that recapitulates bacterial secretion (Hartl and Neupert, 1990), the Rieske Fe/S protein and cytochrome b2, were found to reach only intermediate-sized matrix forms. For all of the polypeptides examined, import to the matrix and proteolytic processing proceeded normally, but there was no further biogenesis. We noted also that *mif4* mutant cells became petite at very high frequency, indicating that one or more macromolecular processes involved with producing or maintaining a functional repiratory chain was defective.

The gene able to rescue the mutant strain proved to be a nuclear gene, *hsp60*. We found that this was the same gene that Hallberg and coworkers identified in a λ gt11 library using the antiserum they had raised against the heat-induced mitochondrial protein of *T. thermophila* (Cheng et al., 1989; Reading et al., 1989). HSP60 is encoded by a nuclear gene. We found by disruption studies that the gene is essential. It encodes a precursor protein with a characteristic cleavable mitochondrial signal peptide at its NH2-terminus, rich in basic residues (6 arginines) and devoid of acidic residues (Reading et al.,

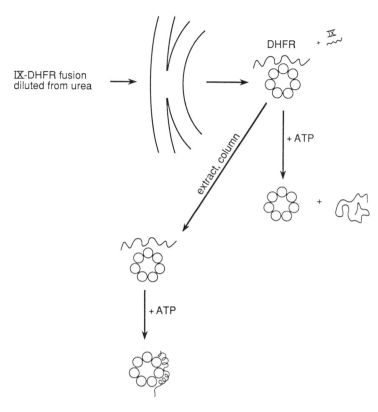

Figure 1. *In vitro* folding of DHFR. Summary of experiments showing that an imported monomeric protein is folded at the surface of the 14mer HSP60 complex in a reaction requiring ATP hydrolysis (Ostermann et al., 1989). Folding and/or release are suggested to require a second component, likely to be a homologue of the *E. coli* groES protein (Lubben et al., 1990).

1989). HSP60 precursor is cytosolically synthesized, recognized by mitochondria and translocated to the matrix, then processed to a mature form that is 58% identical to groEL and to RUBISCO subunit binding protein. The close similarity extends along the entire primary sequence of the proteins. The three related proteins, termed chaperonins (Ellis, 1987), are also similar at a level of higher-order structure. groEL and RUBISCO binding protein had been previously observed on electron microscopy to be comprised in "double-donut" 14mer complexes, consisting of two stacked rings each containing 7 radially arranged subunits (Hendrix, 1979; Puchkin et al., 1982). Electron microscopic examination of HSP60-containing fractions of mitochondrial matrix extracts revealed the identical 14mer appearance (McMullen and Hallberg, 1988).

HSP60 complex binds imported polypeptides and carries out ATP-dependent protein folding

To establish whether newly-imported proteins become associated with the HSP60 complex, and to address whether the complex is involved not only with overall assembly of subunits into multimeric complexes but also with the *folding* of individual polypeptides, a chimeric precursor joining a leader peptide with the monomeric enzyme dihydrofolate reductase (DHFR) was examined (see Fig. 1) (Ostermann et al., 1989). When this precursor was imported into mitochondria in the absence of ATP, the precursor became associated with the HSP60 complex. It remained in an unfolded state, judging from the observation that when extracts of the mitochondria were incubated with protease, the associated DHFR, but not the HSP60 complex, was found to be exquisitely sensitive. When ATP was added back after DHFR had been imported in the absence of ATP, the DHFR was found to be released from the HSP60 complex and was now found in a form resistant to proteinase K, resembling native DHFR. In a further experiment, if the DHFR-HSP60 complex was first partially purified from mitochondrial matrix and ATP then added, the DHFR remained associated, but assumed a partial degree of protease resistance. The collective of observations indicated that DHFR is folded at the surface of the HSP60 complex in a process that requires ATP hydrolysis. The final step(s) of release apparently requires a second component, which presumably became separated from the HSP60-containing fraction during partial purification. This component in *E. coli* is the groES protein, a 10K protein that is itself found as a higher-order structure, a ring of six to eight subunits (Chandrasekhar et al., 1986). Recent studies of bovine mitochondrial extracts indicate the presence of a similar factor in eukaryotic mitochondria, a 9K polypeptide that sediments under native conditions as a 45K complex. This component appears to be capable of functioning in the same manner as groES to mediate release (Lubben et al., 1990).

Behaviour of α143 mutant HSP60 complex - both the complex and "substrate" polypeptides are insoluble - α143 mutant complex may fail to recognize unfolded polypeptides

In the *mif4* mutant at 37°C the HSP60 complex becomes insoluble. A single amino acid substitution in the midportion of the protein is responsible, altering both charge and predicted secondary structure (S. Caplan, unpublished). We asked whether the solubility of newly-imported mitochondrial proteins was also affected in *mif4* cells. Wild-type and *mif4* cells were first grown at 23°C and then pulse-radiolabeled with ^{35}S-methionine 2 hr after shift to 37°C. Mitochondria were isolated, extracted with Triton X-100, and the extracts were separated by centrifugation (15,000 x g for 30 min) into soluble and insoluble fractions which were TCA-precipitated and counted. In the wild-type extract approximately 35% of the radiolabeled protein was found in the soluble fraction and 65% in the insoluble fraction. In striking contrast, in mitochondria from *mif4* cells virtually all of an equivalent amount of total ^{35}S-radiolabeled mitochondrial protein was found in the insoluble fraction. Thus it appears that not only the HSP60 complex but also its normal substrates for folding and assembly become insoluble at 37°C.

Have substrate polypeptides become insoluble by virtue of coaggregation with insoluble complex? This seems unlikely, based on our observations of heterozygous diploid cells. There, we have discovered that mutant subunits assemble independently of wild-type subunits: at 37°C the mutant complexes become insoluble whereas the wild-type complexes remain soluble (Cheng et al., 1990). This explains the recessive behaviour of the *mif4* allele. The relevant observation is that in this heterozygous setting imported proteins did not become universally insoluble as they had in *mif4* cells. Rather, their solubility resembled that in wild-type cells (Cheng, unpublished), suggesting that a dominant effect of aggregation of imported proteins by the mutant complex is not operative. It seems rather more likely that imported proteins are simply not recognized by the mutant HSP60 complex. In a haploid mutant cell their apparent fate is to misfold, aggregate, and become insoluble (Cheng et al., 1989). In a heterozygous diploid cell, while they would fail to be recognized by *mutant* complexes, they *are* recognized and subsequently folded by *wild-type* complexes, permitting assumption of biological function and enabling normal cell growth. Clearly, *in vitro* experiments using purified mutant complex will need to be performed in order to establish whether the *mif4* mutation produces a recognition defect.

Biogenesis of the HSP60 complex - preexistent complex is required in order to produce new complex

Recently we have examined the biogenesis of the HSP60 complex itself. As a further inquiry into the question of whether self-assembly is operative in the intact cell we asked: does a *bone fide* folding catalyst self-assemble, or does it require assistance for its own folding/assembly? More particularly, do imported HSP60 monomers self-assemble, or does folding and assembly of newly-imported HSP60 subunits require preexistent functional complex? The strategy for answering these questions was to import wild-type HSP60 subunits in the absence of HSP60 function, i.e., into *mif4* mutant mitochondria at 37°C, and ask whether they could form soluble wild-type complex (Cheng et al., 1990). In one set of experiments wild-type HSP60 was programmed in *mif4* cells from an inducible Gal operon promoter; in a second set of studies radiolabeled wild-type HSP60 precursor was synthesized *in vitro* and imported into isolated *mif4* mitochondria. Both in intact cells and *in vitro*, newly-assembled wild-type complex could not be produced in mif4 mitochondria at 37°C. While the HSP60

Table 1

Strain	Medium	Temp.	Sp.Act.
Galßgal/wt	EG	37°	0.3
	Gal	23°	5.6
	Gal	37°	5.8
Galßgal/mif4	EG	37°	0.3
	Gal	23°	2.6
	Gal	37°	5.0

Cells were grown at 23°C in ethanol glycerol (EG) medium, shifted where indicated to 37°C, and then expression of the *E. coli* ß-galactosidase gene induced by adding galactose (2%) (Gal) to the medium. Cells were harvested, treated with zymolase, and the spheroplasts were solubilized with Triton X-100 (Cheng et al., 1989). ß-galactosidase activity was measured in the soluble fraction. Activity is expressed as units per mg total soluble protein.

subunits had been normally imported into the mitochondria and proteolytically processed, in the absence of *preexisting functional* complex, no new HSP60 complex was produced. We conclude that even the folding machine itself cannot self-assemble.

Given the requirement for preexistent HSP60 complex in order to produce new complex, the question arises as to how the original mitochondrial HSP60 complex arose. We assume that the complex was originally carried in as an ancient version of groEL in the prokaryotic cell that became engulfed at the time of endosymbiosis. As a further extension of the question of origin, one could ask how the original ancestral groEL complex arose in prokaryotes? Here, the answer seems much less obvious. It seems possible that the original complex could have arisen by an event of self-assembly, inaugurating an essentially autocatalytic process; alternatively, the original complex could have been assembled with the assistance of some other molecule, another protein, or perhaps even a nucleic acid.

HSP60 does not play an identifiable role in folding/assembly of proteins in the cytosol

Reports that mammalian cells treated with the microtubule inhibitor podophyllotoxin exhibit an altered version of HSP60 suggested that perhaps HSP60 plays a role in microtubule assembly, an event presumed to be cytosolic (Jindal et al., 1989). To examine whether HSP60 could play a general role in folding/assembly of proteins in the cytosolic compartment, several cytosolic proteins were examined in

Table 2

Strain	Medium	Temp.	Sp.Act.
GalARG3/arg3	Gal	23°	274
		37°	281
	EG	23°	2
GalARG3/mif4(arg3)	Gal	23°	287
		37°	267
	EG	23°	3
GalOTC/arg3	Gal	23°	29
		37°	16
GalOTC/mif4	Gal	23°	4
		37°	1

Cells were grown at 23°C in ethanol glycerol (EG) medium, shifted where indicated to 37°C, and then expression of either the yeast OTC gene (ARG3) or the human OTC precursor (OTC) was induced by adding galactose (2%)(Gal) to the medium. Cells were harvested, treated with zymolase, and the spheroplasts solubilized with Triton X-100. OTC activity was measured in the soluble fraction (Cheng et al., 1989). Activity is expressed as units per mg total soluble protein.

mif4 cells. Two proteins, E. coli ß-galactosidase, a homotetrameric enzyme, and yeast ornithine transcarbamylase, a cytosolic homotrimer, were programmed for inducible expression from a galactose operon promoter in both "wild-type" and mif4 cells that were devoid of ß-galactosidase and OTC activity, the latter by virtue of mutation at the yeast arg 3 locus. Cells were first grown at 23°C, then shifted to 37°C and induced with galactose. Cell extracts were then prepared and the respective enzyme assays performed. As shown in Tables 1 and 2, identical levels of both ß-galactosidase and OTC enzymatic activities were observed in wild-type and mif4 cells at 37°C. In contrast with this lack of effect of the mif4 mutation on cytosolic enzymes, when we examined the mitochondrial version of OTC, the structurally related human enzyme (targeted via its additional NH2-terminal leader peptide), no activity was observed (Table 2).

As a further test for a role of HSP60 in cytosolic protein assembly, mif4 cells were examined for microtubule assembly using immunofluorescense analysis of α-tubulin. Cells grown at permissive temperature were shifted to 37°C and after 2 hr were fixed and stained

172

with a rat monoclonal anti-α tubulin antiserum (YolI/34)(Kilmartin et al., 1982). In several experiments cells were first synchronized in the cell cycle at permissive temperature using the mating pheromone α factor, were shifted to 37°C, and then 4 hr later were fixed and stained with anti-tubulin antiserum. The fluorescense patterns observed in wild-type and *mif4* mutant cells were identical.

Conclusions

The studies discussed here indicate that in intact cells proteins entering mitochondria require a machinery in order to be folded and assembled. Recent evidence suggests that the same machinery may also be used in the biogenesis of mitochondrial-encoded proteins (Prasad et al., 1990). It seems likely that in the *E. coli* cytosol structurally-related machinery may be utilized for the folding and assembly of newly-translated proteins (Georgopoulos et al., 1983; Gouloubinoff et al., 1989b): similarly, in the chloroplast stroma the related machinery is likely to be utilized for folding and assembly of both imported and synthesized polypeptides (Barraclough and Ellis, 1980). It remains to be seen how newly-translated proteins in the eukaryotic cytosol are folded - the same *hsp60* gene that is involved with mitochondrial folding/assembly does not appear to be involved. Both biochemical and genetic approaches should enable identification of apparently distinct components that are involved. These same approaches can also be applied to a further understanding of how the class of HSP60 proteins carries out its functions.

References

Anfinsen, CB, (1973) Principles that govern the folding of protein chains. Science, 181: 223-230.
Barraclough, R and Ellis, RJ, (1980) Protein synthesis in chloroplasts. IX Assembly of newly-synthesized large subunits into ribulose bisphosphate carboxylase in isolated pea chloroplasts. Biochem. Biophys. Acta, 608: 1931.
Chandrasekhar, GN, Tilly, K, Woolford, C, Hendrix, R and Georgopoulos, C, (1986) The E. coli dnaK gene product, the hsp70 homolog, can reactivate heat-inactivated RNA polymerase in an ATP hydrolysis-dependent manner. J. Biol. Chem., 261: 12414-12419.
Cheng, MY, Hartl, FU, Martin, J, Pollock, RA, Kalousek, F, Neupert, W, Hallberg, EM, Hallberg, RL and Horwich, AL, (1989) Mitochondrial heat-shock protein hsp60 is essential for assembly of proteins imported into yeast mitochondria. Nature, 337: 620-625.
Cheng, MY, Hartl, F-U and Horwich, AL, (1990) The mitochondrial chaperonin hsp60 is required for its own assembly. Nature, 348: 455-458.
Ellis, RJ, (1987) Proteins as molecular chaperones. Nature, 328: 378-379.
Fayet, O, Ziegelhoffer, T and Georgopoulos, C, (1989) The groES and groEL heat shock gene products of *Escherichia coli* are essential for bacterial growth at all temperatures. J. Bact., 171: 1379-1385.
Garel, J-R, Nall, BT and Baldwin, RL, (1976) Guanine-unfolded state of

ribonuclease A contains both fast- and slow-refolding species. Proc. Natl. Acad. Sci. USA, 73: 1853-1857.

Georgopoulos, CP, Tilly, K and Casjens, SR, (1983) In "Lambda II". Hendrix, RW, Roberts, JW, Stahl, FW and Weisberg, RA, eds, Cold Spring Harbor Laboratory Press, pp. 279-304.

Gouloubinoff, P, Gatenby, AA and Lorimer, G, (1989a) GroE heat-shock proteins promote assembly of foreign prokaryotic ribulose bisphosphate carboxylase oligomers in *Escherichia coli* Nature, 337: 44-47.

Gouloubinoff, P, Christeller, JT, Gatenby, AA and Lorimer, GH, (1989b) Reconstitution of active dimeric ribulose bisphosphate carboxylase from an unfolded state depends on two chaperonin proteins and Mg-ATP. Nature, 342: 884-889.

Hartl, F-U and Neupert, W, (1990) Protein sorting to mitochondria: evolutionary conservations of folding and assembly. Science, 247: 930-938.

Hendrix, RW, (1979) Isolation and characterization of the host protein groE involved in bacteriophage lambda assembly. J. Mol. Biol., 129: 359-373.

Jindal, S, Dudani, AK, Singh, B, Harley, CB and Gupta, RS, (1989) Primary structure of a human mitochondrial protein homologous to the bacterial and plant chaperonin and to the 65kd mycobacterial antigen. Mol. Cell. Biol., 9: 2279-2283.

Kilmartin, JV, Wright, B and Milstein, CJ, (1982) Rat monoclononal antibodies derived by using a new non secreting rat cell line. J. Cell Biol., 93: 576-582.

Lubben, TH, Gatenby, AA, Donaldson, GK, Lorimer, GH and Viitanen, PV, (1990) Identification of a groES-like chaperonin in mitochondria that facilitates protein folding. Proc. Natl. Acad. Sci. USA, 87: 7683-7687.

McMullen, TW and Hallberg, RL, (1987) A normal mitochondrial protein is selectively synthesized and accumulated during heat shock in *Tetrahymena thermophila*. Mol. Cell. Biol., 7: 4414-4423.

McMullen, TW and Hallberg, RL, (1988) A highly evolutionary conserved mitochondrial protein is structurally related to the protein encoded by the *Escherichia coli* groEL gene. Mol. Cell. Biol., 8: 371-380.

Ostermann, J, Horwich, AL, Neupert, W and Hartl, F-U, (1989) Protein folding in mitochondria requires complex formation with hsp60 and ATP hydrolysis. Nature, 341: 125-130.

Prasad, TK, Hack, E and Hallberg, RL, (1990) Function of the maize mitochondrial chaperonin hsp60: specific association between hsp60 and newly synthesized F1-ATPase alpha subunits. Mol. Cell. Biol., 10: 3979-3986.

Puchkin, AV, Tsuprun,VL, Solovjeva, NA, Shubin, VV, Evstgneeva, ZG and Kretovich, WL, (1982) High molecular weight pea leaf protein similar to the groE protein of *E.coli* Biochim. Biophys. Acta, 704: 379-384.

Reading, DS, Hallberg, RL and Myers, AM, (1989) Characterization of the yeast hsp60 gene coding for a mitochondrial assembly factor. Nature, 337: 655-659.

Rothman, JE, (1989) Polypeptide chain binding proteins: catalysts of protein folding and related processes in cells. Cell, 59: 591-601.

Sela, M, White, FH and Anfinsen, CB, (1957) Reductive cleavage of disulfide bridges in ribonuclease. Science, 125: 691-695.

Role of Heat Shock Proteins in Mitochondrial Protein Import

N. Pfanner
Institut für Physiologische Chemie
Universität München
Goethestr. 33
W-8000 München 2
FRG

Introduction

Most proteins destined for cell organelles are synthesized on cytosolic polysomes and must cross one or more organelle membranes to reach their functional destination (Wickner and Lodish, 1985). With mitochondria, over 95% of the proteins are made as precursor proteins in the cytosol and are mainly post-translationally imported into the four mitochondrial subcompartments (outer membrane, intermembrane space, inner membrane, and matrix) (Attardi and Schatz, 1988; Hartl and Neupert, 1990; Pfanner and Neupert, 1990). Heat shock proteins (hsps) were found to play an important role at various stages of mitochondrial protein import. This includes the maintenance of a transport-competent conformation of precursor proteins by cytosolic hsp70s (Deshaies et al., 1988; Murakami et al., 1988; Pfanner et al., 1990), the involvement of mitochondrial HSP70 in translocation of precursor proteins through contact sites between mitochondrial outer and inner membranes (Kang et al., 1990), and the refolding of imported proteins at HSP60 in the mitochondrial matrix (Ostermann et al., 1989).

Before discussing these functions of hsps in detail, I will shortly summarize the principles of mitochondrial protein uptake (Attardi and Schatz, 1988; Hartl and Neupert, 1990; Pfanner and Neupert, 1990). The precursor proteins carry targeting signals that are often found in amino-terminal extension sequences (presequences). Upon recognition by specific receptor proteins on the mitochondrial surface, the precursor proteins are imported through membrane contact sites into the mitochondrial inner membrane or matrix. The precursor proteins are translocated in an unfolded state, probably as linear polypeptide chains that are in an extended conformation. The movement of the positively charged presequences through the inner membrane requires the electrical potential $\Delta\Psi$ across this membrane (negative inside). In the matrix, the presequences are proteolytically cleaved off by the enzyme processing peptidase. The proteins are refolded, sorted to their final destination and assembled into functional complexes.

Cytosolic HSP70s and transport-competence of precursor proteins

How are mitochondrial precursor proteins kept in a transport-competent state in the cytosol? On the one hand, misfolding and aggregation of precursor proteins has to be prevented. On the other hand, the precursor proteins must be unfolded prior to or during insertion into the mitochondrial membranes.

Cytosolic heat shock proteins of 70 kd (HSP70s) were shown to stimulate protein translocation into mitochondria *in vivo* and *in vitro* (Deshaies et al., 1988; Murakami et al., 1988; Randall and Shore, 1989). It is generally assumed that HSP70s interact with not fully folded proteins and thereby stabilize a loosely folded conformation of proteins (summarized in Pelham, 1988; Rothman, 1989; Ellis and Hemmingsen, 1989). Hydrolysis of ATP is needed to release proteins from HSP70. It was indeed found that protein import into mitochondria required ATP in the cytosol and the ATP-requirement was correlated to conferring transport-competence to precursor proteins (Pfanner et al., 1987, 1988; Chen and Douglas, 1987; Verner and Schatz, 1987). These results suggest that cytosolic HSP70s and ATP are involved in conferring transport-competence to precursor proteins (a putative role of additional cytosolic cofactors is unclear so far).

By studying the import of precursor proteins that carried a tightly and correctly folded domain (that is also present in the mature protein), we obtained a surprising result. The unfolding and membrane translocation of those domains did not require the addition of ATP in contrast to the ATP-dependence of import of most precursor proteins (Pfanner et al., 1990). Thus, the actual unfolding of precursor proteins seems to be independent of ATP and probably also independent of cytosolic HSP70s. We conclude that ATP (and cytosolic HSP70s) are involved in preventing the misfolding and aggregation of those precursor proteins that may form improper interactions in the cytosol. Fully folded domains have no tendency for misfolding and most likely do not expose binding sites for HSP70s. Therefore they should not interact with cytosolic HSP70s and there is no need for ATP for releasing them (Pfanner et al., 1990). The unfolding of precursor proteins in transit is obviously mediated by the membrane-bound import machinery and involves mitochondrial HSP70 in the matrix (see below).

Mitochondrial HSP70 and protein translocation through contact sites

Contact zones between both mitochondrial membranes represent the major site for import of precursor proteins (Schleyer and Neupert, 1985; Pfanner and Neupert, 1990). We accumulated hybrid proteins carrying a tightly folded carboxyl-terminal domain in contact sites. The amino-terminal portion of the hybrid proteins was inserted into and across both membranes and the presequence was cleaved off by the processing peptidase in the matrix, whereas the folded domain remained on the cytosolic side. It was found that about 50 amino acid residues were sufficient to span both membranes (Rassow et al., 1991). This result implies that the precursor polypeptides are translocated as a linear chain and suggests a high degree of unfolding of the polypeptide in transit.

What provides the driving force for the extensive unfolding and vectorial movement of precursor proteins? The membrane potential $\Delta\Psi$ is needed for the movement of the presequences across the inner membrane. The rest of the polypeptides can be translocated in the absence of $\Delta\Psi$ (Schleyer and Neupert, 1985; Pfanner and Neupert, 1985, 1990). Moreover, as described above cytosolic ATP is not directly required for unfolding of the precursor proteins.

Recent results indicate that the energy for translocation of the major portion of the precursors is derived from binding of the precursors to HSP70 (Ssc1p) in the mitochondrial matrix (Kang et al., 1990). In a yeast mutant, that was defective in mitochondrial HSP70 in a temperature-sensitive manner, mitochondrial precursor proteins accumulated at the non-permissive temperature. The precursor proteins were able to insert into contact sites such that the presequence reached the matrix and was cleaved off by the processing peptidase. The major portion of the polypeptides, however, remained on the cytosolic side of the mutant mitochondria. An artificially unfolded precursor protein circumvented the translocation defect and accumulated at the mutant HSP70 in the matrix. Therefore, functional HSP70 in the matrix seems to be required to promote the unfolding of precursor proteins on the cytosolic side of the import machinery (Kang et al., 1990).

We propose the following model. Mitochondrial HSP70s bind with high affinity to the extended polypeptide chain of a precursor emerging on the matrix side of the inner membrane (while the binding of cytosolic HSP70s to the partially folded precursor portions on the cytosolic side should occur with lower affinity [Rothman, 1989]). Thereby, in a series of successive steps the polypeptide is pulled across the membrane. This includes the unfolding of the precursor entering the membranes in a step-wise fashion (Kang et al., 1990). ATP is probably indirectly required in this process to release the precursor proteins from mitochondrial HSP70s, setting HSP70s free for new rounds of transport.

HSP60 and folding of imported proteins

During or after translocation, the polypeptides seem to be transferred from mitochondrial HSP70 to HSP60. HSP60 forms a 14-mer of identical 60 kd subunits in the matrix and apparently represents an essential component of the machinery responsible for the refolding of imported proteins (Cheng et al., 1989; Ostermann et al., 1989). At low levels of ATP or in the presence of non-hydrolyzable ATP-analogues, imported proteins accumulate at the surface of HSP60 in an unfolded conformation. ATP-hydrolysis is required to promote folding of the proteins in association with HSP60. ATP and additional matrix factors are required to release the proteins from HSP60. It is suggested that HSP60 and associated proteins catalyze the (re)folding of proteins imported into the mitochondrial matrix (Ostermann et al., 1989).

Precursor proteins imported into mitochondria defective in mitochondrial HSP70 accumulate at the mutant HSP70 in an unfolded conformation, implying that HSP70 is also needed for refolding of the imported precursors (Kang et al., 1990). We suggest that HSP70 keeps the incoming precursors in an unfolded conformation and transfers them to HSP60. Mitochondrial HSP70 thus would be indirectly involved in the (re)folding of proteins as it probably prevents the misfolding of imported proteins and thus allows their interaction with HSP60 in a conformation that is conducive to the folding process.

Not all proteins interact with HSP60 on their assembly pathway. The precursor of ADP/ATP carrier, the most abundant protein of the inner mitochondrial membrane, does not depend on functional HSP60 for folding and assembly (Mahlke et al., 1990). In contrast, several other precursor proteins destined for the inner membrane or the intermembrane space are first imported via contact sites into the matrix, interact with HSP60 and are then retranslocated into or across the inner membrane (Hartl and Neupert, 1990). In the latter cases, the interaction with HSP60 probably does not lead to a complete refolding of the proteins, but to the generation of a conformation that is compatible with the retranslocation across the inner membrane. Interestingly, both ADP/ATP carrier and the proteins interacting with HSP60 seem to require mitochondrial HSP70 for import, emphasizing a general role of mitochondrial HSP70 in translocation of precursor proteins (N. Pfanner, J. Ostermann, E.A. Craig and W. Neupert, unpublished results).

Conclusions

Future studies will be directed towards a molecular characterization of the interactions between heat shock proteins and precursor proteins. This includes the definition of structures in the precursor proteins that form the binding sites recognized by hsps and the identification of the active sites of the HSPs. Furthermore, proteins that may influence the functions of HSP70s or HSP60 in eukaryotic cells shall be identified. Among these components may be the putative homologues to the prokaryotic DnaJ and GrpE proteins that are known to interact with DnaK, the prokaryotic HSP70 (Kawasaki et al., 1990), and the putative homologue to the prokaryotic groES protein that cooperates with groEL, the prokaryotic HSP60 (Goloubinoff et al., 1989).

Due to the high evolutionary relationship of mitochondria and chloroplasts it can expected that HSP70s and HSP60 (termed Rubisco subunit-binding protein in chloroplasts) are of comparable function and importance in protein transport into mitochondria and chloroplasts (Ellis and Hemmingsen, 1989; Rothman, 1989). Protein translocation into the endoplasmic reticulum probably involves HSP70s in the cytosol and within the organelle in a manner very similar to that described for mitochondrial protein import (Deshaies et al., 1988; Chirico et al., 1988; Meyer, 1988; Rothman, 1989; Vogel et al., 1990). Moreover, it is tempting to speculate that components with a function equivalent to HSP60, i.e., acting as a "foldase", may exist also in the endoplasmic reticulum and even in the cytosol of eukaryotic cells.

Acknowledgements

The author thanks Drs. Betty Craig, Ulrich Hartl, Art Horwich, Walter Neupert and Joachim Ostermann for stimulating discussions. Work of the author's laboratory was supported by the Deutsche Forschungsgemeinschaft (Sonderforschungsbereich 184, project B1).

References

Attardi, G and Schatz, G, (1988) Biogenesis of mitochondria. Ann. Rev. Cell Biol., 4: 289-333.

Chen, W-J and Douglas, MG, (1987) Phosphodiester bond cleavage outside mitochondria is required for the completion of protein import into the mitochondrial matrix. Cell, 49: 651-658.

Cheng, MY, Hartl, F-U, Martin, J, Pollock, RA, Kalousek, F, Neupert, W, Hallberg, EM, Hallberg, RL and Horwich, AL, (1989) Mitochondrial heat-shock protein hsp60 is essential for assembly of proteins imported into yeast mitochondria. Nature, 337: 620-625.

Chirico, WJ, Waters, MG and Blobel, G, (1988) 70K heat shock related proteins stimulate protein translocation into microsomes. Nature, 332: 805-810.

Deshaies, RJ, Koch, BD, Werner-Washburne, M, Craig, E. A and Schekman, R, (1988) A subfamily of stress proteins facilitates translocation of secretory and mitochondrial precursor polypeptides. Nature, 332: 800-805.

Ellis, RJ and Hemmingsen, SM, (1989) Molecular chaperones: proteins essential for the biogenesis of some macromolecular structures. Trends Biochem. Sci., 14: 339-342.

Goloubinoff, P, Christeller, JT, Gatenby, AA and Lorimer, GH, (1989) Reconstitution of active dimeric ribulose bisphosphate carboxylase from an unfolded state depends on two chaperonin proteins and Mg-ATP. Nature, 342: 884-889.

Hartl, F-U and Neupert, W, (1990) Protein sorting to mitochondria: evolutionary conservations of folding and assembly. Science, 247: 930-938.

Kang, P-J, Ostermann, J, Shilling, J, Neupert, W, Craig, EA and Pfanner, N, Hsp70 in the mitochondrial matrix is required for translocation and folding of precursor proteins. Submitted.

Kawasaki, Y, Wada, C and Yura, T, (1990) Roles of Escherichia coli heat shock proteins DnaK, DnaJ and GrpE in mini-F plasmid replication. Mol. Gen. Genet., 220: 277-282.

Mahlke, K, Pfanner, N, Martin, J, Horwich, AL, Hartl, F-U and Neupert, W, Sorting pathways of mitochondrial inner membrane proteins. Eur. J. Biochem., in press.

Meyer, DI, (1988) Preprotein conformation: the year's major theme in translocation studies. Trends Biochem. Sci., 13: 471-474.

Murakami, H, Pain, D and Blobel, G, (1988) 70-kD heat shock-related protein is one of at least two distinct cytosolic factors stimulating protein import into mitochondria. J. Cell Biol., 107: 2051-2057.

Ostermann, J, Horwich, AL, Neupert, W and Hartl, F-U, (1989) Protein folding in mitochondria requires complex formation with hsp60 and ATP hydrolysis. Nature, 341: 125-130.

Pelham, H, (1988) Heat shock proteins: coming in from the cold. Nature, 332: 776-777.

Pfanner, N and Neupert, W, (1985) Transport of proteins into mitochondria: a potassium diffusion potential is able to drive the import of ADP/ATP carrier. EMBO J., 4: 2819-2825.

Pfanner, N and Neupert, W, (1990) The mitochondrial protein import apparatus. Annu. Rev. Biochem., 59: 331-353.

Pfanner, N, Tropschug, M and Neupert, W, (1987) Mitochondrial protein import: nucleoside triphosphates are involved in conferring import-competence to precursors. Cell, 49: 815-823.

Pfanner, N, Pfaller, R, Kleene, R, Ito, M, Tropschug, M and Neupert, W, (1988) Role of ATP in mitochondrial protein import: conformational alteration of a precursor protein can substitute for ATP requirement. J. Biol. Chem., 263: 4049-4051.

Pfanner, N, Rassow, J, Guiard, B, Söllner, T, Hartl, F-U and Neupert, W, Energy requirements for unfolding and membrane translocation of precursor proteins during import into mitochondria. J. Biol. Chem., in press.

Randall, SK and Shore, GC, (1989) Import of a mutant mitochondrial precursor fails to respond to stimulation by a cytosolic factor. FEBS Lett., 250: 561-564.

Rassow, J, Hartl, F-U, Guiard, B, Pfanner, N and Neupert, W, Polypeptides traverse the mitochondrial envelope in an extended state. Submitted.

Rothman, JE, (1989) Polypeptide chain binding proteins: catalysts of protein folding and related processes in cells. Cell, 59: 591-601.

Schleyer, M and Neupert, W, (1985) Transport of proteins into mitochondria: translocational intermediates spanning contact sites between outer and inner membranes. Cell, 43: 339-350.

Verner, K and Schatz, G, (1987) Import of an incompletely folded precursor protein into isolated mitochondria requires an energized inner membrane, but no added ATP. EMBO J., 6: 2449-2456.

Vogel, JP, Misra, LM and Rose, MD, (1990) Loss of BiP/GRP78 function blocks translocation of secretory proteins in yeast. J. Cell Biol., 110: 1885-1895.

Wickner, WT and Lodish, HF, (1985) Multiple mechanisms of protein insertion into and across membranes. Science, 230: 400-407.

A Role for the 73-kDa Heat Shock Cognate Protein in a Lysosomal Pathway of Intracellular Protein Degradation

J.F. Dice, H.-L. Chiang*, S.R. Terlecky, T.S. Olson
Department of Physiology
Tufts University School of Medicine
136 Harrison Ave
Boston, MA 02111
USA

Introduction

The increased lysosomal degradation of cytosolic proteins during serum withdrawal is stimulated by a member of the 70-kDa heat shock protein (HSP70[1]) family. This HSP70, designated the peptide recognition protein of 73 kDa (prp73), can be isolated by affinity chromatography with RNase S-peptide-Sepharose, and it stimulates lysosomal uptake and degradation of [³H]RNase S-peptide *in vitro*. Prp73 binds to several proteins and peptides that contain peptide regions biochemically related to amino acids 7-11 of RNase S-peptide, Lys-Phe-Glu-Arg-Gln (KFERQ).

Several lines of evidence indicate that prp73 is the heat shock cognate protein of 73 kDa (HSC73). Among four purified HSP70s tested, HSC73 has the greatest capacity to bind RNase S-peptide. Both prp73 and HSC73 bind to RNase S-peptide, RNase A, and aspartate aminotransferase but not to three other proteins. In addition, prp73 and HSC73 promote uptake and degradation of [³H]RNase S-peptide by lysosomes *in vitro* while three other HSP70s are without activity. Finally, two-dimensional gel electrophoresis demonstrates that the major HSP70 in the cytosol of serum-deprived fibroblasts migrates at the position expected for HSC73. We discuss possible mechanisms by which HSC73 might facilitate the transport of proteins containing KFERQ-like peptide regions into lysosomes.

---

* Department of Biochemistry, University of California, Berkeley, CA 94720 USA
[1] The abbreviations used are: HSP70s, heat shock proteins of the 70 kDa family; HSC73, heat shock cognate protein of 73 kDa; grp78, glucose-regulated protein of 78 kDa; prp73, peptide recognition protein of 73 kDa; RNase A, bovine pancreatic ribonuclease A; RNase S-peptide, residues 1-20 of RNase A; RNase S-protein, residues 21-124 of RNase A.

Our laboratory has been interested in pathways of intracellular protein degradation for many years (Goldberg and Dice, 1974; Dice, 1987). Multiple pathways of proteolysis are known to exist within cells (Dice, 1987; Olson et al., 1990), and the molecular determinants within proteins that target them for particular degradative pathways are currently being investigated (Dice, 1987; Olson and Dice, 1989). Our interest in HSP70s arose from studies of a selective pathway for uptake and degradation of specific cytosolic proteins by lysosomes (Neff et al., 1981; Dice, 1982; Backer et al., 1983; McElligott et al., 1985; Backer and Dice, 1986; Dice et al., 1986; Chiang and Dice, 1988; Chiang et al., 1989; Dice, 1990). We introduced radiolabeled ribonuclease A (RNase A) into the cytosol of human diploid fibroblasts and found it to be degraded at an increased rate in response to serum withdrawal (Neff et al., 1981; Dice, 1982; Backer et al., 1983). Several lines of evidence established that its site of degradation is within lysosomes (McElligott et al., 1985; Chiang et al., 1989). The amino terminal 20 amino acids of RNase A (RNase S-peptide) are essential for the enhanced degradation in response to serum withdrawal (Backer et al., 1983), and linkage of RNase S-peptide to heterologous proteins causes their degradation to be increased by serum deprivation (Backer and Dice, 1986). The essential region within RNase S-peptide is amino acids 7-11, KFERQ (Dice et al., 1986), and microinjection of excess unlabeled KFERQ along with [^{125}I]RNase A blocks the enhanced degradation in response to serum deprivation (Chiang and Dice, 1988).

Certain cellular proteins contain peptide regions biochemically related to KFERQ since anti-KFERQ IgGs are able to precipitate approximately 30% of radiolabeled cytosolic proteins from fibroblasts. Only these immunoprecipitable proteins are degraded at an increased rate following serum withdrawal (Chiang and Dice, 1988). Loss of proteins containing KFERQ-like peptide regions is also evident in immunoblots of cytosolic proteins from serum-deprived fibroblasts (Chiang and Dice, 1988). Similar immunoblots of different tissues from fasted rats suggest that this selective pathway of lysosomal protein degradation is also stimulated in liver and kidney (Chiang and Dice, 1988).

The peptide regions within cellular proteins recognized by the anti-KFERQ IgGs are biochemically similar to KFERQ (Table 1) but are not necessarily the exact sequence (Chiang and Dice, 1988; Dice, 1990). Rather, these proteins all contain a peptide motif that consists of a glutamine flanked on either side by a tetrapeptide consisting of only acidic, basic, and very hydrophobic residues (Dice, 1990).

Role of an HSP70

A possible reason for the requirement of a KFERQ-like peptide region in protein substrates of this lysosomal degradation pathway is the existence of a cellular protein that binds to this peptide motif. We purified such a peptide recognition protein of 73 kDa (prp73) by

Table 1

Regulated Protein	Prp73 Sequence	Degradation	Binding
RNase A, RNase S-peptide, KFERQ	K F E R Q	yes	yes
Pyruvate kinase	Q D L K F	yes	yes
Aspartate amino-transferase	R K V E Q Q E K R V	yes	yes
Hemoglobin ß-chain	Q R F F E	yes	?
Clathrin light chain	Q V D R L V D R L Q	?	?
Adaptin	E I L K Q Q E I K V I K E V Q E L R I Q	?	?
Tubulin ß-chain	Q L E R I Q E L F K	?	?
HSC73	Q R D K V Q K I L D	?	?

Table 1. Peptide Sequences Related to KFERQ in Proteins. Regulated degradation refers to increased degradation in response to serum withdrawal, and original citations can be found in Chiang and Dice, 1988.

affinity chromatography using RNase S-peptide-Sepharose followed by elution with RNase S-peptide or KFERQ (Chiang et al., 1989). Amino acid sequences of internal peptides and immunoreactivity indicated that prp73 was an HSP70 (Chiang et al., 1989).

The amount of cytosolic prp73 increases in response to serum withdrawal (Chiang et al., 1989), further suggesting a role for prp73 in enhanced lysosomal proteolysis. Finally, we reconstituted this lysosomal degradative pathway in two different cell-free systems, and uptake and/or degradation of [^{3}H]RNase S-peptide is stimulated by prp73 and ATP (Chiang et al., 1989; Fig. 1).

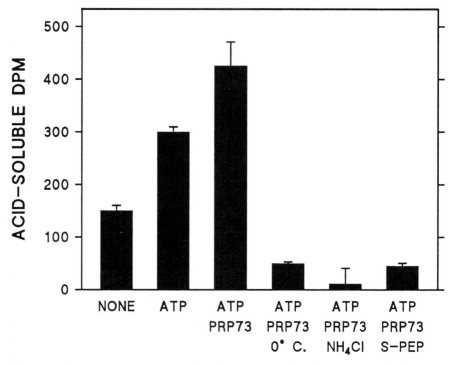

Figure 1. Lysosomes from 2 x 10⁶ cells were incubated with 10,000 dpm of [³H]RNase S-peptide as described by Chiang et al., 1989. Other additions were ATP and an ATP regenerating system (10mM), PRP73 (10 µg/ml), NH₄Cl (10 mM), and unlabeled RNase S-peptide (500 µg/ml).

Binding of Prp73 to Proteins and Peptides

We determined the binding specificity of prp73 for RNase S-peptide by incubating radiolabeled prp73 with RNase S-peptide linked to Sepharose (Terlecky et al., 1990). Prp73 binds to RNase S-peptide while 3 proteins unrelated to prp73 (BSA, RNase A, and RNase S-peptide) do not. The association of prp73 with RNase S-peptide can be competed with unlabeled RNase S-peptide, intact RNase A, or KFERQ but not by unlabeled lysozyme or β-galactosidase (Terlecky et al., 1990), proteins that do not contain KFERQ-related sequences (Chiang and Dice, 1988). Scatchard analysis of radiolabeled prp73 binding to RNase S-peptide indicates that the dissociation constant is approximately 8 µM.

We further evaluated the specificity of the interaction between prp73 and KFERQ by competition experiments using several RNase S-peptide derivatives with single amino acid changes in the KFERQ region (Terlecky et al., 1990). These derivatives were produced by site directed mutagenesis as described (Goff et al., 1987), and were produced in E. coli as fusion proteins with β-galactosidase. Most (5 out of 6) single amino acid substitutions within the KFERQ region of

RNase S-peptide have reduced abilities to compete for binding. This limited analysis suggests that most residues in KFERQ are important for recognition by prp73.

We obtained similar results with a different assay in which prp73 was precoated on microtiter plates and binding of [³H]RNase S-peptide measured (Terlecky et al., 1990). RNase S-peptide and RNase A compete for the binding of [³H]RNase S-peptide, while ovalbumin, lysozyme, and ubiquitin, proteins that do not contain KFERQ-like sequences (Chiang and Dice, 1988), do not. Furthermore, two other purified proteins that are degraded more rapidly during serum withdrawal and contain KFERQ-like sequences, pyruvate kinase and aspartate aminotransferase (Chiang and Dice, 1988; Dice, 1990; Table 1), also compete for RNase S-peptide binding.

Prp73 is HSC73

Several different lines of evidence suggest that the HSP70 family member responsible for stimulating lysosomal degradation of proteins with KFERQ-like peptide regions is HSC73 (Terlecky et al., 1990).

Binding of HSP70s to RNase S-peptide

To determine which HSP70s bind to RNase S-peptide, we precoated microtiter wells with 4 different HSP70s and assayed the specific binding of [³H]RNase S-peptide. HSC73 bound the most [³H]RNase S-peptide of all the HSP70s tested. Binding was 4-10 times less for a heat-inducible HSP70 (SSA1 from yeast), the endoplasmic reticulum glucose-regulated protein (grp78 from bovine brain), and DnaK (an HSP70 from *E. coli*).

Comparison of protein binding specificities of prp73 and HSC73

We precoated prp73 and HSC73 on microtiter wells and determined the ability of various proteins to compete for binding of [³H]RNase S-peptide. Both prp73 and HSC73 interact with RNase S-peptide, RNase A, and aspartate aminotransferase but not with ovalbumin, lysozyme, or ubiquitin.

Stimulation of lysosomal proteolysis by HSP70s

Lysosomes *in vitro* can take up and degrade [³H]RNase S-peptide in an ATP- and prp73-stimulated manner (Chiang et al., 1989). We compared the abilities of different HSP70s to stimulate degradation of [³H]RNase S-peptide by isolated lysosomes. HSC73 and prp73 had approximately equal stimulatory effects on degradation while a heat-inducible HSP70, grp78, and DnaK had little or no activity.

HSP70s in the cytosol of serum-deprived human fibroblasts

Most (>90%) HSP70 in the cytosol of serum-deprived fibroblasts can bind to RNase S-peptide (Chiang et al., 1989). To determine the number of different HSP70s in the cytosol of serum-deprived fibroblasts, we separated cytosolic proteins by two-dimensional electrophoresis (O'Farrell, 1975). Proteins were then transferred to nitrocellulose filters and immunoblotted with mAb 7.10 (kindly provided by Dr. Susan Lindquist, University of Chicago, Chicago, IL) which recognizes all known HSP70s in a wide variety of organisms (Kurtz et al., 1986). Cytosol from serum-deprived fibroblasts contain at least two different HSP70s. The less abundant HSP70 is the major heat inducible HSP70 since its level increases in response to heat shock. The major HSP70 in the cytosol of serum-deprived fibroblasts is slightly larger and more acidic than the heat inducible HSP70. This migration position is consistent with the major HSP70 being HSC73 (Watowich and Morimoto, 1988; Schlesinger, 1990).

Conclusions

HSC73 has been implicated in a variety of subcellular functions including uncoating of clathrin-coated vesicles (Chappell et al., 1986) and binding to microtubules (Napolitano et al., 1987; Green and Liem, 1989). Interestingly, clathrin light chains, clathrin assembly polypeptides (adaptins; Ponnambalam et al., 1990), and tubulin β-chains contain multiple KFERQ-like sequences (Table 1). HSC73 is able to form homodimers (Guidon and Hightower, 1986), and this association may also be through binding to KFERQ-like sequences (Table 1).

On the other hand, certain studies suggest that HSC73 may bind to peptides and proteins at sequences other than KFERQ-like domains. Beckman et al. (Beckman et al., 1990) found that HSC73 transiently associated with most newly-synthesized proteins, and most intracellular proteins do not contain KFERQ-like peptide motifs. In addition, HSC73 binds to two peptides from the vesicular stomatitis virus glycoprotein (Flynn et al., 1990). Neither of these peptides contain KFERQ-like regions, so additional properties of the peptides may also be important for HSC73 binding. For example, peptide folding may allow proper positioning of basic, acidic, equivalent of a KFERQ-like region. Alternatively, HSC73 may contain multiple peptide binding sites, or different forms of HSC73 with different peptide binding properties may exist within cells.

HSC73 appears to have multiple physiological roles. An important role of HSC73 during cell growth has been inferred from the increased expression of HSC73 mRNA in rapidly growing cells (Sorger and

Pelham, 1987; Cairo et al., 1989) This requirement for an increased HSC73 level probably reflects the importance of HSC73 in the posttranslational import of proteins into organelles (Deshaies et al., 1988, Chirico et al., 1988; Sheffield et al., 1990), regulation of clathrin uncoating (Chappell, 1986), and/or regulation of cytoskeletal interactions (Napolitano et al., 1987; Green and Liem, 1989).

We found that the amount of prp73 in the cytosol of confluent fibroblasts increased in response to serum-deprivation (Chiang et al., 1989) and suggested a role for HSC73 in enhanced proteolysis. Our working model is that HSC73 can bind to peptides or peptide regions within proteins and alter the structure of that peptide or protein in an ATP-dependent manner and/or cause the peptide or protein to associate with lysosomes. The interaction between HSC73 and the peptide or protein somehow promotes membrane translocation. How the various protein substrates are targeted for transport across specific cellular membranes is not yet known.

Acknowledgments

We thank Seth Sadis and Larry Hightower for providing bovine HSC73, bovine grp78, and *E. coli* DnaK, Bruce Koch and Randy Schekman for yeast HSP70, and Susan Lindquist for the mAb 7.10. This work is supported by NIH grant AG06116 to J.F.D. T.S.O. is supported by a Damon Runyon-Walter Winchell Cancer Research Fund Fellowship, DRG-960. S.R.T. is supported by NIH Training Grant DK07542.

References

Backer, JM, Bourret, L and Dice, JF, (1983) Regulation of catabolism of microinjected ribonuclease A requires the amino terminal twenty amino acids. Proc. Nat. Acad. Sci. USA, 80: 2166-2170.

Backer, JM and Dice, JF, (1986) Covalent linkage of ribonuclease S-peptide to microinjected proteins causes their intracellular degradation to be enhanced during serum withdrawal. Proc. Nat. Acad. Sci. USA, 83: 5830-5834.

Beckmann, RP, Mizzen, LA and Welch, WJ, (1990) Interaction of HSP 70 with newly synthesized proteins: Implications for protein folding and assembly. Science, 248: 850-853.

Cairo, G, Schiaffonati, L, Rappocciolo, E, Tacchini, L and Bernelli-Zazzera, A, (1989) Expression of different members of heat shock protein 70 gene family in liver and hepatomas. Hepatology, 9: 740-746.

Chappell, TG, Welch, WJ, Schlossman, DM, Palter, KB, Schlesinger, MJ and Rothman, JE, (1986) Uncoating ATPase is a member of the 70 kilodalton family of stress proteins. Cell, 45: 3-13.

Chiang, H-L and Dice, JF, (1988) Peptide sequences that target proteins for enhanced degradation during serum withdrawal, J. Biol. Chem., 263: 6797-6805.

Chiang, H-L, Terlecky, SR, Plant, CP and Dice, JF, (1989) A role for a 70-kilodalton heat shock protein in lysosomal degradation of intracellular

188

proteins. Science, 246: 382-385.

Chirico, W, Waters, MG and Blobel, G, (1988) 70K heat shock related proteins stimulate protein translocation into microsomes. Nature, 332: 805-810.

Deshaies, RJ, Koch, BD, Werner-Washburne, M, Craig, EA and Schekman, R, (1988) 70kd stress protein homologues facilitate translocation of secretory and mitochondrial precursor polypeptides. Nature, 332: 800-805.

Dice, JF, (1982) Altered degradation of proteins microinjected into senescent human fibroblasts. J. Biol. Chem., 257: 14624-14627.

Dice, JF, Chiang, H-L, Spencer, EP and Backer, JM, (1986) Regulation of catabolism of microinjected ribonuclease A: Identification of residues 7-11 as the essential pentapeptide. J. Biol. Chem., 261: 6853-6859.

Dice, JF, (1987) Molecular determinants of protein half-lives in eukaryotic cells. FASEB J., 1: 349-357.

Dice, JF, (1990) Peptide sequences that target cytosolic proteins for lysosomal proteolysis. Trends Biochem. Sci., 15: 305-309.

Flynn, GC, Chappell, TG and Rothman, JE, (1989) Peptide binding and release by proteins implicated as catalysts of protein assembly. Science, 245: 385-390.

Goff, SA, Short-Russell, SR and Dice, JF, (1987) Efficient saturation mutagenesis of a pentapeptide coding sequence using mixed oligonucleotides. DNA, 6: 381-388.

Goldberg, AL and Dice, JF, (1974), Intracellular protein degradation in mammalian and bacterial cells. Ann. Rev. Biochem., 34: 835-869.

Green, LAD and Liem, RKH, (1989) β-Internexin is a microtubule-associated protein identical to the 70-kDa heat-shock cognate protein and the clathrin uncoating ATPase. J. Biol. Chem., 264: 15210-15215.

Guidon, PTJ and Hightower, LE, (1986) The 73 kilodalton heat shock cognate protein purified from rat brain contains nonesterified palmitic and stearic acids. J. Cell Physiol., 128: 239-245.

Kurtz, S, Rossi, J, Petko, L and Lindquist, S, (1986) An ancient developmental induction: Heat shock proteins induced in sporulation and oogenesis. Science, 231: 1154-1157.

McElligott, MA, Miao, P and Dice, JF, (1985) Lysosomal degradation of RNase A and RNase S-protein microinjected into the cytosol of human fibroblasts. J. Biol. Chem., 260: 11986-11993.

Napolitano, EW, Pachter, JS and Liem, RKH, (1987) Intracellular distribution of mammalian stress proteins: Effect of cytoskeletal specific agents, J. Biol. Chem., 262: 1493-1504.

Neff, NT, Bourret, L, Miao, P and Dice, JF, (1981) Degradation of proteins microinjected into IMR-90 human diploid fibroblasts. J. Cell. Biol., 91: 184-194.

O'Farrell, PH, (1975) High resolution two-dimensional electrophoresis of proteins. J. Biol. Chem., 250: 4007-4021.

Olson, TS and Dice, JF, (1989) Regulation of protein degradation rates in eukaryotes. Curr. Opinion Cell Biol., 1: 1194-1200.

Olson, TS, Terlecky, SR and Dice, JF, Pathways of intracellular protein degradation in eukaryotic cells in Stability of Protein Pharmaceuticals: In Vivo Pathways of Degradation and Stratagies for Protein Stabilization. Ahern TJ and Manning MC (eds), Plenum Press, New York, in press.

Ponnambalam, S, Robinson, MS, Jackson, AF, Peiperl, L and Parham, P, (1990) Conservation and diversity in families of coated vesicle adaptins. J. Biol. Chem., 265: 4814-4820.

Schlesinger, MJ, (1990) Heat shock proteins. J. Biol. Chem., 265: 12111-12114.

Sheffield, WP, Shore, GC and Randall, SK, (1990) Mitochondrial precursor protein: Effects of 70-kilodalton heat shock protein on polypeptide folding, aggregation, and import competence. J. Biol. Chem., 265: 11069-11076.

Sorger, PK and Pelham, HRB, (1987) Cloning and expression of a gene encoding hsc 73, the major hsp 70-like protein in unstressed rat cells, EMBO J., 6: 993-998.

Terlecky, SR, Chiang, H-L, Olson, TS and Dice, JF, Protein and peptide binding and stimulation of *in vitro* lysosomal proteolysis by the 73-kDa heat shock cognate protein. J. Biol. Chem., submitted.

Watowich, SS and Morimoto, RI, (1988) Complex regulation of heat shock- and glucose-responsive genes in human cells. Mol. Cell Biol., 8: 393-405.

Immunology

Immune Recognition of Stress Proteins in Infection and in Surveillance of Stressed Cells

S. Jindal, R.A. Young
Whitehead Institute for Biomedical Research
Department of Biology
Massachusetts Institute of Technology
Cambridge, MA 02142
USA

Introduction

Molecular biologists and immunologists have been investigating intensively the major antigens involved in the immune response to infection by mycobacteria, the causative agents of tuberculosis and leprosy. These two infectious diseases are a continuous threat to mankind. Recent WHO data indicates that as much as one third of the total population in the world is infected with *Mycobacterium tuberculosis*, and about 2-3 million people die from tuberculosis each year. About 10 million people in the world are infected with *Mycobacterium leprae* and stand at risk of incurring irreparable nerve damage (Bloom, 1989). The cloning and sequencing of genes coding for major mycobacterial protein antigens revealed that most of these proteins are members of stress protein (or heat shock protein) families (Young et al., 1988). This observation was unexpected because stress proteins occur in highly conserved forms in all organisms, from bacteria to man (Lindquist and Craig, 1988). The discovery raised questions about how immune cells could distinguish bacterial stress proteins and human stress proteins, and why the immune system would treat stress proteins as major targets when such recognition runs the risk of producing autoimmune disease (reviewed in Kaufmann, 1990; Young, 1990; Young and Elliott, 1989).

A variety of T cell types isolated from healthy individuals can recognize one's own stress proteins; this observation, and the finding that viral and bacterial infections can induce stress responses in host cells, has led to the hypothesis that some anti-stress protein T lymphocytes are involved in immune surveillance of autologous cells. If such surveillance mechanisms exists, then occasional over-stimulation of immune cells, perhaps by infection, might lead to autoimmune disease. Indeed, autoimmune responses to stress proteins have been implicated in rheumatoid arthritis and in insulin-dependent diabetes. Here we briefly review the immune response to stress proteins and the role that cellular stress responses may play in an immune defense against infection.

Pathogen's stress proteins are immunodominant antigens

Immunization of mice with whole cell lysates from *M. tuberculosis* and *M. leprae* elicits antibody responses to only a limited set of protein antigens, suggesting that these antigens are major targets of the immune response. Genes coding for these protein antigens were cloned (Young et al., 1985; Husson and Young, 1987; Baird et al., 1988; Nerland et al., 1988; Shinnick et al., 1988; Young et al., 1988; Garsia et al., 1989; Shinnick et al., 1989) and sequence analysis revealed that several are counterparts of well-known stress proteins (Fig. 1). These include HSP70 (dnaK), the 65 kD "common antigen" (a homolog of the *E. coli groEL* product) and a 12 kD species that is homologous to the *E. coli groES* product.

Antibodies and T cells collaborate to manage infections and the T cell response to stress proteins has been intensively investigated. A variety of different types of T lymphocytes have been found to recognize stress proteins, attesting to the importance of stress protein recognition in immune responses. These include T cells with α/β T cell receptors

M. tuberculosis:

Antigens:	Identity:
71 kDa	HSP70 (dnaK)
65 kDa	HSP60 (groEL)
38 kDa	
28 kDa	
19 kDa	
14 kDa	
12 kDa	HSP12 (groES)

M. leprae:

Antigens:	Identity:
71 kDa	HSP70 (dnaK)
65 kDa	HSP60 (groEL)
36 kDa	
28 kDa	superoxide dismutase
18 kDa	HSP17
12 kDa	HSP12 (groES)

Figure 1. Immune targets in tuberculosis and leprosy.

(Emmrich et al., 1988; Mustafa et al., 1986; Boom et al., 1987; Kaufmann et al., 1987; Lamb et al., 1987; Oftung et al., 1987; Oftung et al., 1988; Ottenhoff et al., 1988; Thole et al., 1988; Lamb and Young, 1989) and γ/δ T cell receptors (Haregewoin et al., 1989; Holoshitz et al., 1989; Modlin et al., 1989; O'Brien et al., 1989). Among the α/β T cells that recognize stress proteins, both CD4+ helper cells and CD8+ cytotoxic killer cells have been isolated.

Epitope mapping on HSP60 proteins indicates that antibodies and T cells can recognize many different portions of the molecule (Mehra et al., 1986; Lamb et al., 1987; Thole et al., 1988, Lamb and Young, 1989). Thus, the entire HSP60 protein can be involved in immune responses; no one single epitope is dominant.

Both HSP70 and HSP60 proteins are among the targets of antibodies and T cells in a large variety of bacterial and parasitic infections. Many parasitic infections in humans, including trypanosomiasis (Engman et al., 1988), leishmaniasis (Smith et al., 1988), schistosomiasis (Nene et al., 1986; Hedstrom et al., 1987), filariasis (Selkirk et al., 1988) and malaria (Bianco et al., 1986; Ardeshire et al., 1987) induce antibody responses against HSP70 (Fig. 2). HSP60 appears to be a general target of the immune system in bacterial infections including leprosy and tuberculosis (Young et al., 1988), lyme disease (Hansen et al., 1988), Q fever (Vodkin et al., 1988), syphilis (Hindersson et al., 1987), legionellosis (Plikaytis et al., 1987) and binding trachoma (Morrison et al., 1989). Indeed, before the protein was identified as groEL, it was known as "common antigen".

A number of features of stress proteins may render them particularly important targets of the immune system following bacterial or parasitic infections. For example, bacteria phagocytized by macrophages undergo a stress response (Buchmeier and Heffron, 1990; see Figure 3). Abundant proteins are frequent targets of the immune system. In addition, stress proteins may be very susceptible to processing and presentation by antigen presenting cells. A most important feature of these proteins may be their sequence conservation: they are remarkably similar among pathogens. Thus, infection by a given pathogen during early childhood may be restimulated later in life by other pathogens.

Induction of host stress proteins may permit immune surveillance of stressed autologous cells

Several observations on T lymphocytes that recognize stress proteins have led to the idea that these cells may have a surveillance role that could provide a first line of defense against infection or transformation. Some T lymphocytes from healthy individuals that recognize bacterial stress proteins can equally well recognize self stress proteins (Lamb et al., 1989). Moreover, most healthy individuals have T cells that can recognize self stress protein determinants in vitro (Munk et al., 1989). What kind of T lymphocytes are these and why do they recognize one's own stress proteins? Further, why does this recognition generally not

DISEASE	PATHOGEN	STRESS PROTEIN
Bacteria		
Tuberculosis	Mycobacteria	HSP60 HSP70 HSP12
Leprosy	Mycobacteria	HSP60 HSP70 HSP12
Lyme disease	*Borrelia*	HSP60
Q fever	*Coxiella*	HSP60
Syphilis	*Treponema*	HSP60
Legionaires'disease	*Legionella*	HSP60
Binding trachoma	*Chlamydia*	HSP60
Parasite		
Trypanosomiasis	*Trypanosomes*	HSP70
Leishmaniasis	*Leishmania*	HSP70
Schistosomiasis	*Schistosomes*	HSP70
Filariasis	*Brugia*	HSP70
Malaria	*Plasmodia*	HSP70

Figure 2. Stress protein antigens in bacterial and parasitic infections.

lead to autoimmune disease?
The T lymphocytes that have been demonstrated to be capable of self stress protein recognition include CD4$^+$CD8$^-$, α/β T cell receptor-bearing cells (Munk et al., 1989; Lamb et al., 1989); these "helper" cells produce lymphokines that can stimulate B cells to produce antibodies and can provoke cytotoxic T lymphocytes to more effectively kill their target cells. Cytotoxic T cells that are CD4$^-$CD8$^+$ and have α/β receptors have also been identified that recognize stress proteins (Koga et al.,

STRESS RESPONSE OF PATHOGENS

Bacteria or parasite

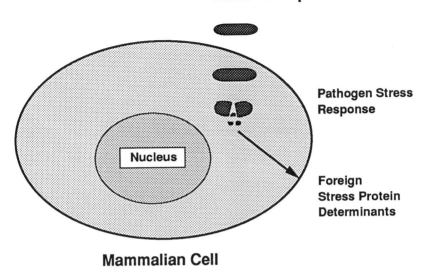

**Pathogen Stress
Response**

**Foreign
Stress Protein
Determinants**

Mammalian Cell

Figure 3. Pathogen's stress response within a host cell. Intracellular bacteria or parasites appear to undergo a stress response upon infection of a host cell. Subsequent killing of a pathogen may result in release of stress proteins which could be processed and presented on the cell surface.

1990). These cells appear to be able to kill autologous cells that express the stress protein antigen on their surface. In addition, foreign or self HSP60 protein is also recognized by a class of T lymphocytes expressing γ/δ T cell receptor (Haregewoin et al., 1989; Holoshitz et al., 1989; Modlin et al., 1989; O'Brien et al., 1989). The γ/δ T cells are localized primarily in the peripheral sites including the epithelial layers of the epidermis, intestine and reproductive tract in the mouse (Raulet, 1989), but their location and specificity is less well defined in humans. These T cells are also capable of cytolysis of target cells expressing HSP60 (Fisch et al., 1990).

Many viruses, including those that can transform their host cells, induce a stress response in infected cells (Hightower and Smith, 1978; Peluso et al., 1978; Collins and Hightower, 1982; Nevins, 1982; Notarianni and Preston, 1982; Garry et al., 1983; Khadjian and Turler, 1983; La Thangue et al., 1984; Macnab et al., 1985; La Thangue and Latchman, 1988). In some cases, as with adenovirus, the virus directly induces HSP70 synthesis by producing a transcriptional activator that stimulates HSP70 gene transcription (Kao and Nevins, 1983; Simmons et al., 1988). In other cases, the response may be due to other mechanisms of induction. We have carefully investigated the effect of infection by different viruses on human macrophages and T cell lines and have found varying levels of response that are both virus and cell type dependent (Jindal and Young, unpublished). Thus, many of these virus-infected cells could become targets of stress protein specific T cells upon cell-surface presentation of stress protein determinants. Indeed, Koga et al. (Koga et al., 1990) have shown that bone marrow derived macrophages are killed by CD8+ cytotoxic T cells that are specific for mycobacterial HSP60 when macrophages are stressed by cytomegalovirus infection, by heat shock or by exposure to IFN-γ treatment. These data support the hypothesis that HSP60 specific T cells can recognize and kill stressed autologous cells.

The HSP60 protein expressed by some cancer cells makes them potential targets for anti-stress protein T lymphocytes (Fisch et al., 1990). Stress protein molecules or their determinants appear to be expressed on the cell surface of some lymphoma cell lines. Fisch et al. (Fisch et al., 1990) investigated the killing of Daudi Burkitt's lymphoma cells by human γ/δ T cells expressing Vγ9 and Vδ2 genes. Using polyclonal antiserum raised against mammalian HSP60, they provided evidence that the ligands recognized by these Vγ9/Vδ2 cells on the cell surface of the lymphoma cells include HSP60 protein or peptides.

Some cytokines and differentiation inducers can induce increased synthesis of some stress proteins in certain human tumor cell lines (Ferm et al., 1990) and this may augment the levels of stress protein that can be recognized by T cells. Monocytic leukemia cell lines which express low levels of HSP60 can be induced to produce moderate to high levels of hsp60 mRNA and protein by heat shock, retinoic acid, IFN-γ and TNF-α treatment. Some of the HSP60 protein can be detected on the cell surface after exposure to cytokines. The production of cytokines by T cells in the vicinity of tumor cells might stimulate enhanced stress protein expression by the tumor cells, possibly making them better targets for T cells that recognize stress proteins.

Conclusions

Why does the presence of T lymphocytes that recognize self stress protein determinants not cause autoimmune disease in the vast majority of people? The immune response is most effective when focusing on local sites of infection; efficient action requires the cooperation of a variety of lymphoid cell types and adequate concentration of lymphokines. Most infections are initiated at a specific site, and the requirement of adequate numbers of lymphoid cells or the accumulation of adequate amounts of cytokines may be necessary for efficient killing by T cells that recognize self stress proteins. T lymphocytes that recognize stress proteins have been implicated in rheumatoid arthritis (van Eden et al., 1988; Lamb et al., 1989; Holoshitz, 1989; Gaston et al., 1989; van der Broek et al., 1989) and in insulin-dependent diabetes (Elias et al., 1989). However, these diseases, if they are a consequence of stress protein recognition, may represent cases in which overstimulation of the immune response to stress proteins has occurred or where regulation of the immune response to stress protein has broken down.

Further studies of stress proteins and the immune response could lead to new strategies for enhancing the body's ability to fight infection, as well as methods for reducing the destructive effects of arthritis and autoimmune diseases.

References

Ardeshir, F, Flint, JE, Richman, J and Reese, RT, (1987) A 75kd merozoite surface protein of *Plasmodium falciparum* which is related to the 70 kd heat-shock proteins. EMBO J., 6: 493-99.

Baird, PN, Hall, LMC and Coates, ARM, (1988) A major antigen from *Mycobacterium tuberculosis* which is homologous to the heat shock proteins groES from *Escherichia coli* and the htpA gene product of *Coxiella burnetii*. Nucl. Acid Res., 16: 9047.

Bianco, AE, Favaloro, JM, Burkof, TR, Culvenor, JG, Crewther, PE, Brown, GV, Anders, RF, Coppel, RL and Kemp, DJ, (1986) A repetitive antigen of *Plasmodium falciparum* that is homologous to heat shock protein 70 of *Drosophila melanogaster*. Proc. Natl. Acad. Sci. USA, 83: 8713-17.

Bloom, BR, (1989) Vaccines for the third world. Nature, 342: 115-20.

Boom, WH, Husson, RN, Young, RA, David, JR and Piessens, WF, (1987) In vivo and in vitro characterization of murine T-cell clones reactive to *Mycobacterium tuberculosis*. Infect. Immun., 55: 2223-29.

Buchmeier, NA and Heffron, F, (1990) Induction of *Salmonella* stress proteins upon infection of macrophages. Science, 248: 730-732.

Collins, PL and Hightower, LE, (1982) Newcastle disease virus stimulates the cellular accumulation of stress (heat shock) mRNAs and proteins. J. Virol., 44: 703-7.

Elias, D, Markovits, D, Reshef, T, van der Zee, R and Cohen, IR (1990) Induction and therapy of autoimmune diabetes in the non-obese diabetic (NOD/Lt) mouse by a 65-KDa heat shock protein. Proc. Natl. Acad. Sci. USA, 87: 1576-1580.

Emmrich, F, Thole, J, van Embden, J and Kaufmann, SHE, (1986) A recombinant 64 kDa protein of *Mycobacterium bovis* BCG specifically stimulates human T 4 clones reactive to mycobacterial antigens. J. Exp. Med., 163: 1024-29.

Engman, DM, Kirchoff, LV, Henkle, K and Donelson, JE, (1988) A novel hsp70 cognate in trypanosomes. J. Cell. Biochem., 12D: Supplement, 290.

Ferm, MT, Soderstrom, K, Jindal, S, Klareskog, L, Gronberg, A, Young, RA and Kiessling, R, Induction and cell surface expression of human hsp60 in monocytic cell lines. Submitted.

Fisch, P, Malkovsky, M, Kovats, S, Sturm, E, Braakman, E, Klein, BS, Voss, SS, Morrissey, W, De Mars, R, Welch, WJ, Bolhuis, RL and Sondel, PM, Recognition by human Vγ9/Vδ2 T cells of a GroEL homolog on Daudi Burkitt's Lymphoma cells. Science, in press.

Garry, RF, Ulug, ET and Bose, HR Jr., (1983) Induction of stress proteins in Sindbis virus- and vesicular stomatitis virus-infected cells. Virology, 129: 319-32.

Garsia, RJ, Hellqvist, L, Booth, RJ, Radford, AJ, Britton, WJ, Astbury, L, Trent, RJ and Basten, A, (1989) Homology of the 70-kilodalton antigens from *Mycobacterium leprae* and *Mycobacterium bovis* with the *Mycobacterium tuberculosis* 71-kilodalton antigen and with the conserved heat shock protein 70 of eucaryotes. Infect. Immun., 57: 204-12.

Gaston, JSH, Life, PF, Bailey, LC and Bacon, PA, (1989) In vitro responses to a 65 kilodalton mycobacterial protein by synovial T cells from inflammatory arthritis patients. J. Immunol., 143: 2494-2500.

200

Hansen, K, Bangsborg, JM, Fjordvang, H, Pedersen, NS and Hinderssen, P, (1988) Immunochemical characterisation of, and isolation of the gene for a *Borrelia burgdorferi* immunodominant 60kDa antigen common to a wide range of bacteria. Infect. Immun., 56: 2047-53.

Haregewoin, A, Soman, G, Hom, RC and Finberg, RW, (1989) Human $\gamma\delta^+$ T cells respond to mycobacterial heat-shock protein. Nature, 340: 309-12.

Hedstrom, R, Culpepper, PJ, Harrison, RA, Agabian, N and Newport, G, (1987) A major immunogen in *Schistosoma mansoni* infections is homologous to the heat-shock protein Hsp70. J. Exp. Med., 165: 1430-35.

Hightower, LE and Smith, MD, (1978) Effects of canavanine on protein metabolism in Newcastle disease virus-infected and uninfected chicken embryo cells. In Negative Strand Viruses and The Host Cell, Mahy, BWJ, Barry, RD, eds., London, Academic Press, 395-405.

Hindersson, P, Knudson, JD and Axelsen, NH, (1987) Cloning and expression of *Treponema pallidum* common antigen (Tp-4) in *E. coli* K-12. J. Gen. Microbiol., 133: 587-96.

Holoshitz, J, Konig, F, Coligan, JE, de Bruyn, J and Strober, S, (1989) Isolation of CD4⁻ CD8⁻ mycobacteria-reactive T lymphocyte clones from rheumatoid arthritis synovial fluid. Nature, 339: 226-29.

Husson, R and Young, RA, (1987) Genes for the major protein antigens of *M. tuberculosis*: the etiologic agents of tuberculosis and leprosy share a major protein antigen. Proc. Natl. Acad. Sci. USA, 84: 1679-83.

Kao, H-T and Nevins, JR, (1983) Transcriptional activation and subsequent control of the human heat shock gene during adenovirus infection. Mol. Cell. Biol., 3: 2058-65.

Kaufmann, SHE, (1990) Heat shock proteins and the immune response. Immunol. Today, 11: 129-36.

Kaufmann, SHE, Vath, U, Thole, JER, van Embden, JDA and Emmrich, F, (1987) Enumeration of T cells reactive with *Mycobacterium tuberculosis* organisms and specific for the recombinant mycobacterial 64 kilodalton protein. Eur. J. Immunol., 17: 351-57.

Khandjian, EW and Turler, H, (1983) Simian virus 40 and polyoma virus induce synthesis of heat shock proteins in permissive cells. Mol. Cell. Biol., 3: 1-8.

Koga, T, Wand-Wurttenberger, A, de Bruyn, J, Munk, ME, Schoel, B and Kaufmann, SHE, (1989) T cells against a bacterial heat shock protein recognize stressed macrophages. Science, 245: 1112-1115.

Lamb, JR and Young, DB, (1989) T cell recognition of stress proteins: A link between infectious and autoimmune disease. Mol. Biol. Med., 7: 311-321.

Lamb, JR, Ivanyi, J, Rees, ADM, Rothbard, JB, Holland, K, Young, RA and Young, DB, (1987) Mapping of T cell epitopes using recombinant antigens and synthetic peptides. EMBO J., 6: 1245-49.

Lamb, JR, Bal, V, Mendez-Samperio, P, Mehlert, A, So, A, Rothbard, J, Jindal, S, Young, RA and Young, DB, (1989) Stress proteins may provide a link between the immune response to infection and autoimmunity. Intl. Immunol., 1: 191-196.

La Thangue, NB, Shriver, K, Dawson, C and Chan, WL, (1984) Herpes simplex virus infection causes the accumulation of a heat-shock protein. EMBO J., 3: 267-77.

La Thangue, NB and Latchman, DS, (1988) A cellular protein related to heat-shock protein 90 accumulates during herpes simplex virus infection and is overexpressed in transformed cells. Exp. Cell. Res., 178: 169-79.

Lindquist, S and Craig, EA, (1988) The heat-shock proteins. Ann. Rev. Genet., 22: 631-37.

Macnab, JCM, Orr, A and La Thangue, NB, (1985) Cellular proteins expressed in herpes simplex virus transformed cells also accumulate on herpes simplex virus infection. EMBO J., 4: 3223-28.

Mehra, V, Sweetser, D and Young, RA, (1986) Efficient mapping of protein antigenic determinants. Proc. Natl. Acad. Sci. USA, 83: 7013-17.

Modlin, RL, Pirmez, C, Hofman, FM, Tongian, V, Uyemura, K, Rea, TH, Bloom, BR and Brenner, MB, (1989) Lymphocytes bearing antigen specific gamma-delta T cell receptors accumulate in human infectious disease lesions. Nature, 339: 544-48.

Morrison, RP, Lyng, K and Caldwell, HD, (1989) Chlamydial disease pathogenesis: ocular hypersensitivity elicited by a genus specific 57-KD protein. J. Exp. Med., 169: 663-675.

Mustafa, AS, Gill, HK, Nerland, A, Britton, WJ, Mehra, V, Bloom, BR, Young, RA and Godal, T, (1986) Human T cell clones recognize a major *M. leprae* protein antigen expressed in *E. coli.* Nature, 319: 63-66.

Munk, ME, Schoel, B, Modrow, S, Karr, RW, Young, RA and Kaufmann, SHE, (1989) T lymphocytes from healthy individuals with specificity to self epitopes shared by the mycobacterial and human 65 kDa heat shock protein. J. Immunol., 143: 2844-49.

Nene, V, Dunne, DW, Johnson, KS, Taylor, DW, and Cordingley, JS, (1986) Sequence and expression of a major egg protein from *Schistosoma mansoni.* Homologies to heat shock proteins and alpha crystallins. Mol. Biochem. Parasitol., 21: 179-88.

Nerland, AH, Mustafa, AS, Sweetser, D, Godal, T and Young, RA, (1988) A protein antigen of *Mycobacterium leprae* is related to a family of small heat shock proteins. J. Bacteriol., 170: 5919-21.

Nevins, JR, (1982) Induction of the synthesis of a 70,000 dalton mammalian heat shock protein by the adenovirus E1A gene product. Cell, 29: 913-19.

Notarianni, EL and Preston, CM, (1982) Activation of cellular stress protein genes by Herpes Simplex virus temperature-sensitive mutants which overproduce immediate early polypeptides. Virology, 123: 113-22.

O'Brien, RL, Happ, MP, Dallas, A, Palmer, E, Kubo, R and Born, WK, (1989) Stimulation of a major subset of lymphocytes expressing T cell receptor γδ by an antigen derived from *Mycobacterium tuberculosis.* Cell, 57: 667-74.

Oftung, F, Mustafa, AS, Husson, R, Young, RA and Godal, T, (1987) Human T-cell clones recognize two abundant *M. tuberculosis* protein antigens expressed in *E. coli.* J. Immunol., 138: 927-31.

Oftung, F, Mustafa, AS, Shinnick, TM, Houghten, RA, Kvalheim, G, Degre, M, Lundin, KEA and Godal, T, (1988) Epitopes of the *Mycobacterium tuberculosis* 65-kilodalton protein antigen as recognized by human T cells. J. Immunol., 141: 2749-54.

Ottenhoff, THM, Ab, BK, van Embden, JDA, Thole, JER and Kiessling, R, (1988) The recombinant 65-kD heat shock protein of *Mycobacterium bovis* Bacillus Calmette-Guerin/*M. tuberculosis* is a target molecule for CD4+ cytotoxic T lymphocytes that lyse human monocytes. J. Exp. Med., 168: 1947-52.

Peluso, RW, Lamb, RA and Choppin, PW, (1978) Infection with paramyxoviruses stimulates synthesis of cellular polypeptides that are also stimulated in cells transformed by Rous sarcoma virus or deprived of glucose. Proc. Natl. Acad. Sci. USA, 75: 6120-24.

Plikaytis, BB, Carlone, GM, Pau, C-P and Wilkinson, HW, (1987) Purified 60kD *Legionella* protein antigen with *Legionella*-specific and nonspecific epitopes. J. Clin. Microbiol., 25: 2080-84.

Raulet, DH, (1989) The structure, function, and molecular genetics of the γ/δ T cell receptor. Ann. Rev. Immunol., 7: 175-207.

Selkirk, ME, Rutherford, PJ, Denham, DA, Partono, F and Maizels, RM, (1987) Cloned antigen genes of *Brugia* filarial parasites. Biochem. Soc. Symp., 53: 91-102.

Shinnick, TM, Vodkin, MH and Williams, JC, (1988) The *Mycobacterium tuberculosis* 65-kilodalton antigen is a heat shock protein which corresponds to common antigen and to the *Escherichia coli* GroEL protein. Infect. Immun., 56: 446-51.

Shinnick, TM, Plikaytis, BB, Hyche, AD, Van Landingham, RM and Walker, LL, (1989) The *Mycobacterium tuberculosis* BCG-a protein has homology with the *Escherichia coli* GroES protein. Nucl. Acids Res., 17: 1254.

Simon, MC, Fisch, TM, Benecke, BJ, Nevins, JR and Heintz, N, (1988) Definition of multiple, functionally distinct TATA elements, one of which is a target in the *hsp70* promoter for E1A regulation. Cell, 52: 723-29.

Smith, DF, Searle, S, Campo, AJR, Coulson, RMR and Ready, PD, (1988) A multigene family in *Leishmania major* with homology to eukaryotic heat shock protein 70 genes. J. Cell. Biochem., 12D: Supplement, 296.

Thole, JER, van Schooten, WCA, Keulen, WJ, Hermans, PWM, Janson, AAM, de Vries, RRP, Kolk, AHJ and van Embden, JDA, (1988) Use of recombinant antigens expressed in *Escherichia coli* K-12 to map B-cell and T-cell epitopes on the immunodominant 65-kilodalton protein of *Mycobacterium bovis* BCG. Infect. Immun., 56: 1633-40.

van der Broek, MF, van Bruggen, MCJ, Hogervorst, EJM, van Eden, W, van der Zee, R and van der Bert, WB, (1989) Protection against streptococcal cell wall induced arthritis by pretreatment with the mycobacterial 65kD heat shock protein. J. Exp. Med., 170: 449-66.

van Eden, W, Thole, JER, van der Zee, R, Noordzij, A, van Embden, JDA, Hensen, EJ and Cohen, IR, (1988) Cloning of the mycobacterial epitope recognized by T lymphocytes in adjuvant arthritis. Nature, 331: 171-73.

Vodkin, MH and Williams, JC, (1988) A heat shock operon in *Coxiella burnetii* produces a major antigen homologous to a protein in both mycobacteria and *Escherichia coli.* J. Bacteriol., 170: 1227-34.

Young, RA, (1990) Stress proteins and immunology. Ann. Rev. Immunol., 8: 401-20.

Young, RA and Elliott, T, (1989) Stress proteins, infection and immune surveillance. Cell, 59: 5-8.

Young, RA, Mehra, V, Sweetser, D, Buchanan, TM, Clark-Curtiss, J, Davis, RW and Bloom, BR, (1985) Genes for the major protein antigens of the leprosy parasite *Mycobacterium leprae.* Nature, 316: 450-454.

Young, DB, Lathigra, R, Hendrix, R, Sweetser, D and Young, RA, (1988) Stress proteins are immune targets in leprosy and tuberculosis. Proc. Natl. Acad. Sci. USA, 85: 4267-70.

Heat Shock Proteins and Mycobacterial Infection

D. Young, T. Garbe, R. Lathigra
MRC Tuberculosis and Related Infections Unit
Hammersmith Hospital
London W12 0HS
UK

Introduction

Mycobacterial diseases - tuberculosis and leprosy - arise from a complex series of interactions between the microbial pathogen and the host immune system. Mycobacteria themselves are relatively benign organisms which multiply very slowly and are devoid of potent biologically active toxins characteristic of more "virulent" bacterial pathogens. Evolution of the disease occurs rather as a consequence of induction of aberrant - immunopathological - responses of the host immune system. Thus, while it is clear that activation of cellular immunity is essential for controlling mycobacterial infection, immune activation is also central to progression of disease.

Heat shock proteins as antigens

A major focus of research on mycobacterial diseases has therefore been dissection of the immune response and identification of individual mycobacterial components involved in interactions with the immune system. From such studies it emerged that members of heat shock protein families make a major contribution to the antigenic profile of mycobacterial pathogens (Table 1). It is now widely recognised that heat shock proteins are potent inducers of both humoral and cell-mediated immune responses involving multiple lymphocyte subsets in a range of infectious and autoimmune diseases (for reviews, see Young et al., 1990; Lamb and Young, 1990). It seems that these ubiquitous and highly conserved proteins hold a particular fascination for the immune system.

Virulence and the heat shock response

During infection, pathogenic mycobacteria survive within host macrophages in an environment which is lethal to most microbes. It is probable that certain heat shock proteins play a vital role in

Table 1

Mycobacterial antigens belonging to heat shock protein families				
heat shock family	E. coli homologue	M. tuberculosis antigen	M. leprae antigen	references
HSP70	DnaK	71kD	70kD	Young et al., 1988 Garcia et al., 1989
HSP60	GroEL	65kD	65kD	Young et al., 1988 Shinnick et al., 1988
low mol wt			18kD	Nerland et al., 1988
	GroES	12kD	14kD	Shinnick et al., 1989 Hunter et al., 1990

adaptation to this hostile environment and, in the case of virulent *Salmonella* bacteria, it has been demonstrated that macrophage entry is indeed accompanied by elevated synthesis of GroEL and DnaK (Buchmeier and Heffron, 1990). Genetic studies of a range of bacterial pathogens have shown that factors which are important in mediating interactions with host cells during infection (adhesins, toxins, etc) are generally subject to co-ordinate transcriptional regulation in response to environmental signals likely to be encountered *in vivo* (Miller et al.,

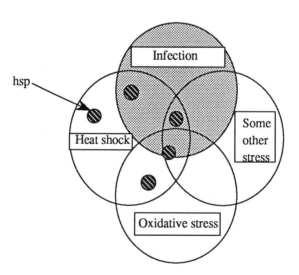

Figure 1. Schematic representation of the relationship between global regulons associated with infection and with laboratory stress stimuli.

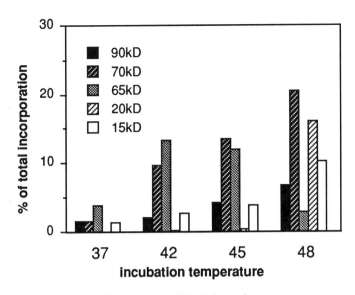

Figure 2. The heat shock response of *M. tuberculosis*.

1989) and it is anticipated that the same pattern will hold true for mycobacteria. Figure 1 provides a schematic illustration of the proposed relationship between the heat shock response and virulence.

A complex range of *in vivo* stimuli results in induction of a set of genes encoding proteins essential for mycobacterial survival within the host (the "infection regulon"). Defined laboratory stress stimuli - such as heat shock - induce partially overlapping regulons with some genes (e.g., *dnaK* and *groEL*) responding to a variety of stresses. We therefore envisage the heat shock response as a convenient model for studying environmentally regulated gene expression in mycobacteria and anticipate that amongst the proteins induced by heat shock we will find some which are also induced during infection.

The heat shock response of *M. tuberculosis*

The heat shock response of *M. tuberculosis* has been analysed by continuous labelling with ^{35}S-methionine or ^{14}C-amino acids for 90 minutes at temperatures ranging from 37 to 48°C. The generation time of *M. tuberculosis* is 24 hr and a 90 min labelling period therefore represents approximately the same proportion of the division cycle as a 90 second pulse of *E. coli*. One and two-dimensional polyacrylamide gel electrophoresis revealed induction of the mycobacterial DnaK, GroEL and GroES proteins in response to a 37° to 42°C heat shock (Fig. 2). At higher temperatures labelling of GroEL decreased sharply while the percentage of total label incorporated into DnaK continued to increase. Four other heat shock proteins with molecular weights of 90kD, 28kD, 20kD and 15kD also became very prominent at temperatures above 42°C. Immunoblot analysis showed that none of these novel heat shock proteins corresponded to known mycobacterial antigens recognised by monoclonal antibodies.

M. tuberculosis cultures were labelled with ^{35}S-methionine at different temperatures and the pattern of protein synthesis was monitored by SDS-polyacrylamide gel electrophoresis followed by autoradiography and densitometric scanning. Results are presented as the percentage of total label incorporated into individual protein bands. The heat shock response of GroES in M. tuberculosis was monitored in separate experiments using mixed amino acids for labelling.

On the basis of such experiments, it is proposed that the heat shock response of M. tuberculosis comprises at least two overlapping components. It is attractive to suggest that regulation at low temperature (induction of groE and dnaK) could be mediated by a factor analogous to E. coli σ^{32}, with high temperature regulation of dnaK and the unidentified heat shock genes being "σ^E" dependent. The existence of two forms of heat shock regulation was also indicated by analysis of the time course of the response, with a delay of approximately 20 minutes between induction of GroEL and the 15kD heat shock protein. In view of the very slow growth rate of M. tuberculosis it was of interest to monitor the kinetics of induction of heat shock protein synthesis. Following a temperature rise from 37° to 42°C, a 10 min delay was seen before the induction of a heat shock response, but this delay was reduced to 1 min at higher temperatures. It is proposed that transcriptional activation is a relatively rapid process in M. tuberculosis, and that the slow response seen at 42°C reflects a delay in "sensing" of the stress stimulus. If the heat shock response is mediated by accumulation of abnormal polypeptides, for example, the lower rate of protein synthesis in a slow growing organism might account for the delay in onset of the heat shock response.

Heat shock genes in M. tuberculosis

Genes corresponding to five of the major E. coli heat shock proteins - the GroE and DnaK families - have been characterised in M. tuberculosis (Shinnick et al., 1988, 1989; Young et al., 1988; Lathigra et al., manuscript in preparation). While groEL and groES form an operon in E. coli, the corresponding genes do not appear to be linked on the mycobacterial chromosome (Fig. 3). In the case of the DnaK family, the grpE gene of M. tuberculosis is located between the dnaK and dnaJ genes, in contrast to E. coli grpE which is unlinked to the dnaK-dnaJ operon (Fig. 3). It remains to be seen whether or not these differences are reflected in regulatory or functional differences in the mycobacterial proteins. Scanning of upstream sequences of the mycobacterial heat shock genes allows identification of potential heat shock promoters corresponding to those recognised by σ^{32} and σ^E in E. coli. While it has been demonstrated that upstream regions from mycobacterial groEL and dnaK can activate transcription in E. coli (Shinnick et al., 1988;

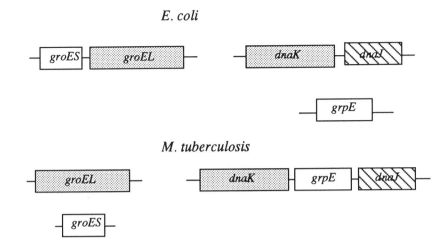

Figure 3. Schematic comparison of linkages between heat shock genes in *E. coli* and in *M. tuberculosis*

Lathigra, unpublished), it remains to be established whether or not the consensus sequences do in fact correspond to functionally active promoters.

Function of mycobacterial heat shock proteins

In common with other members of the *hsp70* gene family, purified DnaK from *M. tuberculosis* has the ability to bind to synthetic peptides *in vitro*. This has been investigated using a series of biotin-labelled peptides corresponding to HLA-DR1 restricted T cell epitopes from a variety of protein antigens. Different peptides show different binding affinities with substitution of single amino acid residues altering binding in some cases. Analysis of peptide-DnaK complexes by non-denaturing polyacrylamide gel electrophoresis indicates that peptides preferentially bind to a multimeric form of the protein (probably a dimer). In the presence of ATP the electrophoretic mobility of the heat shock protein changes to a form (probably monomeric) with decreased affinity for peptide. Preliminary studies suggest that subunit interactions may also be important in GroEL-peptide interactions. In this case we have found increased peptide binding to the dissociated form of *E. coli* GroEL produced by incubation with ATP. Addition of GroES prevents ATP-dependent dissociation leading to a decrease in peptide affinity. In contrast to the stable 14-mer form of *E. coli* GroEL, the mycobacterial homologue is generally found in a proteolytically-sensitive dimeric or tetrameric form which is unaffected by ATP and fails to bind to mycobacterial or *E. coli* GroES.

Conclusions

The recent development of gene transfer systems for mycobacteria (Jacobs et al., 1987; Snapper et al., 1988) allows initiation of a molecular genetic approach to analysis of these important pathogens. The heat shock genes currently represent the best characterised genetic elements in mycobacteria and analysis of the heat shock response will play a central role in establishing fundamental principles of gene regulation in these organisms. Promoter elements for heat shock genes can be used to activate transcription in both E. coli and mycobacteria and are convenient tools for development of "shuttle" vectors allowing expression of foreign genes in mycobacteria. The possibility of in vivo induction of heat shock promoters lends further attraction to their use in heterologous expression systems for multivalent vaccines.

References

Buchmeier, NA and Heffron, F, (1990) Induction of Salmonella stress proteins upon infection of macrophages. Science, 248: 730-732.

Garsia, LJ, Hellqvist, L, Booth, RJ, Radford, AJ, Britton, WJ, Astbury, L, Trent, RJ and Basten, A, (1989) Homology of the 70-kilodalton antigens from Mycobacterium leprae and Mycobacterium bovis with the Mycobacterium tuberculosis 71-kilodalton antigen and with the conserved heat shock protein 70 of eukaryotes. Infect. Immun., 57: 204-212.

Hunter, SW, Rivoire, B, Mehra, V, Bloom, BR and Brennan, PJ, (1990) The major native proteins of the leprosy bacillus. J. Immunol., 265: 14065-14068.

Jacobs, WR, Tuckman, R and Bloom, BR, (1987) Introduction of foreign DNA into mycobacteria using a shuttle phasmid. Nature, 327: 532-535.

Lamb, JR and Young, DB, (1990) T cell recognition of stress proteins. A link between infectious and autoimmune disease. Mol. Biol. Med., 7: 311-321.

Miller, JF, Mekalanos, JJ and Falkow, S, (1989) Coordinate regulation and sensory transduction in the control of bacterial virulence. Science, 243: 916-922.

Nerland, AH, Mustafa, AS, Sweetser, D, Godal, T and Young, RA, (1988) A protein antigen of Mycobacterium leprae is related to a family of small heat shock proteins. J. Bacteriol., 170: 5919-5921.

Shinnick, TM, Vodkin, MH and Williams, JL, (1988) The Mycobacterium tuberculosis 65kDa antigen is a heat shock protein which corresponds to common antigen and to the E. coli GroEL protein. Infect. Immun., 56: 446-451.

Shinnick, TM, Plikaytis, BB, Hyche, AD, van Landingham, RM and Walker, LL, (1989) The Mycobacterium tuberculosis BCG-a protein has homology with the Escherichia coli GroES protein. Nucl. Acids Res., 17: 1254.

Snapper, SB, Lugosi, L, Jekkel, A, Melton, RE, Kieser, T, Bloom, BR and Jacobs, WR, (1988) Lysogeny and transformation in mycobacteria: stable expression of foreign genes. Proc. Natl. Acad. Sci. USA, 85: 6987-6991.

Young, D, Lathigra, R, Hendrix, R, Sweetser, D and Young, RA, (1988) Stress proteins are immune targets in leprosy and tuberculosis. Proc. Natl. Acad. Sci. USA, 85: 4267-4270.

Young, DB, Mehlert, A and Smith, DF, (1990) Stress proteins and infectious diseases. In Stress Proteins in Biology and Medicine. Morimoto, RI, Tissieres, A, Georgopoulos, C, eds., Cold Spring Harbor Press, N.Y., pg 131-165.

Heat Shock Proteins and Antimicrobial Immunity

M.E. Munk, S.H.E. Kaufmann
Department of Immunology
Ulm University
Albert-Einstein-Allee 11
7900 Ulm
FRG

Introduction

The immune system has the important task to recognize antigens, and to discern between the antigenic molecular world of self and non-self. It is composed of different cells which protect the host from infective microorganims and eliminate damaged or altered cells. The immune system possesses a cellular hierarchy whith T lymphocytes making the decisions as to the type of the response. The T cell antigen receptor (TCR) recognizes peptide fragments in association with the major histocompatibility complex (MHC) expressed on the surface of accessory cells (Zinkernagel and Doherty, 1979). The TCR is a heterodimer composed of α/β or γ/δ chains, respectively. The former one which is expressed by the majority of T cells can be classified into two distinct T cell sub-populations: CD4 T cells recognize peptide fragments associated with MHC class II molecules and induce the immune response by secreting interleukins, CD8 T cells recognize peptides bound to MHC class I molecules and have the capacity to lyse target cells. γ/δ T cells also possess cytolytic activities and produce various interleukins; segregation into distinct subsets, however, has not been observed. Mononuclear phagocytes possess the machinery to process and to present antigens thereby informing T cells about the ongoing antigenic situation in the host.

Heat shock proteins (HSP) participate in the normal cellular metabolism, and gain particular importance when the cell is under stress. The infection of mononuclear phagocytes by bacteria is an example of a mutal situation of stress induction (Buchmeier and Hefron, 1990; Kaufmann, 1989; Polla et al., 1987; Young, 1990). On the one hand, microorganisms - in an attempt to escape the defense mechanims of the phagocytic cell and to maintain their viability - synthetize HSPs. On the other hand, phagocytic cells must cope and coexist with infectious agents which persist intracellularly if they want to survive. Thus, it can be envisaged that HSP synthetized by both host cell and pathogen are processed and presented to specific T cells. Our interest is directed towards the antigenic role of mycobacterial HSP65 in the host immune response since this molecule seems to be involved in resistance against mycobacteria (Young et al., 1988).

T cells and mycobacterial HSP

T cells with reactivity towards mycobacterial protein antigens have been characterized and a significant number of T cells has been found to be directed against HSP. In mice immmunized with killed *M.tuberculosis* frequency analyses show that about 20% of T cells with reactivity to *M.tuberculosis* organisms are specific for HSP65 (Kaufmann et al., 1987). Other studies revealed that HSP65 stimulates T cell lines and clones from leprosy and tuberculosis patients (Emmrich et al., 1986; Lamb et al., 1986; Oftung et al., 1987). Even healthy individuals possess T cells capable of responding to different microbial HSPs *in vitro* (including the intact molecule and peptides of *M.tuberculosis* HSP65 (Munk et al., 1988; Munk et al., 1990b). HSPs expressed by various apathogenic bacteria involved in subclinical infections, could stimulate HSP specific T cells leading to crossreactive immunity which may be involved in early resistance against infections with virulent pathogens (Kaufmann, 1990).

Stress protein recognition in the healthy host

The activation of T cells by exogenous antigens resembling self molecules may lead to reactions that are detrimental for the host. Since HSP65 comprises several conserved regions, the possibility that HSP specific T cells may recognize epitopes shared by the infective microorganism and self was investigated. Peripheral blood mononuclear cells from healthy individuals were stimulated *in vitro* with killed *M.tuberculosis* for 7 days. Activated T cells (74% CD4; 12% CD8) were able to lyse autologous target cells pulsed with the homologous agent or with tryptic fragments of HSP65. In order to determine the cross-reactive epitopes more precisely, synthetic peptides consisting of 10 or more amino acids were prepared which corresponded to four distinct regions which are homologous between human and mycobacterial HSP65. Autologous target cells or mouse transfectants expressing HLA-DR molecules compatible with the effector cells were lysed after pulsing with these synthetic peptides. These data suggest that auto-reactive T cells directed against HSP65 evade thymic deletion in normal individuals and that they recognize in a MHC class II restricted fashion common epitopes of the HSP65 shared by the infective microorganism and the host (Munk et al., 1989). It remains unclear whether these autoreactive T cells possess a potentially harmful or protective effect for the host.

We questioned whether stressed host cells could serve as targets for HSP65 specific B and T cells. A cross-reactive B cell epitope shared between HSP65 of mycobacteria and a host molecule of similar size is constitutively expressed in murine bone marrow macrophages (BMMø) (Koga et al., 1989). Murine CD8 α/ß cytotoxic T lymphocytes (CTL) activated *in vitro* with tryptic fragments of *M.tuberculosis* HSP65 lysed BMMø pulsed with the homologous peptides and also targets stimulated with IFN-γ or infected with cytomegalovirus; in contrast, unprimed or unstressed macrophages remained unaffected (Koga et al., 1989). Furthermore, IFN-γ stimulated Schwann cells, the major habitat of *M.leprae*, were also destroyed by α/ß CTL activated with tryptic fragments of HSP65. This mechanism could contribute to nerve damage in tuberculoid leprosy (Steinhoff et al., 1989). These data indicate that host cells process and present cross-reactive HSP65 structures in the context of MHC class I molecules. HSP could play a role in autoimmunity, since clinical and experimental studies indicate its involvement in rheumatoid arthritis (Holoshitz et al., 1989), systemic lupus erythematosus (Minota et al., 1988) and diabetes (Elias et al., 1990).

γ/δ T cells and microbial antigens

There is convincing data showing that mycobacterial antigens are able to stimulate γ/δ T cells (Raulet, 1989). Several features of γ/δ T cells remain unclear inclunding their recognition specificity, activation and MHC restriction. Nevertheless, an epitope of *M.tuberculosis* HSP65 was shown to be frequently seen by human thymus derived γ/δ T cell hybridomas (Born et al., 1990). In contrast, only a limited number of negatively selected γ/δ T cells from peripheral blood of healthy individuals is able to respond to *M.tuberculosis* HSP65, in contrast to an increased frequency of T cells responding to killed *M.tuberculosis* organisms (Kabelitz et al., 1990). Furthermore, human γ/δ T cells activated by different microbes seem to react with common and unique antigens presented by target cells pulsed with different bacteria (Munk et al., 1990a).

Conclusions

T cells - by virtue of their independent and unique antigen repertoire - seem to be responsible for maintaining a functional connectivity of the immune system. By eliminating infected or altered cells, they protect the host. Amongst these antigens, HSP may play a particular role since they contain highly conserved regions which are shared by host cells and infective microorganisms. The immune system seems to be aware of the expression of homologous HSP regions and the emergence of crossreactive T cells could result in the destabilization of this system and the activation of autoreactivity. Hence, immunity to HSP may have protective or injurious consequences. Alternatively, it may merely be a by-product of the immune response to infection.

Acknowledgments

S.H.E. Kaufmann gratefully acknowledges financial support from: UNDP/World Bank/WHO special Program for Research and Training in Tropical Diseases; Sonderforschungsbereich 322; German Leprosy Relief Association; EC-India Science and Technology Cooperation Program; Landesschwerpunkt 30; the A. Krupp award for young professors. M.E. Munk is supported by Conselho Nacional de Desenvolvimento Cientifico & Tecnologico (CNPq) Brazil.

References

Born, W, Hall L, Dallas, A, Boymel, J, Shinnick, TM, Young, DB, Brennan, P and O'Brien, R, (1990) Recognition of a peptide antigen by heat shock-reactive γδ T lymphocytes. Science, 249: 67-69.

Buchmeier, NA and Hefron, F, (1990) Induction of *Salmonella* stress proteins upon infection of macrophages. Science, 248: 730-732.

Elias, D, Markovits, D, Reshef, T, Van der Zee, R and Cohen, IR, (1990) Induction and therapy of autoimmune diabetes in the non-obese diabetic (NOD/Lt) mouse by a 65-kDa heat shock protein. Proc. Natl. Acad. Sci. USA, 87: 1576-1580.

Emmrich, F, Thole, JER and Kaufmann, SHE, (1986) A recombinant 64 kilodalton protein of *Mycobacterium bovis* bacillus Calmette-Guerin specifically stimulates human T4 clones reactive to Mycobacterial antigens. J. Exp. Med., 163: 1024-1029.

Holoshitz, J, Koning, F, Coligan, JE, De Bruyn, J and Strober, S, (1989) Isolation of CD4⁻ CD8⁻ mycobacteria-reactive T lymphocytes clónes from rheumatoid arthritis synovial fluid. Nature, 339: 226-229.

Kabelitz, D, Bender, A, Schondelmaier, S, Schoel, B and Kaufmann, SHE, (1990) A large fraction of human peripheral blood γδ + T cells is activated by *Mycobacterium tuberculosis* but not by its 65kDa heat shock protein. J. Exp. Med., 171: 667-679.

Kaufmann, SHE, Väth, U, Thole, JER, Van Embden, JDA and Emmrich, F, (1987) Enumeration of T cells reactive with *Mycobacterium tuberculosis* organisms and specific for the recombinant mycobacterial 64 kilodalton protein. Eur. J. Immunol., 17: 351-357.

Kaufmann, SHE, Schoel, B, Wand-Württenberger, A, Steinhoff, U, Munk, ME and Koga, T, (1989) T cells, stress proteins and pathogenesis of mycobacterial infections. Curr. Top. Microbiol. Immunol., 155: 125-141.

Kaufmann, SHE, (1990) Heat shock proteins and the immune response. Immunol. Today, 11: 129-135.

Koga, T, Wand-Württenberger, A, De Bruyn, J, Munk, ME, Schoel, B and Kaufmann, SHE, (1989) T cells against a bacterial heat shock protein recognize stressed macrophages. Science, 245: 1112-1115.

Lamb, JR, Ivanyi, J, Ress, A, Young, RA and Young, DB, (1986) The identification of T cell epitopes in *Mycobacterium tuberculosis* using human T lymphocyte clones. Lepr. Rev., 57 (suppl.2): 131-137.

Minota, S, Cameron, B, Welch, WJ and Winfield, JB, (1988) Autoantibodies to the constitutive 73-kD member of the hsp70 family of heat shock proteins in systemic lupus erythematosus. J. Exp. Med., 168: 1475-1480.

Munk, ME, Schoel, B and Kaufmann, SHE, (1988) T cell responses of normal individuals towards recombinant protein antigens of *Mycobacterium tuberculosis*. Eur. J. Immunol., 18: 1835-1838.

Munk, ME, Schoel, B, Modrow, S, Karr, RW, Young, RA and Kaufmann SHE (1989) T lymphocytes from healthy individuals with specificity to self epitopes shared by the mycobacterial and human 65-kilodalton heat shock protein. J. Immunol., 143: 2844-2849.

Munk, ME, Gatrill, AJ and Kaufmann, SHE, (1990a) Target cell lysis and interleukin-2 secretion by γδ T lymphocytes after activation with bacteria. J. Immunol., 145: 2434.

Munk, ME, Shinnick, TM and Kaufmann, SHE, (1990b) Epitopes of the mycobacterial heat shock protein 65 for human T cells comprise different structures. Immunobiology, 180: 272-277.

Oftung, F, Mustafa, AS, Husson, R, Young, RA and Godal, T, (1987) Human T cell clones recognize two abundant *Mycobacterium tuberculosis* protein antigens expressed in *E.coli.* J. Immunol., 138: 927-931.

Polla, BS, Healy, AM, Wojno, WC and Krane, SM, (1987) Hormone 1-α, 25-dihydroxyvitamin D3 modulates heat shock reponse in monocytes. Am. J. Physiol. (Cell Physiol.), 21: C640-C649.

Raulet, DH, (1989) The structure, function, and molecular genetics of the γδT cell receptor. Annu. Rev. Immunol., 7: 175207.

Steinhoff, U, Schoel, B and Kaufmann, SHE, (1989) Lysis of interferon-γ activated Schwann cell by cross-reactive CD8+ a/ß T cells with specificity for the mycobacterial 65 kd heat shock protein. Intl. Immunol., 2: 279-284.

Young, DB, Lathigra, RB, Hendrix, RW, Sweetser, D and Young, RA, (1988) Stress proteins are immune targets in leprosy and tuberculosis. Proc. Natl. Acad. Sci. USA, 85: 4267-4270.

Young, RA, (1990) Stress proteins and immunology. Annu. Rev. Immunol., 8: 401-420.

Zinkernagel, RM and Doherty, PC, (1979) MHC restricted cytotoxic T cells: studies on the biological role of polymorphic major transplation antigens determining T-cell restriction - specificity function and responsiveness. Adv. Immunol., 27: 51-77.

Heat Shock Proteins and Antigen Processing and Presentation

S.K. Pierce, D.C. DeNagel
Northwestern University
Dept. of Biochem., Mol. Biol. and Cell Biol.
Evanston, IL 60208
USA

Introduction

T lymphocytes recognize antigen through their cell surface immune receptor, the T cell receptor (TCR). In several respects the TCR is similar to the B cell immune receptor, antibody. Both are composed of two different protein chains which associate to form a combining site for antigen. These chains have variable and constant regions encoded in separate genes which rearrange during the development of the individual cells. Indeed, TCR and antibody are members of the same super gene family and based on amino acid sequence similarity the TCR is predicted to have a similar three dimensional structure as antibody (reviewed in Davis and Bjorkman, 1988). Despite such similarities the T cell and B cell recognition of antigen differs fundamentally. Antibody binds to protein antigens in their native conformations in an exquisitely specific, lock and key fashion as revealed by recent determinations of the crystal structures of antibodies bound to protein antigens (Amit et al., 1986). In contrast, the TCR has no apparent affinity for native protein antigens but requires that the antigen be denatured, most likely by proteolysis, yielding a peptide or peptide-like fragment containing the T cell antigenic determinant. In general, antigenic peptides alone do not bind to TCRs but require an association with an MHC molecule on the surface of an antigen presenting cell (APC) to activate antigen-specific T cells. This appears to be the case both for helper T cells which recognize antigen in association with MHC class II molecules and for cytolytic T cells which recognize antigen bound to MHC class I molecules. T cells express one of two types of TCR. The vast majority of mature adult T cells express a TCR composed of α and β chains while a minority (5-10%) express a receptor composed of γ and δ chains (reviewed in Matis, 1990). It is well documented that T cells expressing $\alpha\beta$ TCR recognize antigenic peptides when presented by MHC molecules of the class I or class II type. At present it is controversial whether $\gamma\delta$ T cells recognize antigen in a similar fashion and whether antigenic peptides are presented on the MHC class I or II molecules for $\gamma\delta$ T cell recognition. With regard to the role of heat shock proteins in immune responses, the nature of the antigen recognized by γδ T cells and whether heat shock proteins are ligands for γδ T cells is of considerable interest and is addressed by Born and coauthors in this volume. Here we will restrict our discussion to the T cell recognition of antigenic peptides presented by the MHC class II molecules and the potential role for heat shock proteins in this process. Although the overlap in mechanisms by which antigen is processed and presented via class I and class II is not known, we would leave open the possibility of similar roles for the heat shock proteins in both pathways.

The molecular mechanisms by which native antigens are converted to peptides and presented by the MHC class II molecules is, at present, only poorly delineated. Antigen processing is a function of class II expressing APC and involves the internalization of antigen by the APC, transport of the antigen to an acidic compartment, degradation or proteolysis of the antigen and then subsequent association of the resulting peptide fragments with the MHC class II molecules (Harding et al., 1988). All MHC class II expressing cells appear competent to some degree to process and present antigen, including cells which do not normally express MHC class II for example fibroblasts transfected with the genes encoding the class II molecule (Malissen et al., 1984). Thus, antigen processing would not appear to be a specialized function of the differentiated cells which express MHC class II molecules, the most common being B cells and macrophages. Rather, the minimal machinery necessary to process antigen and to present it with the MHC class II would appear to be present in most, if not all cell types. Whether cells which normally express MHC class II have specialized mechanisms which make antigen processing and presentation more efficient remains to be determined.

Antigen can enter the APC through fluid phase pinocytosis, by phagocytosis, or in the cases where the APC express a receptor for antigen, as for antigen-specific B cells, by receptor mediated endocytosis (Lanzavecchia, 1990). The receptor mediated pathway for antigen processing is extremely efficient as compared to pinocytosis. Indeed, specific B cells activate T cells when provided with $1/1000^{th}$ - $1/10,000^{th}$ the antigen concentration necessary for nonspecific B cells. Antigen receptors on B cells may not be the only APC surface structures capable of enhancing antigen processing. In macrophages, the Fc receptor may function to internalize antigen-containing immune complexes or the phagosome itself may represent a specialized means of antigen internalization. At present it is somewhat controversial whether proteins which are synthesized intracellularly have access to the MHC class II processing pathway and under which condition such a pathway might function. Early studies of the processing of influenza viral proteins indicate that intracellularly synthesized viral proteins cannot be presented with MHC class II even though such proteins can be processed and presented with class II when the antigen is provided as killed virus (Morrison et al., 1986). However, studies with endogeneously synthesized Hepatitis B surface antigen (Jin et al., 1988), measles virus (Jacobson et al., 1989), the influenza A virus M_1

matrix protein (Jaraquemada et al., 1990) and B cell antibody (Weiss and Bogen, 1989) clearly indicate that intracellularly synthesized antigen is readily presented with MHC class II. If proteins synthesized in the cytoplasm do indeed become associated with MHC class II molecules, there must be mechanisms to transport the antigenic proteins or peptides from the cytoplasm across a membrane to a post-golgi compartment where MHC class II-peptide association occurs. At present, no such transport mechanisms have been described for MHC class II. Recent intriguing results suggest that proteins related to a large family of transmembrane transporters may be involved in transport of peptide from the cytoplasm to the endoplasmic reticulum for binding to MHC class I (reviewed in Parham, 1990). It is interesting to speculate that other members of this family or of the heat shock family of proteins play a similar role for the transport of peptides from the cytoplasm to the endosomes for binding to MHC class II.

The exact intracellular compartment(s) in which processing occurs and where processed antigen associates with MHC class II has not been identified. Whether newly synthesized class II and/or class II recycling into the cell from the surface associates with processed antigen has not been resolved. Brodsky and coworkers (Guagliardi et al., 1990) have recently described a potential intracellular site for antigen processing. Using immunoelectron microscopy, they demonstrated the endosomal co-localization of Ig, MHC class II, invariant chain and the proteases cathepsin B and D in a human lymphoblastoid B cell line which had been allowed to internalize anti-immunoglobulin for two minutes. While these results offer a good candidate for a processing compartment, no data currently links this endosomal compartment with the site of processing. Studies (Roch and Cresswell, 1990) indicate that newly synthesized MHC class II cannot bind peptide until released of its accompanying invariant chain. This release is believed to occur in a post-golgi compartment immediately prior to transport of MHC class II to the plasma membrane (Cresswell and Blum, 1988; Neefjes et al., 1990). Newly synthesized MHC class II molecules have been shown to be retained for up to two hours in a post-golgi endosomal compartment before transport to the cell surface. The compartment has access to pinocytosized material and is a potential site of MHC class II-antigenic peptide association. The compartment in which processing occurs is of significant interest but its identification will require further studies.

Presumably the final step in the processing pathway involves the binding of an antigenic peptide to the MHC class II molecules. The binding of peptides representing T cell antigenic determinants has been demonstrated *in vitro*. Initial studies were carried out with purified MHC class II in detergent solution (Babbitt et al., 1985; Buus et al., 1986) and more recently with purified MHC class II incorporated into synthetic membranes (Sadegh-Nasseri and McConnel, 1989). These studies revealed highly unusual characteristics of the peptide-MHC class II interaction. Firstly, only a small fraction of the total MHC class II isolated from cells was able to bind peptide. This portion ranged from 1-10%. Although there are several possible explanations for this observation, one frequently cited is that the majority of the MHC class II purified from cells contains antigenic peptides resulting from the processing of either endogenous antigens or exogenous proteins from the cells' environment. Indeed, the crystal structure of a MHC class I molecule showed unresolvable electron density in a groove between two α-helices (Bjorkman et al., 1987). This was presumed to represent peptides which remained bound to the MHC molecules through rigorous purification procedures. Recent studies from our laboratory (Srinivasan and Pierce, 1990) show that MHC class II molecules purified from cells which have processed a foreign antigen do indeed contain antigenic peptides as shown by the ability of the complex, when incorporated into a synthetic membrane, to stimulate specific T cells. This finding indicates that an antigenic peptide resulting from processing, once associated with MHC class II remains bound in a nearly irreversible fashion. Indeed, kinetic analysis of the binding of synthetic peptides to purified class II support a long lived complex. These studies showed an average equilibrium constant of approximately 10^{-6}M with an extraordinary slow association ($t_{1/2}$=5hr) and dissociation rate ($t_{1/2}$=30hr) (Buus et al., 1986). Furthermore, recent studies by Watts and coworkers indicate that extreme conditions are required to release synthetic peptide bound to purified MHC class II *in vitro* (pH 2.0 for several hours) (Lee and Watts, 1990).

Such observations from *in vitro* systems are somewhat puzzling in that they would not be predicted from the biology of antigen processing and presentation. APC can rapidly associate with synthetic peptides within minutes, resulting in functional complexes capable of stimulating specific T cells (Lakey et al., 1988; Roosnek et al., 1988). Once peptides associate with APC and free peptide is washed away the rate of loss of complexes, as measured by loss of the ability to stimulate T cells, is also relatively rapid, complete within 4 hr. The observed loss of peptide from the cell surface cannot be accounted for by turnover of MHC class II. Recent studies in our laboratory studying the assembly of MHC class II-processed antigen complexes during antigen processing show that complexes are formed 1-2 hr after antigen is internalized, remain for an addition 2-3 hr and then disappear. Thus, this a direct biochemical assay clearly shows that APC have the capability of assembling and disassembling peptide MHC class II complexes more rapidly than the rates measured *in vitro* would indicate. Although there may be several explanations for this, one possibility is that APC have mechanisms which facilitate the loading and/or unloading of MHC class II molecules with processed antigen. We have described one protein whose characteristics suggest it as a candidate to participate in such processed. This is a peptide binding protein which is a new member of the heat shock protein (hsp) family. Our evidence for this protein having a role in antigen processing will be reviewed in the following section.

Isolation of a peptide binding protein which plays a role in antigen processing

The puzzling aspects of the reported binding of peptides to Ia lead to a search for peptide binding proteins which might play a role in antigen processing by facilitating the binding of peptide to MHC class II and/or to the TCR. Detergent solubilized cell lysates were chromatographed on a peptide affinity column containing a known antigenic peptide of the soluble globular protein antigen pigeon cytochrome c (Pc), residues 81-104 (Pc 81-104). Two or three proteins in the 72-74 kDMr range are eluted from the column under acid conditions. We refer to these as peptide binding proteins of 72/74 Mr (PBP72/74) (Lakey et al., 1987).

With regard to peptide specificity, PBP72/74 can be competitively eluted from the Pc 81-104 column with the Pc 81-104 peptide but not with intact Pc and thus appears to be specific for some feature of the peptide not found in the native protein. This is the case even though the Pc 81-104 peptide is on the surface of the native protein, exposed to the solvent,and is accessible to binding by specific antibodies. Additional studies indicate that PBP72/74 binding to peptide is not discriminatory of the primary amino acid sequence as other peptides are able to elute PBP72/74 from Pc 81-104, including the corresponding peptide of mouse cytochrome c, and an unrelated peptide of Pc, residues 66-80. In addition, PBP72/74 binds to peptides of the unrelated protein lactate dehydrogenase C4 (LDH-C4). However, the secondary structure of the peptide appears to influence binding to PBP72/74. In preliminary studies it was observed that although a linear peptide of LDH-C4 (residues 315-327) bound PBP72/74, a peptide representing this same region engineered to take on a stable α-helical structure (described in Kaumaya et al., 1990) did not bind. Thus, there does not appear to be a simple algorithm which predicts peptide binding to PBP72/74 and binding may depend on a characteristic of the peptides free from their native proteins.

To determine if PBP72/74 plays a role in antigen processing and/or presentation, antibodies were raised in rabbits to the affinity purified PBP72/74 and tested for their effect on APC function (Lakey et al., 1987). PBP72/74-specific antisera block, to the 80-90% level, the processing and presentation of Pc to a Pc-specific T cell hybrid and of ovalbumin (OVA) to an OVA-specific T cell, AODH. The antisera tested are not nonspecifically toxic as these have no effect on the response of the T cell hybrids to the nonspecific mitogen concanavalin A. Blocking appears to be at the level of the APC as T cells incubated with the antisera and washed are able to respond to antigen presented by APC, while the same treatment of APC blocks presentation. The blocking activity is not MHC specific in that the activity is absorbed by spleen cells of mice of different MHC haplotypes. Subsequently, monoclonal antibodies which showed the same blocking activity were generated from spleens of rats immunized with the purified PBP72/74. However, such lines have proven to be unusually unstable and failed to secrete antibody after a brief time in culture. We have recently obtained direct evidence for PBP72/74's involvement in antigen processing. PBP72/74 purified from APC which have processed a radiolabeled antigen have radiolabeled processed antigen bound to it, as does purified I-E^k. The binding appears highly specific in that other proteins (MHC class I and B220) purified from the same APC contain no radiolabeled antigen.

The PBP72/74-specific rabbit antiserum immunoprecipitates PBP72/74 from [35S]-labeled cell lysates and from [^{125}I]-surface labeled cells and shows a cross reactivity with monkey proteins of similar molecular weight. By flow cytometry, the PBP72/74-specific antiserum stains the surfaces of B cells and macrophages and does not stain T cells, fibroblasts or NK cells (VanBuskirk et al., 1990). The staining of cells does not correlate with the expression of MHC class II. Indeed, L cells show no cell surface expression of PBP72/74 and L cell lines transfected with the genes encoding the α and β chain of either the I-A or the I-E molecules (Germain et al., 1985; Germain and Quill, 1986) are also negative for PBP72/74 cell surface expression. Moreover, mutations in B cell lymphomas, characterized by Glimcher and coworkers (Glimcher et al., 1985) which abrogate the cell surface expression of MHC class II do not affect the expression of PBP72/74. Taken together, such analyses indicate the PBP72/74 surface expression is a function of B cells and macrophages, independent of MHC class II expression.

In collaboration with F. Brodsky and coworkers (University of California, San Francisco) the PBP72/74-specific antisera was used to label PBP72/74 for immunoelectron microscopy in thin sections of the human B cell lymphoma IM-9 (VanBuskirk et al., 1990). The B cells had been incubated with gold labeled antibodies specific for Ig and allowed to internalize for 15 min. As discussed above, surface Ig is taken into vesicles which contain newly synthesized class II, invariant chain and cathepsin B and D, and such vesicles are suggested to represent an intracellular processing site. As compared to control nonspecific immune sera, the PBP-specific antisera stained 32% of the Ig-containing vesicles as detected using gold labeled secondary antibodies. Additional staining of plasma membrane and small cytoplasmic structures was observed. These results indicate that PBP72/74 has a restricted expression within the cell, which is consistent with its playing a role in antigen processing and/or presentation.

Concerning the membrane association of PBP72/74, preliminary studies show that a portion of PBP72/74 is intragrally associated with the membrane and partitions into the detergent phase after Trition 141 extraction. A portion of PBP72/74 is water soluble and partitions into the aqueous phase. It is not known how PBP72/74 becomes associated with the cell's plasma membrane. PBP72/74 could have a membrane anchoring domain, which, if so would be unique among the known HSP70 proteins. Alternatively, PBP72/74 could be membrane associated via a fatty acid tail. However, preliminary studies indicate that lipase treatment of APC does not alter PBP72/74 cell surface expression. Lastly, PBP72/74 may become membrane associated by binding to another membrane protein. Studies in progress to isolate the gene(s) encoding PBP72/74 may clarify the relationship between the membrane and soluble forms.

Identification of PBP72/74 as a member of the heat shock protein 70 (HSP70) family

Our initial analysis of PBP72/74 lead us to conclude that we had identified a protein which played a role in the processing and/or presentation of antigen by binding to processed antigenic peptides and facilitating their interaction with MHC class II and/or the TCR. We considered that PBP72/74 may belong to a family of proteins, members of which function to bind to denatured or newly synthesized proteins, and subsequently to transport these to an appropriate functional site. One candidate protein was the immunoglobulin binding protein (BIP) which had been described to bind to newly synthesized, unfolded μ heavy chains prior to their folding with the appropriate light chains in the endoplasmic reticulum (Hendershot et al., 1987). However, a monoclonal antibody specific for BIP (provided by J. Kearny, University of Alabama, Birmingham) did not recognize purified PBP72/74 in immunoprecipitation although it did precipitate the slightly larger (78kDMr) BIP protein from cell lysates. Subsequently, BIP was shown to be a member of the HSP70 family. Using a monoclonal antibody to HSP70 proteins (mAb7.10) (Velazquez et al., 1983) which recognizes a number of the HSP70 proteins from a variety of species (provided by S. Lindquist, University of Chicago), PBP72/74 was shown to be serologically related to the HSP70 family. Further serological analysis showed PBP72/74 is recognized in Western blot by the mAb N27 (Vass et al., 1988) (provided by W. Welsh, University of California, San Francisco) specific for a constitutive member of the HSP70 family but not with mAb N15 (Welch and Suhan, 1986) (also provided by W. Welsh), which recognizes a stress-induced HSP70 family member.

PBP72/74 shares another characteristic feature of the HSP70 proteins, which is the ability to bind ATP (Munro and Pelham, 1986). The binding of ATP causes the release of the HSP70 proteins from their substrates. Similarly, PBP72/74 binds ATP which causes the release of PBP72/74 from a peptide column (VanBuskirk et al., 1989). The relationship between peptide binding and ATP binding is not known. However, it has been shown that PBP72/74 is eluted from an ATP column by peptide.

Recent N-terminal microsequence analysis of PBP74 shows that it is indeed a member of the hsp family, sharing an invariant six amino acid sequence involved in ATP binding. Significantly, PBP74 is not identical to any known HSP70 protein in the regions outside the conserved sequence. The 72kD protein has a blocked N terminus and consequently has not yet been sequenced.

Because PBP72/74 is related to the hsp family, it was of interest to determine if its expression is regulated by stress or alternatively by activation which has been shown in lymphocytes to induce transcription of HSP70 (Spector et al., 1989). The cell surface expression of PBP72/74, measured by flow cytometry using the PBP72/74-specific antisera and a fluoreseinated secondary antibody, does not change following incubation of cells at 42°C for 30 min. PBP72/74 cell surface expression is similar in B cells activated with lipopolysaccharide or with the lymphokine IL-4. Kaufmann and coworkers report that in mouse macrophages the mAb N27 recognizes a structure present on the surfaces of stressed but not on untreated macrophages (Kaufmann et al., 1990). Although mAb N27 recognizes PBP72/74 in Western blot, we do not know if it reacts with native surface PBP72/74. It is also not known if the regulation of PBP72/74 is similar in B cells and in macrophages. Winfield and coworkers have also described the expression of HSP70 proteins on activated human HL60 cells (Jarjour et al., 1989). In this case surface expression was detected using human antibodies to HSP70 proteins found in the serum of autoimmune individuals. We observed that the same human antisera recognized PBP72/74 in Western blot but have not tested them for staining of mouse B cell surfaces (unpublished observation).

Conclusions

Our results indicate that a heat shock protein expressed on the cell surface and in endosomes is linked to the processing and presentation of antigen by the observation that antibodies specific for it block APC function and antigen is found associated with it during processing. Current knowledge of the function of the heat shock proteins indicates that they are chaperons binding to newly synthesized polypeptides before folding occurs so as to prevent inappropriate interactions among chains. In some cases hsp proteins serve to transport these to a functional site in the cell. In this context, one could envision that members of this family function in binding newly produced peptides in the processing compartments. The function of heat shock proteins, such as PBP72/74, may be to scavenger peptides from degradative compartments thereby preventing complete proteolysis, and concentrating these for binding to MHC class II. It is possible that PBP72/74 binds the released peptides and transports these, from the degradative compartment, to a compartment where MHC class II binds peptide if these are not one and the same. A portion of PBP72/74 associated with peptide may be transported to the cell surface in class II-containing vesicles to display peptides for TCR binding. PBP72/74 could also play a chaperone function in receiving antigens from the cytoplasm in cases where transport must occur. Dice and coworkers this volume described a function for an HSP70 in transporting degraded proteins from the cytoplasm to lysosomes or to late endosomes. PBP72/74 could facilitate this transport by receiving proteins on the luminal side of the membrane and transport these to the site of peptide-MHC class II assembly. In each of these cases HSP70 proteins would play a scavenging and concentrating role in antigen processing. It is also possible that the HSP70 proteins play a more fundamental role in facilitating class II-peptide binding. In this regard, the observation (Cresswell and Blum, 1990) that MHC class II bound to its invariant chain is not accessible for peptide binding is of interest. The invariant chain is proteolytically cleaved from the MHC class II molecules at a late stage in synthesize prior to cell surface expression. During such a process, surfaces of the class II molecule which were bound to invariant chain must be exposed to the environment. Heat shock proteins may function to bind to the exposed sites on class II, allowing for appropriate folding processes which could include peptide binding. Such a function for heat shock proteins could be in addition to the simpler scavenger/concentration function. It is intriguing that the *hsp70* gene is located in the MHC complex in man and thus is genetically linked to the MHC proteins (Sargent et al., 1989). Future studies may clarify whether there is a functional link between the MHC and the HSP70 proteins.

References

Amit, AG, Mariuzza, RA, Phillips, SEV and Poljak, RJ, (1986) Three-dimensional structure of an antigen-antibody complex at 2.8 A resolution. Science, 233: 747-753.

Babbitt, BP, Allen, PM, Matsueda, G, Haber, E and Unanue, ER, (1985) Binding of immunogenic peptides to Ia histocompatibility molecules. Nature, 317: 359-360.

Bjorkman, PJ, Saper, MA, Samraoui, B, Bennett, WS, Strominger, JL and Wiley, DC, (1987) Structure of the human class I histocompatibility antigen, HLA-A2. Nature, 329: 506-512.

Buus, S, Sette, A, Colon, SM, Jenis, DM and Grey, HM, (1986) Isolation and characterization of antigen-Ia complexes involved in T cell recognition. Cell, 47: 1071-1077.

Cresswell, P and Blum, JS, (1988) Intracellular transport of class II HLA antigens. In Processing and presentation of antigens. Pernis, G, Silverstein, SC and Vogel, HJ, eds., Acad. Press., 43-51.

Davis, MM, Dansereau, JQ, Johnstone, RM and Bennett, V, (1986) Selective externalization of an ATP-binding protein structurally related to the clathrin-uncoating ATPase/heat shock protein in vesicles containing terminal transferrin receptors during reticulocyte maturation. J. Biol. Chem., 261: 15368-15371.

Davis, MM and Bjorkman, PJ, (1988) T-cell antigen receptor genes and T-cell recognition. Nature, 334: 395-402.

Germain, RN, Ashwell, JD, Lechler, RA, Margulies, DH, Nickerson, KM, Suzuki, G and Tou, JYL, (1985) "Exon-shuffling" maps control of antibody and T cell recognition sites to the NH2 terminal domain of the class II major histocompatibility polypeptide A_β. Proc. Natl. Acad. Sci. USA, 82: 2940-2944.

Germain, R and Quill, H, (1986) Unexpected expression of a unique mixed type class II MHC molecule by transfected L cells. Nature, 320: 72-75.

Glimcher, LH, McKean, DJ, Choi, E and Seidman, JG, (1985) Complex regulation of class II gene expression analysis with class II mutant cell lines. J. Immunol., 135: 3542-3550.

Guagliardi, LE, Koppelman, B, Blum, JS, Marks, MS, Cresswell, P and Brodsky, FM, (1990) Co-localization of molecules involved in antigen processing and presentation in an early endocytic compartment. Nature, 343: 133-139.

Harding, CV, Layva-Cobian, F and Unanue, ER, (1988) Mechanisms of antigen processing. Immunol. Rev., 106: 77-92.

Hendershot, L, Bole, D, Kohler, G and Kearney, JF, (1987) Assembly and secretion of heavy chains that do not associate posttranslationally with immunoglobulin heavy chain-binding protein. J. Cell Biol., 104: 761-767.

Jacobson, S, Sekaly, RP, Jacobson, CL, McFarland, HF and Long, EO, (1989) HLA class II-restricted presentation of cytoplasmic measles virus antigens to cytotoxic T cells. J. Virol., 63: 1756-1762.

Jaraquemada, D, Merce, M and Long, EO, (1990) An endogenous processing pathway in vaccinia virus-infected cells for presentation of cytoplasmic antigens to class II-restricted T cells. J. Exp. Med., 172: 813-820.

Jarjour, W, Tsai, V, Woods, V, Welch, W, Pierce, S, Shaw, M, Mehta, H, Dillmann, W, Zvaifler, N and Winfield, J, (1989) Cell surface expression of heat shock proteins. Arthrit. Rheum., 32: S44.

Jin, Y, Shih, JWK and Berkower, I, (1988). Human T cell response to the surface antigen of hepatitis B virus (HBsAG). Endosonal and nonendosomal processing pathways are accessible to both endogenous and exogenous antigen. J. Exp. Med., 168: 293-300.

Kaufmann, SHE, (1990) Heat shock proteins and the immune response. Immunol. Today, 11: 129-136.

Kaufmann, SHE, Schoel, B and Wand-Wurttenberger, A, Curr. Top. Microbiol. Immunol., in press.

Kaumaya, PTP, Berndt, KD, Heidorn, DB, Trewhella, J, Kezdy, FJ and Goldberg, E, (1990) Synthesis and biophysical characterization of engineered topographic immunogenic determinants with αα topology. Biochemistry, 29: 13-23.

Lakey, EK, Casten, LA, Niebling, WL, Margoliash, E and Pierce, SK, (1988) Time dependence of B cell processing and presentation of peptide and native protein antigens. J. Immunol., 140: 3309-3314.

Lakey, EK, Margoliash, E and Pierce, SK, (1987) Identification of a peptide binding protein that plays a role in antigen presentation. Proc. Natl. Acad. Sci. USA, 84: 1659-1663.

Lanzavecchia, A, (1990) Receptor-mediated antigen uptake and its effect on antigen presentation to class II restricted T lymphocytes. Ann. Rev. Immunol., 8: 773-793.

Lee, JM and Watts, TH, (1990) On the dissociation and reassociation of MHC class II-foreign peptide complexes. J. Immunol., 144: 1829-1834.

Malissen, B, Peele, P, Goverman, JM, McMillan, M, White, S, Kappler, J, Marrack, P, Pierres, A, Pierres, M and Hood, G, (1984) Gene transfer of H-2 class II genes: antigen presentation by mouse fibroblast and hamster B-cell lines. Cell, 36: 319-327.

Matis, LA, (1990) The molecular basis of T cell specificity. Ann. Rev. Immunol., 8: 65-82.

Morrison, LA, Lukacher, AE, Braciale, VL, Fan, DP and Braciale, TJ, (1986) Differences in antigen presentation to MHC class I-and class II-restricted influenza virus-specific cytolytic T lymphocyte clones. J. Exp. Med., 163: 903-921.

Munro, S and Pelham, HRB, (1986) An hsp70-like protein in the ER: identity with the 78 Kd glucose-regulated protein and immunoglobulin heavy chain binding protein. Cell, 46: 291-300.

Neefjes, JJ, Stollorz, V, Peters, PJ, Geuze, HJ and Ploegh, HL, (1990) The biosynthetic pathway of MHC class II but not class I molecules intersects the endocytic route. Cell, 61: 171-183.

Parham, P, (1990) Transporters of delight. Nature, 348: 674-675.

Roche, PA and Cresswell, P, (1990) Invariant chain associated HLA-DR molecules do not bind immunogenic peptide. Nature, 345: 615-618.

Roosnek, E, Demotz, S, Corradin and Lanzavecchia, A, (1988) Kinetics of MHC-antigen complex formation on antigen-presenting cells. J. Immunol., 140: 4079-4082.

Sadegh-Nasseri, S and McConnell, HM, (1989) A kinetic intermediate in the reaction of an antigenic peptide and I-E^k. Nature, 337: 274-276.

Sargent, CA, Dunham, I, Trowsdale, J and Campbell, RD, (1989) Human major histocompatibility complex contains genes for the major heat shock protein HSP70. Proc. Natl. Acad. Sci. USA, 86: 1968-1972.

Spector, NL, Freedman, AS, Freeman, G, Segil, J, Whitman, JF, Welch, WJ and Nadler, LM, (1989) Activation primes human B lymphocytes to respond to heat shock. J. Exp. Med., 170: 1763-1768.

Srinivasan, M and Pierce, SK, (1990) Isolation of a functional antigen-Ia complex. Proc. Natl. Acad. Sci. USA, 87: 919-922.

VanBuskirk, A, Crump, BL, Margoliash, E and Pierce, SK, (1989) A peptide binding protein having a role in antigen presentation is a member of the hsp70 heat shock family. J. Exp. Med., 170: 1799-1809.

VanBuskirk, AM, DeNagel, DC, Guagliardi, LE, Brodsky, FM and Pierce, SK, Cellular and subcellular distribution of PBP72/74, a peptide binding protein which plays a role in antigen processing. J. Immunol., in press.

Vass, K, Welch, WJ and Nowak, TS, (1988) Localization of 70 kDa stress protein in gerbil brain after ischemia. Acta Neuropathol., 77: 128-132.

Velazquez, JM, Sonoda, S, Bugaisky, G and Lindquist, S, (1983) Is the major drosophila heat shock protein present on cells that have not been heat shocked? J. Cell Biol., 96: 286-290.

Weiss, S and Bogen, B, (1989) B-lymphoma cells process and present their endogenous immunoglobulin to major histocompatibility complex-restricted T cells. Proc. Natl. Acad. Sci. USA, 86: 282-286.

Welch, WJ and Suhan, JP, (1986) Cellular and biochemical events in mammalian cells during and after recovery from physiological stress. J. Cell Biol., 103: 2035-2052.

Young, RA, (1990) Stress proteins and immunology. Ann. Rev. Immunol., 8: 401-420.

Recognition of Heterologous and Autologous Stress Protein Sequences by Heat Shock Reactive γδ T Lymphocytes

W. Born[*#], R. Cranfill[#], R. O'Brien[#]
#Department of Medicine
National Jewish Center for Immunol. and Resp. Medicine
Denver, Colorado
USA

Introduction

Stress proteins of a number of pathogens have been identified as immune targets (Young, 1990). Many lymphocytes bearing γδ T cell receptors (γδ cells) appear to recognize both heterologous and autologous stress proteins (O'Brien et al., 1989). This reactivity of γδ cells may not be a classical immune response to foreign antigen but may rather represent a previously unrecognized surveillance function of the immune system (Born et al., 1990a).

Clonal analysis of γδ cell specificities

Using a modified version of the AKR thymoma BW5147 as a fusion partner (White et al., 1989), we have, without antigen selection, generated several collections of γδ T cell receptor (TCR) bearing hybridomas, derived from fetal (Born et al., 1987), newborn (O'Brien et al., 1989), and adult mouse thymus (unpublished) as well as mouse spleen (unpublished), mammary glands (Reardon et al., in press) and epidermis (Reardon, unpublished). Since the fusions were carried out directly, without antigen stimulation or prolonged tissue culture, these collections of γδ+ hybridomas probably closely reflect the composition of normal γδ cell populations at these sites.

For studies on possible antigen specificities, we have initially chosen newborn thymus-derived hybridomas because thymic γδ populations at this stage of development express a greater variety of TCRs than some of the peripheral γδ cell subsets and are less likely to have undergone antigen-driven clonal expansion in vivo. Therefore, they probably constitute a broader panel of detectable specificities.

---

* Department of Microbiology and Immunology, University of Colorado, Health Sciences Center, Denver, Colorado, USA

Most, if not all, T lymphocytes expressing αβ TCRs (αβ cells) respond to allogeneic antigens and this reactivity is thought to be the primary cause for tissue graft rejection. The molecular basis for this response is probably a close resemblance between the allogeneic ligands and self major histocompatibility complex (MHC) encoded molecules in association with antigenic peptides. Since γδ TCRs structurally resemble αβ TCRs, we attempted to stimulate γδ⁺ hybridomas derived from C57BL/10 mice with allogeneic cells from a number of different mouse strains. In these and all following stimulation experiments, antigen receptor engagement and subsequent responder cell activation was measured indirectly, via determinations of lymphokine release in an HT-2 cell line dependent bioassay (Mosmann, 1983). In approximately 1000 combinations of stimulators and responders, not a single γδ cell response to allogeneic antigens was found (O'Brien et al., 1989, and unpublished results). In contrast, about 20% of αβ⁺ hybridomas generated in the same fusion experiments reacted with one or several allogeneic stimulators. Although alternative possibilities are not excluded, these data suggest that γδ cells are less likely than αβ cells to react with allogeneic antigens and thus may recognize their ligands in a manner altogether different from αβ cells. It should be noted, however, that one can obtain alloreactive γδ clones using appropriate selective conditions (Matis et al., 1987; Matis et al., 1989). We speculate that these clones could be "falsely" alloreactive due to antigenic mimicry or may be representatives of a more "αβ-like" subset among γδ populations.

γδ cells reactive with autologous and mycobacterial antigens

During attempts to stimulate with allogeneic antigens, we noticed that many γδ⁺ hybridomas secreted lymphokines without addition of stimulator cells (O'Brien et al., 1989). This rather unusual reactivity was not found with αβ⁺ hybridomas. Further experimentation provided strong evidence that this "spontaneous" reactivity was TCR-dependent and probably ligand-specific: Firstly, testing of TCR loss-mutants showed that the response was dependent on the presence of TCRs. Secondly, blocking of TCRs with monoclonal antibodies directed against the γδ heterodimer or the TCR-associated complex of transmembrane proteins (CD3) abrogated the response. Thirdly, sequencing analyses revealed that only cells expressing certain γδ TCRs were reactive. The "spontaneous" reactivity was not directed against heterologous serum determinants in the culture medium because it could not be induced by addition of the serum to serum-free cultures. Thus, we speculate that the reactivity is directed against determinants expressed by the hybridoma cells themselves, either bona fide "self" antigens, antigens introduced by the fusion partner, or antigens induced by the fusion process.

In a second screening with a number of bacterial antigens, including several staphylococcal enterotoxins ("superantigens" for certain αβ cells), every single hybridoma that had shown "spontaneous" reactivity, but no others, could be stimulated to increased lymphokine release by adding mycobacterial purified protein derivative (PPD). In these experiments, no other antigen for γδ cells was found, but it should be noted that others have been able to stimulate human γδ clones with tetanus toxoid (Kozbor et al., 1989) and staphylococcal enterotoxin A (Rust et al., 1990). In attempts to fractionate PPD in open columns, most of the stimulatory activity was found in fractions with apparent high molecular weight (> 70 kD).

Taken together, our data suggested that both eukaryotic (mouse) and prokaryotic (mycobacterial) cells could provide ligands for the same γδ TCRs, implying antigenic similarity in the two systems. Assuming that γδ cells, like αβ T lymphocytes, may be specialized to interact with proteins, we considered proteins highly conserved in evolution as candidate antigens. Among these, stress proteins and particularly Gro-EL homologs are known to be present in PPD. Therefore, we tested the possibility that PPD-reactive γδ+ hybridomas might be capable of recognizing Gro-EL homologs.

The response of γδ cells to GroEL homologs

Using one hybridoma that strongly responded to PPD (BNT-19.8), we tried to inhibit the response with monoclonal antibodies (mAbs) directed against mycobacterial HSP-65. Three mAbs, WTB78-HI, IVD8 and IIIC8 (kindly provided by Dr. T. Shinnick), partially inhibited the response when added to the cultures, but did not have any effect on a response by another hybridoma against an allogeneic antigen that served as a control. This result suggested that HSP-65 was the stimulatory component in PPD, at least for hybridoma BNT-19.8.

In a second experiment, we stimulated approximately 40 PPD-reactive and -nonreactive γδ hybridomas with purified recombinant HSP-65 from M. bovis (a kind gift from Drs. J. van Embden and R. van der Zee). The majority of PPD-reactive hybridomas was stimulated by the heat shock protein whereas the cells that did not respond to PPD were not (O'Brien et al., 1989). In all cases, however, the responses to HSP-65 were weaker than to PPD. This result suggested two possibilities: Either HSP-65 was not the antigen contained in PPD and stimulated the γδ clones through an accidental crossreaction or the purified molecule could not efficiently be made available in a stimulatory form. In order to circumvent possible problems in our culture system with antigen processing, we tried to stimulate the same hybridomas with small synthetic peptide antigens corresponding to different portions of mycobacterial HSP-65 (kindly donated by Drs. D. Young and T. Shinnick). One peptide, covering amino acids 180-196 of M. leprae HSP-65, was found to be stimulatory for most cells (Born et al., 1990b). Some clones responded better to the peptide antigen than to PPD. These findings, together with a sequence analysis of all PPD/HSP-65 reactive γδ TCRs (Happ et al., 1989) established HSP-65 as an antigen for γδ cells.

Several other groups have reported that GroEL homologs can stimulate γδ cells. One HSP-65 reactive clone has been isolated from the synovial fluid of a patient with rheumatoid arthritis (Holoshitz et al., 1989), and another cell line from a normal, tuberculin-positive individual (Haregewoin et al., 1989). These cells are also reactive with lysates of mycobacteria or PPD. The frequency of HSP reactive γδ cells in humans remains controversial. Whereas a limiting dilution analysis with uncloned peripheral blood lymphocytes suggested that the frequency is low if compared with the frequency of all mycobacteria-reactive γδ cells (Kabelitz et al., 1990), a more recent study suggests that essentially all Vδ2/Vγ9-positive clones, i.e., about 50% of all γδ cells in human peripheral blood, can recognize the human homolog of this protein and many of these cells are stimulated by mycobacterial antigen as well (Fisch et al., in press).

Mechanism of GroEL recognition

As antigens for γδ cells, Gro-EL homologs could be "seen" as native molecules or as processed peptide fragments. They could be presented by specialized antigen presenting cells (APC), like classical T cell antigens or like the recently discovered "superantigens", or be otherwise surface membrane associated. Alternatively, they could be soluble antigens. The structural similarities between αβ and γδ TCRs and the fact that γδ clones can be isolated that recognize allogeneic or syngeneic MHC molecules suggest that the ligand recognition might be similar between these two major T cell populations. Our data, showing that γδ cells are stimulated by small synthetic protein fragments, are compatible with this notion (Born et al., 1990b). Peptides corresponding to amino acids 180-196 of M. leprae HSP-65 strongly stimulated about 10% of the PPD-reactive γδ hybridomas, and virtually all Vδ6/Vγ1-positive cells that were tested showed at least a weak response. The smallest peptide that could stimulate strong responses covered 15 amino acids. Interestingly, two hybridomas with identical TCR V genes but different junctional sequences showed different requirements for the terminal amino acids: One cell, BNT-9.12.18, responded more strongly to peptide 180-196 than to a peptide with two fewer amino acids (181-195), whereas the other, BNT-19.8.7, was stimulated equally well by both peptides. Based on these data, we would expect that processed and presented peptides are also natural antigens for γδ cells, and that junctional TCR sequences might influence fine specificities. The nature of a possible antigen presenting molecule in this system in not clear, however. Our data exclude MHC class II molecules as a necessary requirement and further suggest that presenting molecules may be non-polymorphic (O'Brien et al., 1989). Thus, unlike αβ T cells, γδ cells might "see" protein fragments in the "context" of the oligomorphic class Ib molecules, i.e., TL, Qa and CD1. It remains possible, however, that our synthetic peptides come close enough to the native configuration of the GroEL homologs that some of the cells, specialized to "see" the native proteins, can be stimulated, perhaps because their affinity for the antigenic sequence is particularly high. We therefore cannot formally exclude that γδ cells can interact with GroEL homologs in their native configuration (Fisch et al., in press).

Conclusions

Clearly, we would like to know if the response to GroEL homologs is only an isolated example of a variety of γδ cell responses or whether these cells perhaps tend to recognize stress proteins. We would also like to know whether γδ cells distinguish "self" from "non-self" similarly to αβ T lymphocytes or whether other criteria need to be applied to define a typical γδ cell ligand. Finally, the biological role of cells specialized to recognize stress proteins is of considerable interest. Some clues to these problems are provided by recent observations: Data by Rajasekar et al. (Rajasekar et al., 1990) and Havran et al., (personal communication) indicate that there must be several distinct populations of γδ cells with the ability to recognize stress-induced antigens. Epidermal γδ cells which are apparently stimulated by stressed keratinocytes express γδ TCRs that are entirely different from those expressed by GroEL-reactive cells, and therefore probably recognize a different antigen. The case of lung associated γδ cells is less clear, although preliminary data suggest that they express a γ chain that is different from the γ chains expressed by GroEL-reactive cells in our collection. At this point, we would like to conclude that there are several stress-reactive γδ populations but that it is not yet clear whether only some or most γδ cells have this specificity.

In every case where γδ cell responses to heterologous heat shock proteins have been observed, at least some evidence for reactivity with the autologous homologs also exists (O'Brien et al., 1989; Born et al., 1990b; Holoshitz et al., 1989; Fisch et al., in press; Rajasekar et al., 1990; Haregewoin et al., in press). We have recently shown that the same clones that recognize a defined mycobacterial protein sequence (amino acids 180-196 of *M. leprae* HSP-65) were also stimulated by the mouse equivalent of this peptide. "Self" stress has been shown to induce responses of γδ cells derived from lung, lymph nodes and epidermis (Rajasekar et al., 1990, and Havran et al., personal communication). One human γδ cell line has been shown to react with the purified human GroEL homolog (Haregewoin et al., in press) and many human peripheral blood γδ cells appear to recognize GroEL determinants on a human Burkitt's lymphoma, DAUDI (Fisch et al., in press). Therefore, it seems likely that γδ cells can recognize autologous heat shock proteins, and that they may not primarily distinguish "self" from "non-self" but rather "stressed" from "not stressed". Why several of our clones preferred the mycobacterial over all other sequences is not clear. Such a (heteroclitic?) preference, however, might explain why mycobacterial infections can occasionally initiate autoaggressive immune responses.

Based on the data that are presently available, stress-reactive γδ cells could play a role as an early warning system, regardless of whether they primarily recognize autologous or heterologous stress. It has also been suggested that γδ cells might be involved in the termination of inflammatory responses, after the pathogen has been cleared (Carding et al., in press). Apart from pathology, stress-reactive γδ (and other) cells could play a role as immune regulators (Born et al., 1990a) since activated and differentiating cells tend to express heat shock proteins. As more and more connections between the stress response and the immune system are uncovered, γδ cells should be watched as potentially major players in this interaction.

References

Born, W, Miles, C, White, J, O'Brien, R, Freed, JH, Marrack, P, Kappler, J and Kubo, RT, (1987) Peptide sequences of T-cell receptor δ and γ chains are identical to predicted X and γ proteins. Nature, 330: 572-574.

Born,W, Happ, MP, Dallas, A, Reardon, C, Kubo, R, Shinnick, T, Brennan, P and O'Brien, R, (1990a) Recognition of heat shock proteins and γδ cell function. Immunol. Today, 11: 40-43.

Born, W, Hall, L, Dallas, A, Boymel, J, Shinnick, T, Young, D, Brennan, P and O'Brien, R, (1990b) Recognition of a peptide antigen by heat shock reactive γδ T lymphocytes. Science, 249: 67-69.

Carding, SR, Allan, W, Kyes, S, Hayday, A, Bottomly, K and Doherty, PC, Late dominance of the inflammatory process in murine influenza by γδ+ T cells. J. Exp. Med., in press.

Fisch, P, Malkovsky, M, Klein, BS, Morrissey, LW, Carper, SW, Welch, WJ and Sondel, PM, Human Vγ9/Vδ2 T cells recognize a groEL homolog on Dudi Burkitt's lymphoma cells. Science, in press.

Happ, MP, Kubo, RT, Palmer, E, Born, WK and O'Brien, RL, (1989) Limited receptor repertoire in a mycobacteria-reactive subset of γδ T lymphocytes. Nature, 342: 696-698.

Haregewoin, A, Soman, G, Hom, RC and Finberg, RW, (1989) Human γδ+ T cells respond to mycobacterial heat-shock protein. Nature, 340: 309-312.

Haregewoin, A, Singh, B, Gupta, RS and Finberg, RW, A mycobacterial heat shock protein responsive γδ T cell clone responds to homologous human heat shock protein: a possible link between infection and autoimmunity. J. Infect. Dis., in press.

Holoshitz, J, Koning, F, Coligan, JE, De Bruyn, J and Strober, S, (1989) Isolation of CD4- CD8- mycobacteria-reactive T lymphocyte clones from rheumatoid arthritis synovial fluid. Nature, 339: 226-229.

Kabelitz, D, Bender, A, Schondelmaier, S, Schoel, B and Kaufmann, SHE, (1990) A large fraction of human peripheral blood γδ+ T cells is activated by Mycobacterium tuberculosis but not by its 65 kD heat shock protein. J. Exp. Med., 171: 667-679.

Kozbor, D, Trinchieri, G, Monos, DS, Isobe, M, Russo, G, Haney, JA, Zmijewski, C and Croce, CM, (1989) Human TCR-γ+/δ+, CD8+ T lymphocytes recognize tetanus toxoid in an MHC-restricted fashion. J. Exp. Med., 169: 1847-1851.

Matis, LA, Cron, R and Bluestone, JA, (1987) Major histocompatibility complex-linked specificity of γδ receptor-bearing T lymphocytes. Nature, 330: 262-264.

Matis, LA, Fry, AM, Cron, RQ, Cotterman, MM, Dick, RF and Bluestone, JA,

(1989) Structure and specificity of a class II alloreactive γδ T cell receptor heterodimer. Science, 245: 746-749.

Mosmann, T, (1983) Rapid colorimetric assay for cellular growth and survival: application to proliferation and cytotoxicity assays. J. Immunol. Methods, 65: 55-63.

O'Brien, RL, Happ, MP, Dallas, A, Palmer, E, Kubo, R and Born, WK, (1989) Stimulation of a major subset of lymphocytes expressing T cell receptor γδ by an antigen derived from Mycobacterium tuberculosis. Cell, 57: 667-674.

Rajasekar, R, Sim, G and Augustin, A, (1990) Self heat shock and γδ T-cell reactivity. Proc. Natl. Acad. Sci. USA, 87: 1767-1771.

Reardon, C, Lefrancois, L, Farr, A, Kubo, R, O'Brien, R and Born, W, Expression of γδ T-cell receptors on lymphocytes from the lactating mammary gland. J. Exp. Med., in press.

Rust, CJJ, Verreck, F, Vietor, H and Koning, F, (1990) Specific recognition of staphylococcal enterotoxin A by human T cells bearing receptors with the Vγ9 region. Nature, 346: 572-574.

White, J, Blackman, M, Bill, J, Kappler, J, Marrack, P, Gold, D and Born, W, (1989) Two better cell lines for making hybridomas expressing specific T cell receptors. J. Immunol., 143: 1822-1825.

Young, RA, (1990) Stress proteins and immunology. Ann. Rev. Immunol., 8: 401-420.

Expression of Stress Proteins on the Surface of Cells of the Immune System

W.N. Jarjour, W.J. Welch*, J.B. Winfield
Division of Rheumatology and Immunology
University of North Carolina at Chapel Hill
Chapel Hill, North Carolina 27514
USA

Introduction

Localization of stress proteins on plasma membranes is of fundamental biologic and immunologic interest (reviewed in Winfield and Jarjour, 1990). For example, the detection of HSP70 as a peptide-binding protein on the surface of antigen-presenting cells implies a role for at least some members of this family of stress proteins in the recognition of MHC class II/antigen peptide complexes by T cell receptors (TCRs) (Lakey et al., 1987; Vanbuskirk et al., 1989). Furthermore, surface expression of autologous stress proteins, or stress protein epitopes, can render cells susceptible to specific immunological attack (Koga et al., 1989; Ottenhof et al., 1988). Relatively little information is available concerning which stress proteins are expressed on lymphoid cells and under what circumstances, however. In this review, recent immunologic and biochemical data from our laboratories that bear on these issues are summarized briefly.

Experimental Approach

The overall design of our experiments to identify stress protein molecules, or molecules immunologically-related to stress proteins, on the cell surface involved indirect immunofluorescence/flow microfluorimetry (Mimura et al., 1990) and complement-dependent microcytotoxicity assays (Winfield et al., 1975) using specific antibody probes and various lymphoid or non-lymphoid cells as targets. Control immunoglobulins, including mouse myeloma proteins and normal rabbit sera which are devoid of antibody activity and irrelevant rabbit antibodies were used to assess the specificity of positive reactions with anti-stress protein antibodies. When positive staining and cytotoxicity were encountered, immunoblotting and

* The Lung Biology Research Center, University of California, San Francisco, California 94143, USA

236

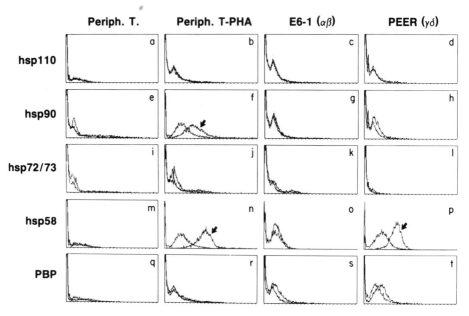

Figure 1. Reactivity of T cells with anti-stress protein antibodies. Peripheral blood T cells and T cells lines stained with anti- stress protein antibodies and fluorochrome-conjugated anti- immunoglobulin reagents were examined by flow microfluorimetry, as described in Materials and Methods. Mouse myeloma immunoglobulin or normal rabbit serum were used as controls. PEER (TCR γδ+) and PHA-stimulated peripheral T cells were stained with anti-HSP58 (panels p and n, respectively). Anti-HSP90 stained PHA-stimulated peripheral T cells (panel f). Both antibody and control staining are shown in each histogram. Arrows indicate antibody-specific staining.

immunoprecipitation experiments using ^{125}I-surface-labeled cells (Minota and Winfield, 1987; Hamada and Tsuruo, 1987) were performed in an attempt to define the nature of the reactive antigen.

A variety of cell types were studied, both in a resting or unstimulated state and following heat-shock (43°C for 60 min followed by culture for 18 hr at 37°C) or stimulation with mitogens, soluble antigens, anti-CD3 plus phorbol myristate acetate (PMA), PMA plus the Ca++ ionophore A23187, or PMA alone. Cells included: normal human peripheral blood mononuclear cells (PBMC); E6-1, a Jurkat T cell line expressing TCR αβ; PEER, a leukemia T cell line expressing TCR γδ; a human thymocyte cell line expressing TCR γδ; JRT3-T3.5, a mutant Jurkat line which does not express TCRs (Saito et al., 1987); MOLT-4, a leukemia cell line which does not transcribe TCR α chain genes (Hara et al., 1987); HSB-2, a primitive T cell line which lacks TCRs and most of the other antigens expressed by peripheral T cells; Raji, a B cell line; Wil-2, a B cell line; and the non-lymphoid lines HL60, THP-1 and U937.

Antibodies to members of the HSP70 family of stress proteins included mouse monoclonals N4, N6 F3-3, N15, N21, N27, C92, H, and N (constitutive and highly-induceable 72-73 kD proteins) and rabbit

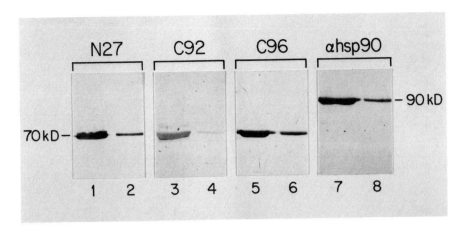

Figure 2. Reactivity of Wil-2 and HL60 cells with anti-stress protein antibodies. A B cell line (Wil-2) and a monomyelocytic cell line (HL60), with or without heat-shock (HS) or stimulation with phorbol myristate acetate (PMA), were stained and analyzed as described in the legend to Figure 1. Except for anti-HSP58 staining of Wil-2 cells (arrow in panel m), these cell lines did not react with anti-stress protein antibodies when cell death was minimized.

antiserum specific for the 75kD glucose-regulated protein (4442,6147) (Welsh, unpublished data). Rabbit antiserum to mouse peptide binding proteins (PBP 72/74 (Lakey et al., 1987) was a gift from Dr. Sue Pierce (Northwestern University, Evanston, IL). Rabbit antisera to HSP90 (Koyasu et al., 1986) and HSP110 (Subjeck et al., 1983) were gifts from Drs. Ichiro Yahara, Tokyo Metropolitan Institute of Medical Science, Tokyo, Japan, and Dr. John Subjeck, Roswell Park Memorial Institute, Buffalo, NY, respectively. Preparation and characterization of rabbit anti-HSP58 has been described previously (Mizzen et al., 1989). Rabbit antiserum to P1, another mammalian homologue of groEL isolated from chinese hamster ovary cells (Gupta et al., 1985), was provided by Dr. R. Gupta, McMaster University, Hamilton, Ontario.

Surface expression of members of the HSP70 family, HSP90, and HSP110 on T cells (representative data are illustrated in Figure 1)

Unstimulated or heat-shocked T cells (peripheral blood and cell lines) were consistently non-reactive by indirect immunofluorescence and C-dependent microcytotoxicity with antibodies to HSP90, HSP110, and members of the HSP70 family. PBMC cultured with tetanus

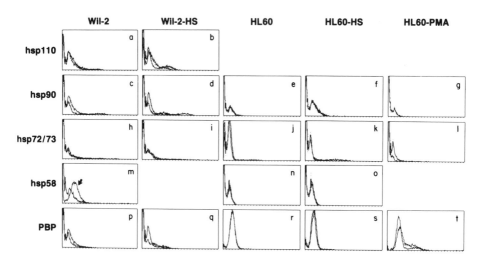

Figure 3. Localization of stress proteins in enriched plasma membranes. One-dimension blots of detergent whole cell lysate (odd-numbered lanes) or ~32-fold enriched plasma membranes (even-numbered lanes) of E6-1 cells were stained with anti-HSP70 mAbs N27, C92, or C96 and rabbit anti-HSP90, as described in Materials and Methods.

toxoid and PPD were non-reactive as well. Staining with some anti-HSP70 mABs (weak) and with anti-HSP90 (strong) was observed when peripheral T cells were stimulated with mitogens, but only anti-HSP90 killed mitogen-stimulated peripheral T cells in the presence of complement. None of the antibody probes immunoprecipitated ^{125}I-labeled molecules from T cells, however.

When one-dimension (1D) blots of protein from a T cell line (E6-1) were examined for reactivity with rabbit anti-HSP90 and mAbs specific for members of the HSP70 family, strongly stained bands were apparent in both the whole cell lysate (as expected for these intracellular molecules) and in ~32-fold-enriched plasma membranes (Fig. 2). While it is possible that the membrane staining simply reflects contamination with intracellular stress protein, the morphology of the stained bands suggested the presence of at least two distinct proteins, only one of which localized to (or co-localized with) the membrane fraction.

Surface expression of members of the HSP70 family, HSP90, and HSP110 on B cell and non-lymphoid cell lines (representative data are illustrated in Figure 3)

Staining of two B cell lines (Raji, Wil-2) with antibodies to HSP90, HSP110, members of the HSP70 family (including PBP 72/74) was absent or only weakly and inconsistently present, and was not induced by heat-shock. In contrast to our own preliminary data suggesting that the monomyelocytic line HL60 expressed HSP90 after heat-shock or stimulation with phorbol esters (Jarjour et al., 1989), positive staining was observed in only 2 of 12 separate experiments when cell death was reduced to 5% or less. Nor was reactivity with anti-HSP90 evident when U937 and THP-1 were studied in this fashion. An extensive series of immunoprecipitation experiments using these and other B and non-lymphoid cell lines also failed to demonstrate unequivocally the presence of surface-labeled stress proteins (Winfield, unpublished observations).

Surface expression of a protein cross-reactive with HSP58 on T cells bearing γδ receptors

As reported in detail elsewhere (Jarjour et al., 1990), polyclonal rabbit antisera against HSP58 and P1 stained two unrelated γδ T cell lines, but did not react with resting peripheral blood T cells, T cell lines expressing αβ receptors, T cells lacking TCRs altogether, or non-lymphoid cell lines. Identical results were obtained when this panel of cells was examined in complement-dependent microcytotoxicity assays for reactivity with anti-HSP58 or anti-P1. In Western blot experiments utilizing whole cell lysates or enriched plasma membranes as substrate and immunoprecipitation experiments with ^{125}I-surface-labeled cells, anti-HSP58 reacted specifically with a ~77 kD surface membrane protein only on T cells with γδ TCRs. Reactivity with anti-HSP58 was not increased on γδ cells or induced on αβ cells following brief heat shock or stimulation with OKT3 + PMA or PMA + A23187. In more recent experiments, certain B cell lines, e.g., Wil-2, and mitogen-activated peripheral T cells also were found to express an anti-HSP58-reactive protein.

Conclusions

Except for an anti-HSP90-reactive protein on mitogen-activated T cells, little convincing evidence was obtained in indirect immunofluorescence and complement-dependent microcytotoxicity experiments for the expression of HSP90, HSP110, and members of the HSP70 family of stress proteins on the surface of T cells, B cells, and several monomyelocytic cell lines. Although the possibility of localization (or co-localization) of HSP70 and HSP90 in plasma membranes was raised by unusual staining patterns in immunoblotting experiments, immunoprecipitation experiments with ^{125}I-surface-labeled cells were inconclusive. In contrast to these generally negative data was the detection of a ~77 kD anti-HSP58 cross-reactive antigen on the surface of several γδ T cell lines, but not on T cell lines that express αβ receptors or which lack TCRs (Jarjour et al., 1990). Some evidence was obtained to suggest that this molecule, or another anti-

Table 1

Evidence for expression of stress proteins or stress protein epitopes on the cell surface
HSP60-specific CTL lyse naive monocytes (Koga et al., 1989; Ottenhof et al., 1988; Munk et al., 1989)
HSP90 functions as tumor specific transplantation antigen (Ullrich et al., 1986)
"self-reactivity" of neonatal γδ thymocyte hybridomas is related to expression of HSP60 determinants (O'Brien et al., 1989)
brief heat-shock of murine T cells induces selective proliferation of γδ cells (Rajasekar et al., 1990)
surface membrane expression of stress proteins can be induced by virus infection of the cell, certain drugs, and cytokines (Koga et al., 1989; La Tangue and Latchman, 1988; Martin and Regan, 1988)
HSP70 can be detected as a peptide-binding protein on the surface of murine antigen-presenting B cells (Lakey et al., 1987; Vanbuskirk et al., 1989)
grp100 is located in the pericellular matrix and is spontaneously shed (McCormick et al., 1979; McCormick et al., 1982)
cultured rat embryo cells selectively release HSP110 and several members of the HSP70 family (Hightower and Guidon, 1989)
certain proteins homologous to grp94 and HSP90 are integral membrane proteins (La Tangue and Latchman, 1988; Mazzarella and Green, 1987)

HSP58 cross-reactive molecule, also is expressed on some B cells and on peripheral T cells after stimulation with the mitogen, phytohemagglutinin.

Considerable previous data suggest that certain stress proteins, or peptides of stress proteins, may be localized on the cell surface (summarized in Table 1). Some possible mechanisms by which this might occur are listed in Table 2. Although special efforts were made to exclude trivial explanations for our results, e.g., non-specific Fcγ receptor binding, precise definition of the nature and significance of our data and of the phenomena listed in Table 1 will require additional investigation. A major difficulty in such experiments is the cell death and consequent release of intracellular stress proteins that occurs when cells are subjected to heat-shock, prolonged culture *in vitro*, or

Table 2

Mechanisms by which stress proteins, or peptides thereof, could be expressed on the cell surface
presentation of stress protein peptides by MHC class I and class II molecules
presentation of autologous stress protein peptides by class Ib molecules
translocation of stress proteins to the cell surface as they chaperon MHC or other nascent integral membrane proteins to the plasma membrane
direct binding of circulating extra-cellular stress proteins to surface membrane proteins
expression of stress proteins, or closely-related proteins, as integral plasma membrane proteins

other types of manipulations. Indeed, the authors suspect that cell death and secondary binding of released intracellular stress protein may have been responsible for our previous observation (Jarjour et al., 1989) that HL60 cells expressed HSP90 when heat-shocked or stimulated with phorbol esters.

Nevertheless, in the case of anti-HSP58 staining of γδ cells, some evidence was obtained for the presence on the plasma membrane of a novel protein distinct from, but related immunologically to, intracellular HSP58. It remains to be determined whether the 77 kD target is a glycosylated version of mitochondrial HSP58 (Mizzen et al., 1989; Dudani and Gupta, 1989; McMullin and Hallberg, 1988; Jindal et al., 1989) or cytoplasmic TCP-1 in the mouse (Gupta, 1990), or, indeed, even a stress-inducible protein. Of interest in this regard are data discussed at this meeting from Dr. R. A. Young's laboratory at the Whitehead Institute in Cambridge, MA, which suggest that there are at least five genes related to groEL in mammalian cells. Further experiments to better define the biochemistry of this molecule(s) and the HSP90-related molecule on mitogen-activated T cells are in progress.

The fact that stress protein epitopes are expressed at the cell surface has important implications for immunology and autoimmunity. A high proportion of murine neonatal γδ thymocyte hybridomas recognize mycobacterial antigens, including HSP60 and a conserved HSP60 peptide, and exhibit unusual, TCR-dependent "self-stimulatory" activity. This is probably due to recognition of mycobacterial HSP60 cross-reactive autologous stress protein peptides (O'Brien et al., 1989; Born et al., 1990a). The 77 kD HSP58 cross-reactive protein described herein may be relevant to such "self-stimulatory" activity. Born and colleagues have developed an interesting schema in which γδ cells that recognize stress proteins represent a "rapid-response" first line of defense against infection (Born et al., 1990b). With stress or other stimuli, autologous stress protein peptides are "presented" on cell surfaces to special subsets of γδ cells, which respond by mediating cellular activation, immune surveillance, and regulation of lymphocyte growth and differentiation. Cross-reactions of autologous stress proteins with bacteria may lead to a break in tolerance, expansion of autoreactive T cells that recognize stress proteins, and autoimmune-mediated cell and tissue injury. Consistent with this idea are data implicating immunodominant groEl-related microbial stress proteins as both "triggers" and autologous targets in various disorders characterized by T cell autoreactivity, including arthritis (Van eden et al., 1988; Res et al., 1988; Gaston et al., 1988; Gaston et al., 1989; Gaston et al., 1990), trachoma (Morrison et al., 1989), and insulin-dependent diabetes (Elias et al., 1990). Cells expressing stress proteins on their surface may be subject to complement-dependent antibody-mediated cell lysis as well, as suggested by the occurrence of anti-stress protein autoantibodies in systemic lupus erythematosus and other diseases (Bahr et al., 1988b; Bahr et al., 1988a; Minota et al., 1988a; Minota et al., 1988b; Tsoulfa et al., 1988) and by an association of anti-lymphocyte and anti-stress protein autoantibody activities in patient sera (Minota and Winfield, 1987).

Acknowledgments

Research in the authors' laboratories is supported by National Institutes of Health grants RO1 AM30863, T32 AR7416, and P60 AR30701, and a Biomedical Research Center grant from the Arthritis Foundation.

References

Bahr, GM, Rook, GAW, Al-Saffar, M, van Embden, J, Stanford, JL and Behbehani, K, (1988a) Antibody levels to mycobacteria in relation to HLA type: evidence for non-HLA-linked high levels of antibody to the 65 kD heat shock protein of M. bovis in rheumatoid arthritis. Clin. Exp. Immunol., 74: 211-215.
Bahr, GM, Rook, GAW, Shahin, A, Stanford, JL, Sattar, MI and Behbehani, K, (1988b) HLA-DR-associated isotype-specific regulation of antibody levels to mycobacteria in rheumatoid arthritis. Clin. Exp. Immunol., 72: 26-31.
Born, W, Hall, L, Dallas, A, Boymel, J, Shinnick, T, Young, D, Brennan, P and O'Brien, R, (1990a) Recognition of a peptide antigen by heat shock reactive γδ T lymphocytes. Science, 249: 67-69.
Born, W, Happ, MP, Dallas, A, Reardon, C, Kubo, R, Shinnick, T, Brennan, P and O'Brien, R, (1990b) Recognition of heat shock proteins and γ/δ cell function. Immunol. Today, 11: 40-43.
Dudani, AK and Gupta, RS, (1989) Immunological characterization of a

human homolog of the 65-kilodalton mycobacterial antigen. Infect. Immun., 57: 2786-2793.

Elias, D, Markovits, D, Reshef, T, van der Zee, R and Cohen, IR, (1990) Induction and therapy of autoimmune diabetes in the non-obese diabetic (NOD/Lt) mouse by a 65-kDa heat shock protein. Proc. Natl. Acad. Sci. USA, 87: 1576-1580.

Gaston, JSH, Life, PF, Bailey, L and Bacon, PA, (1988) Synovial fluid T cells and 65 kD heat-shock protein. Lancet, 2: 856-856.

Gaston, JSH, Life, PF, Bailey, L and Bacon, PA, (1989) In vitro responses to a 65-kilodalton mycobacterial protein by synovial T cells from inflammatory arthritis patients. J. Immunol., 143: 2494-2500.

Gaston, JSH, Life, PF, Jenner, PJ, Colston, MJ and Bacon, PA, (1990) Recognition of a mucobacteria-specific epitope in the 65-kD heat-shock protein by synovial fluid-derived T cell clones. J. Exp. Med., 171: 831-841.

Gupta, RS, Venner, TJ and Chopra, A, (1985) Genetic and biochemical studies with mutants of mammalian cells affected in microtubule-related proteins other than in tubulin: mitochondrial localization of a microtubule-related protein. Can. J. Biochem. Cell Biol., 63: 489-502.

Gupta, RS, (1990) Sequence and structural homology between a mouse T-complex protein TCP-1 and the "chaperonin" family of bacterial (GroEl, 60-65 kDa heat shock antigen) and eukaryotic proteins. Biochem. Intl., 24: 833-841.

Hamada, H and Tsuruo, T, (1987) Determination of membrane antigens by a covalent crosslinking method with monoclonal antibodies. Anal. Biochem., 160: 483-488.

Hara, J, Benedict, SH, Champagne, E, Mak, TW, Minden, M and Gelfand, EW, (1988) Comparison of T cell receptor alpha, beta, and gamma gene rearrangement and expression in T cell acute leukemia. J. Clin. Invest., 81: 989-996.

Hightower, LE and Guidon, PT Jr, (1989) Selective release from cultured mammalian cells of heat- shock (stress) proteins that resemble glia-axon transfer proteins. J. Cell Physiol., 138: 257-266.

Jarjour, W, Tsai, V, Woods, V, Welch, W, Pierce, S, Shaw, M, Mehta, H, Dillmann, W, Zvaifler, N and Winfield, J, (1989) Cell surface expression of heat shock proteins. Arthritis Rheum., 32: S44.

Jarjour, W, Mizzen, LA, Welch, WJ, Denning, S, Shaw, M, Mimura, T, Haynes, BF and Winfield, JB, Constitutive expression of a groEL-related protein on the surface of human γ/δ cells, submitted.

Jindal, S, Dudani, AK, Singh, B, Harley, CB and Gupta, RS, (1989) Primary structure of a human mitochondrial protein homologous to the bacterial and plant chaperonins and to the 65-kilodalton mycobacterial antigen. Mol. Cell. Biochem., 9, No.5: 2279-2283.

Koga, T, Wand-Wurttenberger, A, DeBruyn, J, Munk, ME, Schoel, B and Kaufmann, SHE, (1989) T cells against a bacterial heat shock protein recognize stressed macrophages. Science, 245: 1112-1115.

Koyasu, S, Nishida, E, Kadowaki, T, Matsuzaki, F, Iida, K, Harada, F, Kasuga, M, Sakai, H and Yahara, I, (1986) Two mammalian heat shock proteins, HSP90 and HSP100, are actin-binding proteins. Proc. Natl. Acad. Sci. USA, 83: 8054-8058.

Lakey, EK, Margoliash, E and Pierce, SK, (1987) Identification of a peptide binding protein that plays a role in antigen presentation. Proc. Natl. Acad. Sci. USA, 84: 1659-1663.

La Thangue, NB and Latchman, DS, (1988) A cellular protein related to heat-shock protein 90 accumulates during herpes simplex virus infection and is overexpressed in transformed cells. Exp. Cell Res., 178: 169-179.

Martin, ML and Regan, CM, (1988) The anticonvulsant sodium valproate specifically induces the expression of a rat glial heat shock protein which is identified as the collagen type IV receptor. Brain Res., 459: 131-137.

Mazzarella, RA and Green, M, (1987) ERp99, an abundant, conserved glycoprotein of the endoplasmic reticulum, is homologous to the 90-kDa heat shock protein (HSP90) and the 94-kDa glucose regulated protein

244

(GRP94). J. Biol. Chem., 262: 8875-8883.

McCormick, PJ, Keys, BJ, Pucci, C and Millis, AJT, (1979) Human fibroblast-conditioned medium contains a 100k dalton glucose-regulated cell surface protein. Cell, 18: 173-182.

McCormick, PJ, Millis, AJT and Babiarz, B, (1982) Distribution of a 100k dalton glucose-regulated cell surface protein in mammalian cell cultures and sectioned tissues. Exp. Cell Res., 138: 63-72.

McMullin, TW and Hallberg, RL, (1988) A highly evolutionarily conserved mitochondrial protein is structurally related to the protein encoded by the Escherichia coli groEL gene. Mol. Cell Biol., 8: 371-380.

Mimura, T, Fernsten, P, Jarjour, W and Winfield, JB, (1990) Autoantibodies specific for different isoforms of CD45 in systemic lupus erythematosus. J. Exp. Med., 172: 653-656.

Minota, S and Winfield, JB, (1987) Identification of three major target molecules of IgM antilymphocyte autoantibodies in systemic lupus erythematosus. J. Immunol., 139: 3644-3651.

Minota, S, Koyasu, S, Yahara, I and Winfield, JB, (1988) Autoantibodies to the heat-shock protein hsp90 in systemic lupus erythematosus. J. Clin. Invest., 81: 106-109.

Minota, S, Cameron, B, Welch, WJ and Winfield, JB, (1988) Autoantibodies to the constitutive 73-kD member of the hsp70 family of heat shock proteins in systemic lupus erythematosus. J. Exp. Med., 168: 1475-1480.

Mizzen, LA, Chang, C, Garrels, JI and Welch, WJ, (1989) Identification, characterization, and purification of two mammalian stress proteins in mitochondria, grp 75, a member of the hsp 70 family and hsp 58, a homolog of the bacterial groEL protein. J. Biol. Chem., 264: 20664-20675.

Morrison, RP, Belland, RJ, Lyng, K and Caldwell, HD, (1989) Chlamydial disease pathogenesis. The 57-kD chlamydial hypersensitivity antigen is a stress response protein. J. Exp. Med., 170: 1271-1283.

Munk, ME, Schoel, B, Modrow, S, Karr, RW, Young, RA and Kaufmann, SHE, (1989) T lymphocytes from healthy individuals with specificity to self-epitopes shared by the mycobacterial and human 65-kilodalton heat shock protein. J. Immunol., 143: 2844-2849.

O'Brien, RL, Happ, MP, Dallas, A, Palmer, E, Kubo, R and Born, WK, (1989) Stimulation of a major subset of lymphocytes expressing T cell receptor γδ by an antigen derived from Mycobacterium tuberculosis. Cell, 57: 667-674.

Ottenhoff, TH, Ab, BK, van Embden, JD, Thole, JE and Kiessling, R, (1988) The recombinant 65-kD heat shock protein of Mycobacterium bovis Bacillus Calmette-Guerin/M. tuberculosis is a target molecule for CD4+ cytotoxic T lymphocytes that lyse human monocytes. J. Exp. Med., 168: 1947-1952.

Rajasekar, R, Sim, G-K and Augustin, A, (1990) Self heat shock and γδ T-cell reactivity. Proc. Natl. Acad. Sci. USA, 87: 1767-1771.

Res, PC, Breedveld, FC, Van Embden, JDA, Schaar, CG, Van Eden, W, Cohen, IR and De Vries, RRP, (1988) Synovial fluid T cell reactivity against 65 kD heat shock protein of mycobacteria in early rheumatoid arthritis. Lancet, ii: 478-480.

Saito, T, Weiss, A, Gunter, KC, Shevach, EM and Germain, RN, (1987) Cell surface T3 expression requires the presence of both alpha- and beta-chains of the T cell receptor. J. Immunol., 139: 625-628.

Subjeck, JR, Shyy, T, Shen, J and Johnson, JR, (1983) Association between the mammalian 110,000-dalton heat-shock protein and nucleoli. J. Cell Biol., 97: 1389-1398.

Tsoulfa, G, Rook, GA, van Embden, JD, Young, DB, Mehlert, A, Isenberg, DA, Hay, FC and Lydyard, PM, (1988) Raised serum IgG and IgA antibodies to mycobacterial antigens in rheumatoid arthritis. Ann. Rheum. Dis., 48: 118-123.

Ullrich, SJ, Robinson, EA, Law, LW, Willingham, M and Appella, E, (1986) A mouse tumor-specific transplantation antigen is a heat shock-related protein. Proc. Natl. Acad. Sci. USA, 83: 3121-3125.

Vanbuskirk, A, Crump, BL, Margoliash, E and Pierce, SK, (1989) A peptide binding protein having a role in antigen presentation is a member of the HSP70 heat shock family. J. Exp. Med., 170: 1799-1809.

Van Eden, W, Thole, JE, van der Zee, R, Noordzij, A, van Embden, JD, Hensen, EJ and Cohen, IR, (1988) Cloning of the mycobacterial epitope recognized by T lymphocytes in adjuvant arthritis. Nature, 331: 171-173.

Winfield, JB, Winchester, RJ, Wernet, P, Fu, SM and Kunkel, HG, (1975) Nature of cold-reactive antibodies to lymphocyte surface determinants in systemic lupus erythematosus. Arthritis Rheum., 18: 1-8.

Winfield, JB and Jarjour, WN, (1990) Stress proteins, autoimmunity, and autoimmune disease. Curr. Top. Microbiol. Immunol., 167: 161-189.

Medical Applications
of Heat Shock Response

Quantification of Protein Denaturation with Sulfhydryl Agents: Application to Heat Shock in Mammalian Cells

G.M. Hahn, K.-J. Lee, Y.-M. Kim
Department of Radiation Oncology
Stanford University School of Medicine
Stanford, CA 94305-5468
USA

Introduction

Considerable attention has recently been focused on the role that denatured or otherwise "nonfunctional" proteins play in the induction of the heat shock response and the development of thermotolerance (Ananthan et al., 1986; Carlson et al., 1987; Hightower, 1980; Lee and Hahn, 1988; Magun and Fennie, 1981). The concept of denatured proteins is imprecise: in the context of heat shock, what is usually meant is that cytosolic proteins have parts or all of their hydrophobic regions uncovered. These normally well-hidden parts of the proteins then become reactive and, therefore, readily combine with other cellular components, thus preventing renaturation. To date, no technique has been reported that would permit quantifying the amount of denatured proteins within the cell either before, during, or after exposure to elevated temperatures. For this reason, the quantitative relationship between protein denaturation and induction of stress responses has not been investigated.

We present data that suggest that this aspect of protein denaturation can readily be quantified, and we demonstrate that proteins that cannot be renatured, for whatever reason, may very likely be among the cellular lesions that are responsible for heat-induced cell death as well as being able to induce the heat shock response.

Protein denaturation and heat shock

The idea behind the technique for quantifying protein denaturing that we will describe is based on the following: As discussed earlier, at least one consequence of what is referred to as protein denaturation is the exposure, in aqueous environments, of hydrophobic (or, in membranous portions of the cell, of lipophobic) regions of some proteins. These hydrophobic regions then interact with other similar regions, or with different intracellular compounds and thus become permanently nonfunctional. One role of heat shock proteins, specifically of HSP70,

250

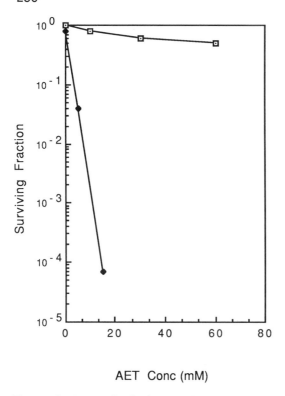

Figure 1. Survival of Chinese hamster ovary (HA-1) cells exposed to AET. Monolayers of HA-1 cells were exposed to graded doses of AET either at 37°C (□) or at 43°C (◆). Cells treated with AET at 43°C were much more sensitive than cells at 37°C.

is thought to bind reversibly to such exposed regions and, thereby, to facilitate renaturation by refolding. Our idea was that any compound, or compounds, that could compete with HSP70 for binding sites (or would inhibit the refolding ability of HSP70) would then inhibit renaturation of the proteins. The presence of such agents during heating would make the cells more sensitive to heat. More importantly, addition of such agents to previously heated cells should result in binding to denatured proteins. Measurement of the bound compound in the form of (protein) high molecular weight complexes (HMW) could then serve as a measure of the amount of denatured protein present in the heated cells. The compounds that we have examined for their possible role as competitors for HSP70-binding sites (or of inhibitors of HSP70 binding) are a series of sulfhydryl agents. These are known to interact with hydrophobic regions of proteins, and, by formation of disulfide bridges, to prevent renaturation. These bridges can be between different proteins, leading to the development of HMW. If our concept is correct, we should first be able to demonstrate that in the presence of such drugs, cells become very sensitive to heat, since HSP70 could then only partially fulfil its protective (or refolding) function. Secondly, in the presence of increased amounts of HSP70, the cytotoxic effects of the sulfhydryl compounds should become progressively inhibited. We first show that indeed such agents, at elevated temperatures, become highly cytotoxic at doses that are nontoxic at 37°C, and that cells with increased amounts of HSP70 become more resistant.

Figure 1 shows survival of Chinese hamster cells exposed to graded doses of AET at 37° or 43°C. The very much enhanced killing at the higher temperature is certainly consistent with the view that the sulfhydryl compound is interfering with renaturing of one or more proteins. The data say nothing, however, about possible competition with HSP70. To examine the latter possibility, we made use of a recent finding in our laboratory, namely, that by proper timing of multiple heat exposures of progressively severe heat doses, it is possible to obtain populations of viable cells that contain vastly increased concentrations of HSP70 (Hahn et al., in preparation). Data on the heating protocols and and the resulting HSP70 concentrations are shown in Table 1. We treated cells according to the protocols shown and then exposed them to diamide, another sulfhydryl agent either at 37°C or at 43°C. At 37°C, there was no measurable cytotoxicity over the range of diamide tested. Even at 43°C, these cells were substantially more resistant to the drug than their counterparts that contained normal amounts of HSP70 (Fig. 2). Further, resistance to diamide increased with progressively increasing concentrations of HSP70, i.e., group 1 cells were somewhat less resistant, while group 3 cells were the most resistant. Cells from all three groups were, however, far more

TABLE 1

	HSP70 in hyperinduced Chinese hamster cells.		
	Treatment	Maximum Concentration of HSP70	Percent of Total Cellular Protein
Control	No treatment	1.0	3.0 + 1.5
Group I	10' 45°C; 10 hr 37°C	5.9 + 2.9	18 + 6
Group II	10' 45°C; 7 hr 37°C 45' 45°C; 4 hr 37°C	12.4 + 3.1	32 + 18
Group III	10' 45°C; 10 hr 37°C 45' 45°C; 7 hr 37°C 60' 45°C; 4 hr 37°C	17.5 + 2.0	52 + 28

Table 1. HSP70 concentrations were determined by ELISA. Details of these experiments will be published elsewhere.

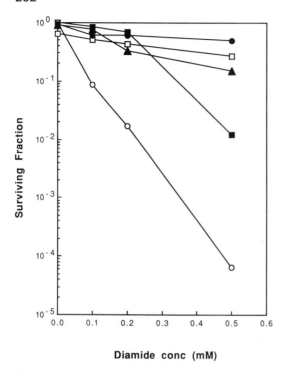

Diamide conc (mM)

Figure 2. Survival of mouse RIF-1 cells exposed to diamide. Cells were pretreated as described in Table 1, followed by 1 h diamide treatment at 43°C. Survival of control cells (with no preheat treatment) at 37°C is shown for comparison (■). Control (o); group I (▲); group II (□) ; group III (●).

resistant to diamide than cells not given any pretreatment. Thus, if the model was correct, the data strongly suggested that HSP70 inhibited diamide reaction on heat-damaged proteins.

Quantitation of binding of diamide to cellular proteins

Our next step was to devise a method to evaluate the binding of the drug diamide to cellular proteins. Diamide was chosen because it is readily available, and it has been used in many cellular studies. The technique that we found convenient was to prelabel cellular proteins with ^{35}S-methionine, and then to expose such cells to diamide. The cells were then lysed and the proteins were examined by SDS PAGE (poly-acrylamide gel electrophoresis) performed under nonreducing conditions (i.e., in the absence of β-mercaptoethanol). One-dimensional gels of proteins from cells from one such experiment are shown in Figure 3. Each lane was loaded with equal amounts of radioactivity. With increasing doses of diamide, we saw increasing amounts of HMW. These are seen at two locations: at the top of the gel, (actually at the bottom of the well) thus not even entering into the stacking gel

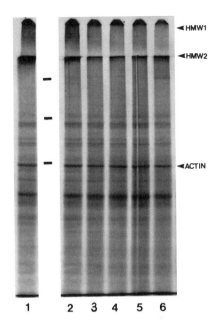

Figure 3. Protein patterns of mouse RIF-1 cells exposed to graded doses of diamide for 30 min at 37°C. Lane 1, 25 mM; lane 2, 10 mM; lane 3, 1 mM; lane 4, 0.5 mM; lane 6, 0 mM.

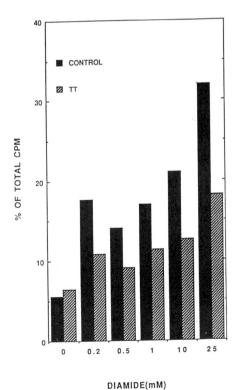

Figure 4. The amount of HMW as a function of diamide concentration. Thermotolerant (TT) cells produced less HMW than control cells.

(5% acrylamide), and another band that accumulates at the interface between the stacking and the resolving (7.5-10% acrylamide) gels. Figure 4 shows the amounts of HMW in these two bands as a function of diamide concentration. Quantification was carried out by counting the radioactivity of the two bands and expressing that as a fraction of the total counts in each lane. Clearly, the amount of HMW increased with increasing dose of drug.

Quantification of heat-induced protein denaturation

The next step was to see if this approach could detect the amounts of denatured proteins that remained in cells that had previously been heated. We performed an experiment in which we heated cells for various lengths of time at 45°C and then exposed one group of cells for 30 min to diamide (0.2 mM). The other half were processed without exposure to diamide. Results are depicted in Figure 5, where we show the amounts of HMW as function of duration of heating time. That pattern also shows the amounts of HMW complexes in heated cells not exposed to the sulfhydryl agent. Cells not exposed to diamide showed no clearly measurable changes in the amount of HMW. Those cells exposed to the sulfhydryl agent all contained increased amount of large molecules. Even the unheated controls contained more HMW than cells not exposed to diamide, consistent with the data presented in Figure 4. This is not surprising, because denaturation is a process that goes on at all temperatures, including 37°C. It is, however, greatly enhanced at elevated temperatures. Cells exposed to increasing durations at the elevated temperatures contained amounts of HMW that increased rapidly until the exposure time exceeded about 30 min. Increasing the duration of heating beyond that time no longer resulted in the appearance of increasing amounts of HMW. The times of heating over which the amounts of HMW did increase cover two important ranges: 1) Induction of thermotolerance, which in these cells takes up to about 20 min of exposure to 45°C, and 2) cell killing which is most pronounced in the interval of 20-40 min. By 45 min at 45°C, only one cell in 104 survives. In other words, our results show that only over the range that the cells show increases in content of HMW does induction of tolerance as well as cell killing occur.
The idea that induction of the heat shock response, induction of thermotolerance and cell killing have their origin in the same molecular events is not new and has considerable supporting data. For example, the Arrhenius plot for the three processes in these cells has considerable similarity, at least above 43°C, and each shows a marked "break" at 43°C (Hahn and Li, 1990). High molecular weight complexes have been implicated in induction of stress responses by agents other than heat (Lee and Hahn, 1988). The results shown here can therefore be considered additional evidence supporting the idea that protein denaturation in modest doses leads to the induction of HSPs and of thermotolerance. Excessive protein denaturation leads to a situation

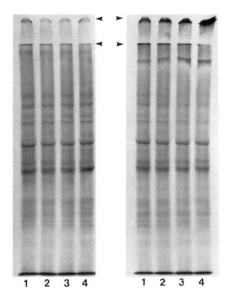

Figure 5. The increase of HMW as a function of heating time. Left panel shows a protein pattern of heat treated cells. Right panel shows a pattern of diamide treated cells followed by heating. Lane 1, control; lane 2, 10 min at 45°C; lane 3, 25 min at 45°C; lane 4, 45 min at 45°C.

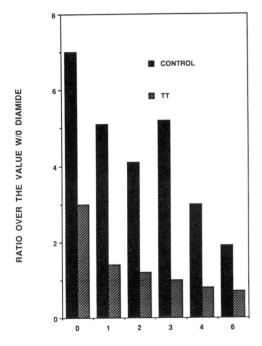

HR AFTER DIAMIDE(.2mM/30MIN)

Figure 6. Disappearance of HMW in HA-1 cells after diamide treatment.

when the cell presumably can no longer deal with the massive amounts of denatured proteins; it then dies.

HMW induction in normal and thermotolerant cells

If these ideas are correct, we would predict that thermotolerant cells should be more resistant to the formation of HMW than cells previously unheated. Therefore, we examined two features: the amount of HMW formed, and their rate of disappearance in tolerant and normal cells. Cells were exposed to diamide (0.2 mM) and proteins from these cells electrophoresed at various times thereafter. The radioactivity associated with the HMW was determined and expressed as a fraction of total counts in each lane. Results are shown in Figure 6. Two features are apparent: 1) the thermotolerant cells had less initial HMW formation, and 2) in both normal and tolerant cells, the amounts of HMW disappeared within a few hours, presumably because of internal degradation.

Conclusions

Our data show that sulfhydryl agents detect a lesion in heated cells; very likely this lesion represents denatured proteins. Excess of HSP70 efficiently prevents cell killing by diamide at elevated temperatures, suggesting that HSP70 protects binding sites of the sulfhydryl agents on the denatured proteins. A comparison of kinetics of induction and decay of HMW formation in normal and tolerant cells suggests that HSP70 acts primarily to prevent their formation, rather than aid in their dissociation and/or degradation.

References

Ananthan, J, Goldberg, AL and Voellmy, R, (1986) abnormal proteins serve as eukaryotic stress signals and trigger the activation of heat shock genes. Science, 232: 522-524.

Carlson, N, Rogers, S and Rechsteiner, M, (1987) Microinjection of ubiquitin: changes in protein degradation in Hela cells subjected to heat shock. J. Cell Biol., 104: 547-555.

Hahn, GM and Li, GC, (1990) Thermotolerance, thermoresistance and Thermosensitization. In Stress Proteins in Biology and Medicine, Morimoto, R, Tissieres, A and Georgopoulos, C, eds., Cold Spring Harbor Press New York, pp. 79-100.

Hightower, LE, (1980) Cultured animal cells exposed to amino acid analogues or puromycin rapidly synthesize several polypeptides. J. Cell Physiol., 102: 407-427.

Lee, K-J and Hahn, GM, (1988) Abnormal proteins as the trigger for the induction of stress responses: Heat, diamide and sodium arsenite. J. Cell Physiol., 136: 411-420.

Magun, BE and Fennie, CW, (1981) Effects of hyperthermia on binding, internalization, and degradation of epidermal growth factor. Radiat. Res., 86: 133-146.

Stable Expression of Human HSP70 Gene in Rodent Cells Confers Thermal Resistance

G.C. Li*, L. Li, R. Liu, J.Y. Mak, W. Lee
MCB-200, Radiation Oncology Research Laboratory,
Dept. of Rad. Oncol., University of California,
San Francisco, CA 94143
USA

Introduction

When exposed to a non lethal heat shock, a variety of organisms and cell lines acquire a transient resistance to one or more subsequent exposures at elevated temperatures (Gerner and Schneider, 1975; Henle and Leeper, 1976). This phenomenon has been termed thermotolerance (Henle and Dethlefsen, 1978). The mechanism for thermotolerance is not well understood. In mammalian systems, several studies suggest that heat shock proteins (HSPs) may be involved in the development of thermotolerance (Lindquist and Craig, 1988; Landry et al., 1982; Li and Werb, 1982; Subjeck et al., 1982). HSP70, a major heat shock protein, is synthesized by cells of many organisms in response to thermal or other environmental stresses (Lindquist and Craig, 1988; Landry et al., 1982; Li and Werb, 1982; Subjeck et al., 1982; Li, 1985; Lasglo and Li, 1985; Hahn and Li, 1990). It has been hypothesized that the transient induction of HSP70 may enable cells to recover from previous thermal stress, and/or to provide cells a transient degree of protection to subsequent heat challenge. A corollary of this hypothesis is that the overexpression of HSP70 in cells confers permanent heat resistance. We have initiated direct tests of the hypothesis that HSP70 protects cells from thermal stress by microinjecting human HSP70 into CHO cells or introducing a cloned human HSP70 gene into rat fibroblasts, and then evaluating the effect of human HSP70 gene expression on transient thermotolerance and permanent thermal resistance.

Microinjected purified HSP70 protects CHO cells from thermal stress

Purified human HSP70 was microinjected directly into CHO cells to

---

* Present address: Dept. of Med. Physics, Memorial Sloan-Kettering Cancer Center, 1275 York Ave., New York N.Y. 10021, USA

258

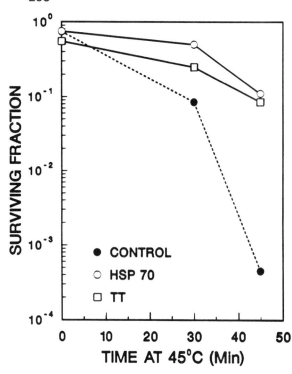

Figure 1: Microinjected HSP70 protects cells from thermal stress.
Chinese hamster HA-1 cells were microinjected with purified HSP70 (20 mg/ml), incubated at 37°C for 4 hr, and then heated at 45°C for 0-45 min. Cellular survivals were determined by colony formation. For comparison, the 45°C curvival for thermotolerant cells are also shown. Control HA-1 cells: •; HA-1 cells microinjected with HSP70: o; thermotolerant cells: ◻, cells were made thermotolerant by a 45°C, 15 min heat treatment followed by 16 hr incubation at 37°C. Data from Li (Li, 1989).

vary its intracellular concentration and the relationship between HSP70 and thermal resistance was examined. We clearly demonstrated that cells microinjected with HSP70 are more resistant to 45°C heat killing (Li, 1989). When the 45°C survivals between thermotolerant cells and cells microinjected with human hsp70 are compared, we found that cells having elevated levels of HSP70, either by mechanical microinjection, or by a 45°C, 15 min pre-heat treatment all become resistant to thermal stress (Fig. 1). Furthermore, there is a log-linear relationship between the survival after a 45°C, 45 min heat treatment and the concentration of HSP70 microinjected into HA-1 cells (Li, 1989).

Human *hsp70* can be stably expressed in transfected Rat-1 cells

A plausible alternative to microinjection is to place a cloned mammalian *hsp70* gene in an expression vector, and to transfect tissue culture cells with the construct. Transfected cells can then be examined for cellular level of HSP70 and thermal resistance. A 2.3 kb DNA fragment of the human *hsp70* was excised from pHHsp70, a plasmid containing the entire transcribed portion of the *hsp70* locus plus 5'-sequences required for heat inducible expression (Wu et al., 1985; Hunt and

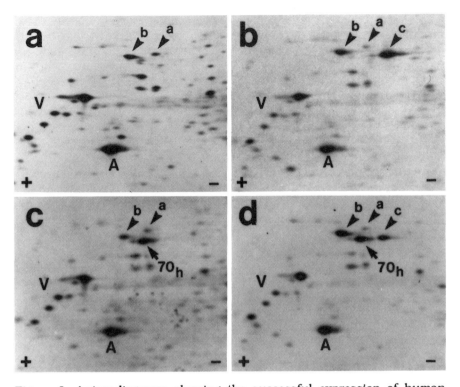

Figure 2. Autoradiograms showing the successful expression of human HSP70 protein in transfected rat HR-24 cells.
Monolayers of cells were heated at 45°C for 10 min, and then labelled with H³-leucine for 8 hr at 37°C. Cellular proteins from control unheated or heat-shocked cells were analyzed by two-dimensional gel electrophoresis and autoradiography. (a) Rat-I cells, at 37°C; (b) Rat-1 cells, 45°C for 10 min; (c) HR-24 cells, at 37°C; and (d) HR24 cells, 45°C for 10 min. The endogenous rat HSP70s are indicated by downward arrowhead. The human HSP70 expressed in HR-24 cells is indicated by arrow (70$_h$). Molecular weight is from top to bottom, isoelectric point is from left to right. A is actin and V is vimentin.

Morimoto, 1985), and subcloned into the polylinker region of pSVSP65. In the resulting plasmid, pSV-hsp70, transcription of the *hsp70* gene in eukaryotic cells is driven by the SV40 early promoter and enhancer, and ends with termination signals in the 3'-region of the *hsp70* locus. Exponential growing rat fibroblasts -Rat-1 cells were co-transfected with pSV-hsp70 and a plasmid containing the gene conferring neomycin resistance. Neomycin-resistant cells were selected in medium containing antibiotic G418 (400 μg/ml), pooled, grown to monolayers, and then subjected to six cycles of heating over a period of 60 days. Individual colonies were then isolated, trypsinized and grown to monolayers. All cell lines (HR-cells) used for this study were derived from individual colonies, and were designated as HR-21, HR-24, etc. The expression of human *hsp70* gene in transfected cells was first verified by Northern hybridization analysis. The identity and integrity of the human HSP70 protein expressed in transfected Rat-1 cells were

260

assessed subsequently by two-dimensional gel electrophoresis. The proteins extracted from Rat-1 cells cotransfected with pSV-hsp70 and a plasmid containing neomycin resistance gene (e.g., HR-24 cells), Rat-1 cells transfected only with neomycin resistance gene, human 293 cells and untransfected Rat-1 cells were analyzed, and the results are shown in Figure 2.

Our analysis of the pattern of HSP70 in Rat-1 cells utilizing two-dimensional gel electrophoresis reveals that the rat HSP70 identified in Rat-1 cells represents at least three major components of approximate molecular weight: 73 kD (rat HSP70-a), 72 kD (rat HSP70-b) and 70 kD (rat HSP70-c). Rat hsp70 b is found in cells under normal growth condition at 37°C, and its expression is enhanced after heat shock (compare Figs. 2a and 2b). Rat HSP70-c is not detectable in Rat-1 cells under normal growth condition at 37°C (Fig. 2a), and is induced upon heat shock (Fig. 2b). Rat HSP70-a is probably not a heat shock-inducible protein (see Figs. 2a and 2b).

When the protein synthesis profile of HR-24 cells was examined, a new protein with molecular weight around 70 kD (7°_h) was found under normal growth condition at 37°C (Fig. 2c), or after a 10 min heating at 45°C (Fig. 2d). This protein is clearly separated from the constitutive rat HSP70-b and the heat inducible rat HSP70-c in Rat-1 cells and its electrophoretic mobility and isoelectric point are identical to those of the human HSP70 in 293 cells (data not shown). All HR-cell lines

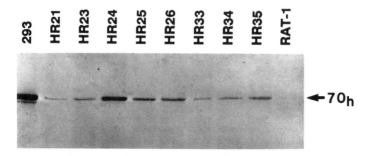

Figure 3. Western blot analysis of cellular proteins from Rat-1 cells and Rat-1 cells transfected with human hsp70 gene.
Cells, grown at 37°C, were lysed in NP-40 lysis buffer, omitting boiling in the presence of SDS and treatment of ß-mercaptoethanol. Equal amounts of cellular proteins were separated by gel electrophoresis, transferred to nitrocellulose membrane and probed with the Amersham monoclonal antibody against HSP70. 293: human 293 line; Rat-1: wild type Rat-1 cells. HR-21, 23, 24, 25, 26, 33, 34, 35: clonal lines derived from Rat-1 cells transfected with human hsp70 gene. It is clearly demonstrated that human hsp70 (indicated as 70_h) is expressed constitutively in the HR-cells, but not in wild type Rat-1 cells. Human 293 line, which constitutively expresses human hsp70 is shown for comparison. Note: McAb-HSP70 does not recognize the rat-HSP70 in control Rat-1 or HR-0 cells at 37°C.

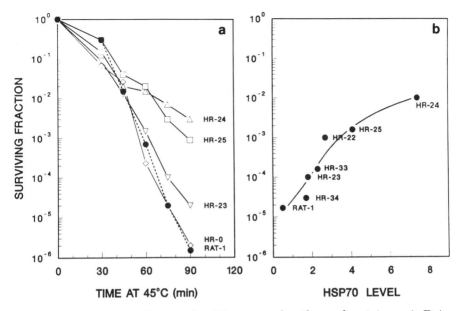

Figure 4. Expression of human *hsp70* gene confers thermal resistance to Rat-1 cells.
Panel A: Monolayers of exponentially growing HR-cells were exposed to 45°C for 0-90 min, and survivals were determined by colony formation assay. Each cell line expresses human *hsp70* in the following order: HR-24 (highest amount) > HR-25 > HR-23 (lowest amount). Rat-1: control untransfected Rat-1 cells. HR-0: Rat-1 cells transfected with only the neomycin resistance gene. Thermal survival for HR-0 cells is indistinguishable from that of the control untransfected Rat-1 cells. The thermal resistance seen in HR-cells is unlikely due to the transfection of a human gene per se. We have found that Rat-1 cells, transfected with human myc oncogene, singly or in combination with ras, were more sensitive to thermal stress (Li et al., 1990).
Panel B: Monolayers of exponentially growing HR-cells were exposed to 45°C for 75 min, and survival was determined by colony formation assay. In parallel experiments, the relative levels of human HSP70 expressed in HR-cells were measured by FCM using monoclonal antibody against HSP70. Thermal survival of various transfected cell line is plotted against the relative level of human *hsp70* in these cell lines. At least 20,000 cells were analyzed per data point for the FCM measurement. Experiments have been done at least twice and yield nearly identical results.

express human HSP70 at 37°C. The levels of human HSP70 protein in various clones are different. Their expression is stable for at least six months in culture. Introduction of the human *hsp70* gene into Rat-1 cells, and the heat selection procedures employed, did not affect the synthesis profiles of other endogenous rat heat shock proteins.

The expression of this human *hsp70* gene product was further confirmed by Western blot analysis using the Amersham monoclonal antibody originally prepared by Welch and Suhan (Welch and Suhan, 1986). Our results demonstrated immunologically that high levels of human HSP70 were expressed in the unstressed HR-cells at 37°C. As shown in Figure 3, the level of expression of human HSP70 protein in the various HR-cell lines follows the order: HR-24 (highest amount) ≥

HR-25, HR-26 > HR-35, > HR-23, HR-33, HR-34 > HR-21 (lowest amount).

Expression of human *hsp70* gene in Rat-1 cells confers thermal resistance

When monolayers of exponentially growing HR-cells were expose to 45°C and survivals were determined by colony formation assay, we found that cells expressing high level of human *hsp70* were more resistant to thermal stress. As shown in Figure 4a, when surviving fraction of HR-cells after a 45°C heat shock is plotted as a function of heating time at 45°C, HR-24 cells, in which the level of human *hsp70* is the highest, are most thermal resistant. The 45°C thermal survival for control HR-0 cells, the pooled population derived from Rat-1 cells transfected with only the neomycin resistance gene, and subjected to the identical heat selection procedure was indistinguishable from that of untransfected Rat-1 cells.

To evaluate the correlation between the level of human *hsp70* and thermal sensitivity, the survival of various transfected cell lines after a 45°C, 75 min heat shock treatment is plotted against their intracellular concentration of human *hsp70*. Results from one set of experiments are shown in Figure 4b. Here, monolayers of exponentially growing HR-cells were exposed to 45°C for 75 min, and survivals were determined. In parallel experiments, the relative levels of human *hsp70* expressed in HR-cells were measured by flow cytometric analysis using monoclonal antibody against human HSP70. Clearly, there appears to be a good correlation between the human *hsp70* level and thermal resistance. The higher the level of expression of human HSP70 protein, the more resistant the cells are to 45°C thermal stress.

Heat shock causes a transient import of cytoplasmic human HSP70 into the nucleus/nucleoli of transfected Rat-1 cells

The cellular localization of the human HSP70 in transfected Rat-1 cells at normal growth temperature or after heat shock were determined by indirect immunofluorescence assay using specific monoclonal antibodies against human HSP70. Figure 5 shows the intracellular distribution of human HSP70 in unheated HR-24 cells, and in heat shocked HR-24 cells at 0, 1, 2, 4, or 16 hr after a 45°C, 10 min heat treatment. At 37°C, human hsp70 is found exclusively in the cytoplasm (Fig. 5a). After a 10 min heating at 45°C, the protein rapidly moves into nucleus (Fig. 6b) and becomes associated with the nucleoli (Fig. 6 b-e). By 16 hr after the heat shock, however, most if not all of the protein has again left the nucleus/nucleoli, and the staining pattern of the cells reverts to that of unstressed HR-cells at 37°C (Fig. 5f). Control experiments with untransfected Rat-1 cells showed that the antibody did not react with the constitutively expressed rat-HSP70 under the identical immunostaining protocol (data not shown).

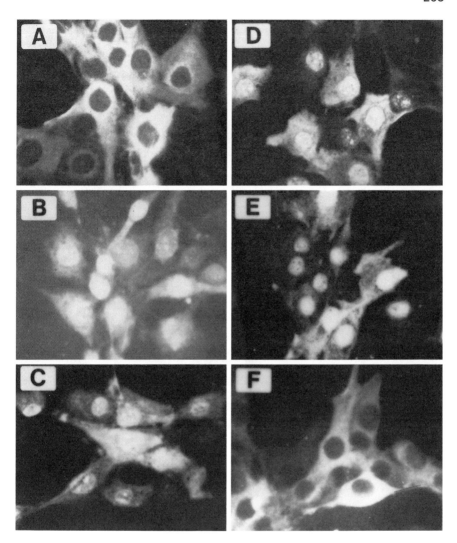

Figure 5. Intracellular distribution of the human HSP70 expressed in rat HR-24 cells before and after a 45°C, 10 min heat shock treatment.
HR-24 cells, grown on glass coverslips, were heated at 45°C for 10 min, and returned to 37°C for 0, 1, 2, 4 or 16 hr. After the indicated recovery times, the cells were fixed and stained with monoclonal antibody against human hsp70. The distribution of human HSP70 was analyzed by indirect immunofluorescence. (A) Cells at 37°C; (B) cells heat-shock treated at 45°C, no recovery; (C) cells heat-shock treated and recovered at 37°C for 1 hr; (D) cells heat-shock treated and recovered at 37°C for 2 hr; (E) cells heat-shock treated and recovered at 37°C for 4 hr; and (F) cells heat-shock treated and recovered at 37°C for 16 hr.
It is shown that at 37°C, human HSP70 is found exclusively in the cytoplasm (A). After 45°C, 10 min heating, it is rapidly moved into nucleus and becomes associated with the nucleoli (B-E). By 16 hr after the heat shock, the protein has mostly left the nucleus/nucleoli, and the staining pattern reverts to that of unstressed HR-cells at 37°C (F).

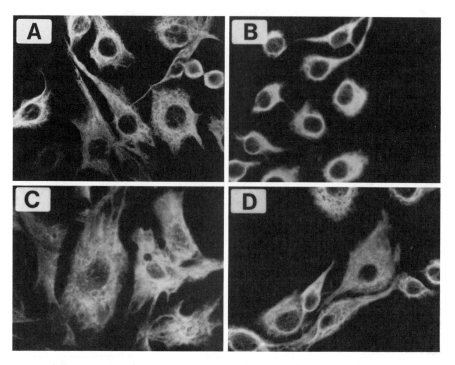

Figure 6. Protection of intermediate filament integrity in HR-cells after a 45°C, 30 min heat shock treatment.
Rat-1 cells or HR-24 cells, growing on glass coverslips, were heat shocked at 45°C for 30 min. The cells were fixed immediately afterwards and analyzed for the distribution of the vimentin-containing intermediate filaments using a monoclonal antibody specific for vimentin. (A) Rat-1 cells, 37°C; (B) Rat-1 cells, 45°C for 30 min; (C) HR-24 cells, 37°C; (D) -HR-24 cells, 45°C for 30 min. Comparing (A) and (B), it is clearly shown that the vimentin-containing intermediate filament have redistributed from their normal well-shaped cytoplasmic array (A) and collapsed in and around the nucleus of Rat-1 cells after a 45°C, 30 min heat shock treatment (B). In contrast, HR-24 cells exhibited little or no collapse of the intermediate filaments after an identical 45°C, 30 min heat shock treatment (compare C and D).

The effects of heat shock on intermediate filament integrity

It has been shown that one of the early events after heat shock is the collapse of the vimentin-containing intermediate filament network (Welch and Mizzen, 1988; Welch et al., 1985). For example, in rat fibroblasts after a 43°C, 90 min heat treatment, the intermediate filaments rapidly redistribute from their normal well-spread cytoplasmic array into a tight cage around the nucleus. In contrast, cells first made thermotolerant exhibited little or no collapse of the intermediate filaments after an identical 43°C, 90 min heat shock treatment.

Using a similar approach, we examined whether intermediate filament integrity would be protected after heat stress in HR-cells overexpressing human HSP70. Monolayers of Rat-1 cells or HR-24 cells were first exposed to 45°C, 30 min heat shock treatment, the cells were then fixed and analyzed for the distribution of the vimentin-containing intermediate filaments using a monoclonal antibody specific for vimentin. Figure 6 shows the distribution of vimentin-containing intermediate filament in Rat-1 cells and HR-24 cells at 37°C or immediately after a 45°C, 30 min heat shock. In Rat-1 cells, after the 45°C heat treatment, the intermediate filaments rapidly redistribute themselves into a tight cage around the nucleus, a finding consistent with that previously reported by Welch and Mizzen (Welch and Mizzen, 1988). In contrast to Rat-1 cells, HR-cells exhibit little or no collapse of the intermediate filament after the 45°C, 30 min heat shock. Expression of human HSP70 appears to protect the intermediate filament integrity in HR-cells after thermal stress.

Conclusions

The major heat shock protein, HSP70, is synthesized by cells of a wide variety of organisms in response to heat shock or other environmental stresses and is assumed to play an important role in protecting cells from thermal stress (Pelham, 1984). We have tested the validity of this hypothesis directly by transfecting a constitutively expressed recombinant human *hsp70* gene into rat fibroblasts and examining the relationship between the levels of human HSP70 expressed and thermal resistance of the stably transfected rat cells. Successful transfection and expression of the human *hsp70* gene were characterized by Northern hybridization analysis. The human HSP70 gene product was further identified by two-dimensional gel electrophoresis and Western blot analysis.

Using monoclonal antibodies against HSP70 protein, we have examined the distribution of human HSP70 expressed from a single cloned gene in stably transfected rat cell line. We found that at 37°C, this protein is mostly cytoplasmic. Immediately after a 45°C, 10 min heat shock, a substantial fraction of the protein rapidly moved into the nucleus and became transiently associated with nucleoli. By 16 hr after the heat shock, the protein had mostly left the nucleus, and intracellular distribution of HSP70 reverted to that of unstressed HR-cells at 37°C. Thus, in response to a heat shock treatment, the intracellular distribution of this constitutively expressed human HSP70 in Rat-1 cells is similar to that of the endogenous rat HSP70 proteins (Welch and Mizzen, 1988; Welch et al., 1985).

When individual cloned cell lines were exposed to 45°C and their thermal survivals were determined by colony formation assay, we found that the expression of this human HSP70 protein confers heat resistance to the rat cells. In addition, the overexpression of human HSP70 protein appears to protect the intermediate filament integrity in the stably transfected rat cells after heat shock treatment. Our results provide strong evidence that HSP70, directly or indirectly, protects cells from thermal stress.

Munro and Pelham (Munro and Pelham, 1984) and Milarski and Morimoto (Milarski and Morimoto, 1989) have successfully transfected mammalian cells with *hsp70* genes. However, the transfection and expression were transient, and may not adequately give a complete and accurate picture of *hsp70* in an equilibrium condition. Stably transfected cell lines that express human *hsp70* should provide a steady-state condition for studying cellular targeting and physiological functions of HSP70 at 37°C, after heat shock or other environmental stresses, or during the development of thermotolerance. In addition, the use of stable cell lines should eliminate the limitations of transient transfection experiments that are imposed by the narrow time window. Detailed thermal survival studies, studies on the kinetics of thermotolerance development, and nucleus/nucleolar localization, etc., can be performed repeatedly and more reproducibly for the same clonal derivatives. These studies should provide further insights on the mechanisms of thermal killing, thermal resistance and transient thermotolerance of mammalian cells.

Acknowledgments

We thank R. Morimoto for providing the plasmid pHHsp70 containing the human *hsp70* gene fragment, P. Krechmer for typing the manuscript. This work was supported in part by Grant CA-31397 and CA-53788 from the National Cancer Institute, NIH, Department of Health and Human Service, USA.

References

Gerner, EW and Schneider, MJ, (1975) Induced thermal resistance to Hela cells. Nature (London), 256: 500-502.

Hahn, GM and Li, GC, (1990) Thermotolerance, thermoresistance and thermosensitization. In "Stress Proteins in Biology and Medicine", Morimoto, R, Tissieres, A and Georgopoulos, C, eds., Cold Spring Harbor Press, N.Y., 79-100.

Henle, KJ and Leeper, DB, (1976) Interaction of hyperthermia and radiation in CHO cells: recovery kinetics. Radiation Res., 66: 505-518.

Henle, KJ and Dethlefsen, LA, (1978) Heat fractionation and thermotolerance: A review. Cancer Res., 38: 1843-1851.

Hunt, C and Morimoto, RI, (1985) Conserved features of eukaryotic *hsp70* gene revealed by comparison with the nucleotide sequence of human hsp70. Proc. Natl. Acad. Sci. USA, 82: 6455-6459.

Landry, S, Bremier, D, Chretien, P, Nicole, LM, Tanquay, RM and Marceau, N, (1982) Synthesis and degradation of heat shock proteins during development and decay of thermotolerance. Cancer Res., 42: 2457-2461.

Laszlo, A and Li, GC, (1985) Heat resistant variants of Chinese hamster fibroblasts altered in heat shock protein expression. Proc. Natl. Acad. Sci. USA, 82: 8029-8033.

Li, GC and Werb, Z, (1982) Correlation between synthesis of heat shock proteins and development of thermotolerance in Chinese hamster fibroblast. Proc. Natl. Acad. Sci. USA, 79: 3219-3222.

Li, GC, (1985) Elevated levels of 70,000 dalton heat shock protein in transiently thermotolerant Chinese hamster fibroblasts and in their stable heat resistant variants. Intl. J. Rad. Onc. Biol. Phys., 11: 165-177.

Li, GC, (1989) Hsp70 as an indicator of thermotolerance. In "Hyperthermic Oncology 1988". Sugahara, T and Saito, M, eds., Taylor & Francis, London, Vol. 2, pp. 256-259.

Li, GC, Ling, CC, Endlich, B and Mak, JY, (1990) Thermal response of oncogene-transfected rat cells. Cancer Res., 50: 4515-4521.

Lindquist, S and Craig, EA, (1988) The heat shock proteins. Ann. Rev. Gen., 22: 631-677.

Milarski, KL and Morimoto, RI, (1989) Mutation analysis of the human hsp70 protein: Distinct domains for nucleolar localization and ATP-binding. J. Cell Biol., 109: 1947-1962.

Munro, S and Pelham, HRB, (1984) Use of peptide tagging to detect proteins expressed from cloned genes: Deletion mapping functional domains of Drosophila hsp70. EMBO J., 3: 3087-3093.

Pelham, HRB, (1984) Hsp70 accelerates the recovery of nucleolar morphology after heat shock. EMBO J., 3: 3095-3100.

Subjeck, JR, Sciandra, JJ and Johnson, RJ, (1982) Heat shock proteins and thermotolerance: Comparison of induction kinetics. Brit. J. Radiol., 55: 579-584.

Welch, WJ and Suhan, JP, (1986) Cellular and biochemical events in mammalian cells during and after recovery from physiological stress. J. Cell. Biol., 103: 2035-2052.

Welch, WJ and Mizzen, LA, (1988) Characterization of thermotolerant cells. II. Effects on the intracellular distribution of heat shock protein 70, intermediate filaments and small ribonucleoprotein complexes. J. Cell Biol., 106: 1117-1130.

Welch, WJ, Feramisco, JR and Blose, SH, (1985) The mammalian stress response and the cytoskeleton: Alterations in intermediate filaments. Ann. N. Y. Acad. Sci., 455: 57-67.

Wu, B, Hunt, C and Morimoto, RI, (1985) Structure and expression of human gene encoding major heat shock protein hsp70. Mol. Cell. Biol., 5: 330-341.

The Role of HSP90 in Tumor Specific Immunity

S.J. Ullrich, S.K. Moore, L.W. Law, W.D. Vieira,
K. Sakaguchi, E. Appella
National Institutes of Health
National Cancer Institute
Laboratory of Cell Biology
Bethesda, MD 20892
USA

Introduction

Tumors induced by chemical carcinogens usually express tumor specific transplantation antigens (TSTAs) (Law et al., 1980; Law, 1985). These antigens are defined by their ability to induce tumor specific immunity when syngeneic hosts are preimmunized with irradiated tumor cells or cellular fractions containing the tumor specific antigens. An important characteristic of TSTAs, is that upon challenge with tumors the induced immunity is confined to the original tumor and not other independently derived syngeneic tumors. The immune response to these antigens is, for the most part, limited to a cellular immune response. However, under certain conditions of hyperimmunization, tumor specific cytotoxic sera have been developed in syngeneic hosts (De Leo et al., 1978). The ability of these proteins to act as antigens is probably due to a molecular change(s) in the protein, such as a missense mutation or altered post translational processing of the tumor antigen. Other mechanisms could potentially induce immunogenicity, such as expression of a normally silent gene or an alternative splicing mechanism. However, it is difficult to envision how these latter mechanisms could generate the extensive diversity of TSTAs. The goal of the research discussed here was to purify and identify these TSTAs and to determine the molecular change which rendered these proteins immunogenic and tumor specific.

The 3-methylcholanthrene-induced fibrosarcoma, Meth A, was chosen for the purification of TSTAs, since it neither expresses alien class I molecules nor MuLV gene products, gp70 or p30, at the cell surface (Law et al., 1980; De Leo et al., 1977). The majority of the TSTA activity was found to reside in the cytosol (Du Bois et al., 1982). TSTA activity was purified from the cytosol by sequential fractionation over hexylamine agarose, Sepharose S300, hydroxylapatite and Mono Q columns to give an apparently homogeneous protein (Ullrich et al., 1986). We found that the purified TSTA actually consisted of two closely related proteins of 84 and 86 kDa. Furthermore, these two proteins were identified as the murine equivalent of the 90 kDa heat shock proteins (HSP90) (Ullrich et al., 1986; Moore et al., 1989) and referred to are hereafter as HSP84 and 86.

Biochemical and immunological analysis of HSP84 and HSP86

The expression of HSP84/86 was found to be elevated 2 to 3 fold in Meth A tumors compared to nontransformed fibroblasts when measured at the mRNA and steady state protein levels and at the rate of protein synthesis (Ullrich et al., 1989). Immunofluorescence and immune electron microscopic analyses revealed that HSP84/86 is found mainly in the cytoplasm (Ullrich et al., 1986). Furthermore, none of the HSP84/86 was found inside any intracellular membrane vesicles in these cells, suggesting that these proteins are not normal secretory proteins. Surprisingly, HSP84/86 was also observed at the cell surface in both nontransformed and transformed cells, even though, neither HSP84 nor 86 contains a signal peptide sequence (Moore et al., 1987; Moore et al., 1989). Cell surface radiolabeling experiments indicate that a significant amount of the cell surface localized HSP84/86 occurs as intact molecule (data not shown). Besides HSP84/86, two other stress proteins, HSC70 and GRP94, are present at the cell surface. One possible mechanism whereby these normally cytoplasmic proteins become associated with the cell surface is by cell lysis followed by binding to the plasma membrane of adjacent cells. This hypothesis is supported by the observation that HSP84/86 is present *in vivo* in the ascites fluid of Meth A tumors and by the fact that ~5% of the *in vivo* Meth A ascites cells are lysed (unpublished observations). Also, the ascites fluid surrounding the tumor cells *in vivo* was found to have extremely low levels of glucose ($\leq 10 \mu$ M), conditions under which Meth A cells are easily lysed *in vitro* (unpublished observation). The significance of the cell surface and extracellular localization of HSP84/86 with respect to immune recognition of these proteins is unclear; however they may act to prime the host's immune system.

To further explore the intrinsic immunogenicity of HSP84/86, the ability of HSP84/86 purified from three other syngeneic tumors to act as a TSTA was examined. Purified preparations of HSP84/86 from two methyl-cholanthrene induced tumors, CI-4 (Du Bois et al., 1982) and CII-7 (Du Bois and Law, 1986), and an SV40-induced tumor, mKSA (Du Bois et al., 1984), also were found to possess TSTA activity. The immune response to these tumors was tumor specific; no cross protection was observed against each other or other syngeneic tumors. However, biochemical analysis of these proteins by 1D and 2D SDS/ PAGE analysis did not indicate any gross differences in their properties from those of Meth A or normal cells. Since the amino acid sequences of HSP84/86 are highly conserved (Moore et al., 1989), one possible explanation for the observation that TSTA activity is associated with HSP90 from several tumors is that the actual antigen is not HSP84/

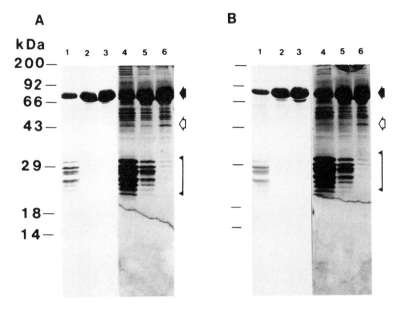

Figure 1. SDS/PAGE analysis of Mono Q purified HSP84/86. HSP84/86 was isolated from Meth A cells and subjected to anion exchange chromatography. The Mono Q peak was divided into three pools: pool A (lanes 1,4), B (lanes 2,5) and C (lanes 3,6). Pools A-C were subject to 10% SDS/PAGE and stained with coomassie blue (lanes 1-3) or silver stain reagent (lanes 4-6). A, HSP84/86 pools as isolated and B, after incubation for 1 hr at 37°C. Intact HSP84/86 is indicated by closed arrowhead.

86 *per se*, but rather a protein or peptide which is bound to or copurifies with HSP84/86. Since, HSP70 has recently been found to bind peptides (Flynn et al., 1989), perhaps HSP84/86 also binds peptides, one or more of which are responsible for TSTA activity.

If a subpopulation of HSP84 and/or 86 was complexed with other proteins or peptides, then its retention time on a Mono Q column may be different from that of uncomplexed or free HSP84/86. In order to investigate this possibility, the HSP84/86 peak from the Mono Q column was subdivided into three pools, A, B and C, in order of their relative elution from the column. In the previous reported study (Ullrich et al., 1986), TSTA activity was determined with Mono Q fractions consisting of HSP84/86 pools B and C, but not pool A, as it contained lower M_r material. Ten µg of each of the individual pools were analyzed by SDS/PAGE and stained with coomassie blue (Fig. 1A, lanes 1-3); pool A was found to be the least pure and contained multiple protein species between 20-30 kDa, whereas pool C appeared to contain only HSP84/86. When these pools were analyzed for TSTA activity, pool C contained the majority of TSTA activity with pool B displaying minimal but statistically significant activity (representative results are shown in Table 1). To further assess for the presence of

Table 1				
Immunogenicity of Mono Q Fractions of HSP84/86				
Fraction	Dose µg	Avg. Tumor Diameter (mm)*	Tumor Incidence	*p* value
A	2.5 20.	15.7 13.6	5/5 4/5	
B	2.5 20.	13.0 ± 3.1 7.5 ± 2.0[1]	4/5 4/5	<0.05
C	2.5 20.	13.7 ± 1.4 2.4 ± 2.6[1]	5/5 1/5	<0.001
none	—	14.6 ± 2.2	14/15	

* 20 days post intradermal challenge with 5×10^4 Meth A ascites cells.
[1] statistically significant difference from control immunizations.

impurities in the different fractions, all three pools were subjected to SDS/PAGE analysis followed by silver staining (Fig. 1A). In addition to the group of proteins in the 20 to 30 kDa range, all three pools had multiple protein species in the range of 40 to 80 kDa (Fig. 1A, lanes 4-6). Thus, all three pools contained protein species other than intact HSP84/86, however, due to the high sensitivity of silver staining, each of these species actually represents a relatively minor impurity (ng quantities). Comparison of the 40 to 80 kDa species among pools A-C revealed that all of the species were present in essentially equal amounts, except for one species of ~50 kDa which was enriched in pool C (Fig. 1A, open arrow head). The presence of peptides bound to HSP84/86 and having M_r below 10 kDa was assessed by Tricine gels (Schagger and von Jagow, 1987). Silver staining revealed that there were no peptides bound to HSP84/86 below 10 kDa; the limit of sensitivity of this assay was 10 ng per peptide band or 1 part per 1000 (data not shown). Thus, Mono Q purified HSP84/86 does not contain detectible peptide species.

The presence of various proteins with Mrs lower than HSP84/86 in pools A-C suggested that either HSP84/86 contains bound proteins or that these lower M_r species represent degradation products of HSP84/86. In order to ascertain whether these lower M_r species represent

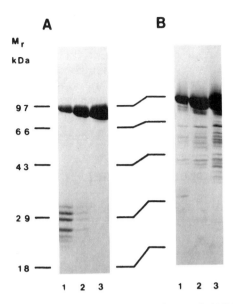

Figure 2. Western blot analysis of HSP84/86. HSP84/86 Mono Q pools A-C, lanes 1-3, respectively, were subjected to 10% SDS/PAGE analysis followed by western blot analysis. A, coomassie blue stained gel, B, western blot developed with affinity-pure rabbit anti-HSP84/86.

proteolytic degradation products, Western blot analysis was performed using affinity purified HSP84/86 antisera (affinity antisera was prepared by immunoadsorption onto the region of a western blot corresponding to intact HSP84/86). The results indicate that the 40-80 kDa species correspond to breakdown products of HSP84/86, including the ~50 kDa band enriched for in pool C (Fig. 2B). This finding was also confirmed using an HSP84/86 C-terminal peptide antiserum (data not shown). In order to see if proteolytic activity is present in purified HSP84/86 fractions, pools A-C were incubated at 37°C for 1 hr. A limited degree of proteolytic degradation of intact HSP84/86 was found in pools B and C, resulting in partial digestion of HSP84/86 to a fragment of ~80 kDa (Fig. 1B, lanes 2,3,5 and 6). The proteolytic activity present in these fractions also appeared to degrade only a subfraction of the total intact HSP84/86 present, since prolonged incubation resulted in only a partial degradation of the intact HSP 84/86 (data not shown).

Interestingly, the only other proteins which showed differential distribution across the Mono Q fractions were a group of proteins between 20 to 30 kDa which were highly enriched in pool A (Fig. 1, see bracketed region). One band (~30 kDa) in this region gave a weak reaction with the affinity pure anti-HSP84/86 sera, whereas, all the other bands were negative (Fig. 2B). When this band was subjected to N-terminal sequence analysis, its sequence did not correspond to either HSP84 or 86. Thus, this group of proteins appears to be unrelated in its primary sequence to HSP84/86. This group of proteins displayed a pattern similar to that of the proteasome, an intracellular multicatalytic proteinase complex (Arrigo et al., 1988). In fact, the N-terminal sequence of one band in this region (data not shown) was found to correspond to the sequence of the C-5 proteasome component (Tokunaga et al., 1990). Thus, the protease activity present in purified HSP84/86 may be due to the copurification of the proteasome with HSP84/86.

Molecular and genetic analysis of *hsp84* and *hsp86*

In order to determine the actual number of *hsp84* and *86* genes and their chromosomal localization in normal BALB/c mice, chromosomal analysis was performed using hamster/mouse somatic cell hybrids. Southern blot analysis showed the presence of at least five *hsp84*-related genes and three *hsp86*-related genes (Moore et al., 1987; Moore et al., 1989). Sequence analysis of *hsp84*-related genomic clones revealed that there is actually one active gene and at least six pseudogenes (Moore et al., 1990). Analysis of *hsp86*-related genomic clones indicated that there is apparently only one active gene and two or more pseudogenes (Moore et al., 1989 and unpublished observations). The active gene for *Hsp84* was assigned to chromosome 17 near the MHC locus (Romano et al., 1989). Recently, the active *hsp86* gene was mapped to chromosome 12 near the *IgH* loci (Moore and Kozak, unpublished observations). Karyotype analysis of Meth A tumors showed that it is subtetraploid with three copies of chromosome 17 and two normal copies of chromosome 12, as well as a translocation of chromosome 12 to both chromosome 16 and the X chromosome (Pravtcheva et al., 1981). The X^{12} chromosome contained a nearly complete copy of chromosome 12. Thus, Meth A potentially may have and express multiple active copies of both *hsp84* and *86*.

The above biochemical analysis of HSP84/86 protein purified from the Meth A tumor suggested that either HSP84 and/or 86 was the actual immunogen. In order to determine whether the *hsp84* or *86* genes contained any mutations which might render them immunogenic, cDNA clones corresponding to both *hsp84* and *86* were isolated from a Meth A cDNA library (Moore et al., 1987; Moore et al., 1987) and compared to the wild-type sequences of syngeneic *hsp84* and *86* (Moore et al., 1990 and unpublished results). No differences in the nucleotide sequence was found in the Meth A clones which would lead to a change in the amino acid sequence of either *hsp84* or *86*. However, since karyotype analysis of Meth A tumors indicated that there are multiple copies of the chromosomes on which the active *hsp84/86* genes are located, we cannot rule out the possibility that potential mutant *hsp84* or *86* alleles exist in the Meth A genome but were not found in our screening of the Meth A cDNA library.

Immunization of BALB/c mice with hamster/Meth A somatic cell hybrids followed by subsequent challenge with Meth A cells indicated that a TSTA activity was associated with those hybrids containing the Meth A X12 chromosome and not those containing the Meth A X chromosome (Law, 1984). Moreover, serological analysis of these chromosomal hybrids using an anti-Meth A cytotoxic sera (unable to immunoprecipitate antigen) also indicated that the Meth A cytotoxic antigen was apparently encoded on the X^{12} chromosome and closely linked to the *IgH* cluster (Pravtcheva et al., 1981). These data are consistent with the notion that Meth A TSTA activity could be encoded by the *hsp86* allele present on the X^{12} chromosome.

Conclusions

Our results indicate that Mono Q purified HSP84/86 can be subdivided into immunogenic (pool C) and non-immunogenic fractions (pool A). The non-immunogenic pool was found to be enriched for a mutlicatalytic protease or proteasome. This finding suggests that an important function of HSP84/86 may be the binding of the proteasome, as these proteins appear to be the only proteins which copurify with HSP84/86. In fact, some purified proteasome preparations have been found to contain a 90 kDa component (Schmid et al., 1984). Our biochemical data indicated that pool C contained essentially pure HSP84/86. If the actual TSTA present in pool C represents a copurified component other than HSP84/86, then it must be present at a concentration of less than 0.1%. The protease activity associated with purified HSP84/86 appeared to digest only a subfraction of the total protein. One reason for this may be that only denatured and/or mutant HSP84/86 is sensitive to digestion. Also, the presence of protease activity may explain the differential TSTA activity of the various pools. If the protease activity preferentially destroys the antigenic epitope of the TSTA, one would expect pool A to have little or no activity due to its higher concentration of proteasomes. Whereas, pool C would display the most TSTA activity. This is exactly what was observed. Alternatively, pool C may be enriched in TSTA activity by virtue of intrinsic chromatographic property of the antigen itself. Whatever the case may be, our data suggest that a subpopulation of Meth A HSP84/86 contains a change rendering it immunogenic and perhaps rendering it more protease sensitive, although, the molecular and genetic analysis of hsp84/86 has not yet revealed the presence of any mutation in these genes. Moreover, the results of TSTA experiments using hamster/Meth A somatic hybrids are consistent with the notion that one of the *Hsp86* alleles present on the X^{12} chromosome may code for the TSTA. This possibility can be tested by analysis of X^{12} hybrids that have lost the *Hsp86* gene and determining whether TSTA activity is also lost. Finally, what is the general mechanism as to why HSP84/86 appear to be common targets in the generation of TSTAs among different tumors? One possibility is that since heat shock genes are in an open transcriptional state which is known to render DNA more susceptible to mutagenic damage, they may be easily mutated.

References

Arrigo, A-P, Tanaka, K, Goldberg, AL and Welch, WJ, (1988) Identity of the 19S 'prosome' particle with the large multifunctional protease of mammalian cells (the proteasome). Nature, 331:192-194.

De Leo, AB, Shiku, H, Takahashi, T, John, M and Old, LJ, (1977) Cell surface antigens of chemically induced sarcomas of the mouse. J. Exp. Med., 146: 720-734.

De Leo, AB, Shiku, H, Takahashi, T and Old, LJ, (1978) In Biological Markers of Neoplasia: Basic and Applied Aspects. Ruddon, RW, ed, pg. 25-34, Elsevier, Amsterdam.

Du Bois, GC, Appella, E and Law, LW, (1984) Isolation of a tumor associated transplantation antigen from an SV40-induced sarcoma. Intl. J. Cancer., 34: 561-566.

Du Bois, GC and Law, LW, (1986) Biochemical characterization and biologic activities of 82- and 86 kDa tumor antigens isolated from a methylcholanthrene-induced sarcoma, CII-7. Intl. J. Cancer., 37: 925-931.

Du Bois, GC, Law, LW and Appella, E, (1982) Purification and biochemical properties of tumor-associated transplantation antigens from methylcholanthrene- induced murine sarcomas. Proc. Natl. Acad. Sci. USA, 79: 7669-7673.

Flynn, GC, Chappell, TG and Rothman, JE, (1989) Peptide binding and release by proteins implicated as catalysts of protein assembly. Science, 245: 385-390.

Law, LW, (1984) Assignment of the gene encoding for Meth A tumor rejection antigen (TATA) to chromosome 12 of the mouse. Br. J. Cancer, 50: 109-111.

Law, LW, (1985) Characteristics of tumour-specific antigens. Cancer Surveys, 4: 3-19.

Law, LW, Rogers, MJ and Appella, E, (1980) Tumor antigens on neoplasms induced by DNA- and RNA-containing viruses: Properties of the Solubilized antigens. Adv. Cancer Res., 32: 201-235.

Moore, SK, Kozak, C, Robinson, EA, Ullrich, SJ and Appella, E, (1987) Cloning, nucleotide sequence and chromosome assignment of cDNA coding for murine hsp84 isoform. Gene, 56: 29-40.

Moore, SK, Kozak, C, Robinson, EA, Ullrich, SJ and Appella, E, (1989) Murine 86- and 84- kDa Heat Shock Proteins, cDNA sequences, chromosomal assignments, and evolutionary origins. J. Biol. Chem., 264: 5343-5351.

Moore, SK, Rijli, F and Appella, E, (1990) Characterization of the mouse 84-kD heat shock protein gene family. DNA and Cell Biol., 9: 387-400.

Pravtcheva, DM, De Leo, A, Ruddle, FH and Old, LJ, (1981) Chromosomal assignments of the tumor-specific antigen of a methylcholanthrene-induced mouse sarcoma. J. Exp. Med., 154: 964-977.

Romano, JW, Seldin, MF and Appella, E, (1989) Linkage of the mouse Hsp84 heat shock protein structural gene to the H-2 complex. Immunogenetics, 29: 142-144.

Schagger, H and Von Jagow, G, (1987) Tricine-sodium dodecyl sulfate-polyacrylamide gel electrophoresis for the separation of proteins in the range from 1 to 100 kDa. Analyt. Biochem., 166: 368-379.

Schmid, HP, Akhayat, O, Martins De Sa, C, Puvion, F, Koehler, K and Scherrer, K, (1984) The prosome: an ubiquitous morphologically distinct RNP particle associated with repressed mRNPs and containing specific ScRNA and a characteristic set of proteins. EMBO J., 3: 29-34.

Tokunaga, F, Aruga, R, Iwanaga, S, Tanaka, K, Ichihara, A, Takao, T and Shimonishi, Y, (1990) The NH_2-terminal residues of rat liver proteasome subunits, C2, C3 and C8, are not N α-acetylated. FEBS Lett., 263: 373-375.

Ullrich, SJ, Robinson, EA, Law, LW, Willingham, M and Appella, E, (1986) A mouse tumor-specific transplantation antigen is a heat shock-related protein. Proc. Natl. Acad. Sci. USA, 83: 3121-3125.

Ullrich, SJ, Moore, SK and Appella, E, (1989) Translational and transcriptional analysis of the murine 84- and 86- kDa heat shock protein. J. Biol. Chem., 264: 6810-6816.

Heat Shock and Oxidative Injury in Human Cells

B.S. Polla, N. Mili, S. Kantengwa
Allergy Unit
University Hospital
CH 1211 Geneva 4
Switzerland

Introduction

For the last few years, we have been interested in the interactions between oxygen free radicals (OFR) and the heat shock, or stress response. Oxidative injury participates in a variety of pathological conditions such as inflammation and ischemia. During inflammation, OFR are generated by the phagocytic cells (polymorphonuclear leukocytes, monocytes-macrophages) infiltrating the inflamed tissues, whereas after ischemia and during reperfusion, OFR are generated by a xanthine-xanthine oxidase system. *In vitro*, we have investigated in human monocytes-macrophages the effects of exogenous addition, as well as of endogenous production (during phagocytosis) of OFR. These cells respond to oxidative injury by the synthesis of the classical heat shock proteins (HSPs) and of oxidation-specific stress proteins such as heme oxygenase or superoxide dismutase (SOD). HSPs may thus be part of a physiological response to injury during inflammation. In animal models, HSPs are induced during reperfusion injury in the ischemic organs.

The simultaneous induction, under certain conditions, of heat shock and oxidation-specific stress proteins suggests that they are part of the same system. They may however be induced by distinct mechanisms and we have shown that in monocytes-macrophages calcium is neither required nor necessary for the induction of HSPs, whereas in the context of oxidative injury, calcium enters the cells and participates to cell death. Preexposure of the human premonocytic line U937 to heat shock partially protects these cells from subsequent exposure to oxidative stress such as hydrogen peroxide (H_2O_2) or the cytotoxic agent bleomycin. Mechanisms by which heat shock/HSPs may protect cells or tissues from oxidative injury are investigated in these models and include inhibition of phospholipase A_2, maintenance of ATP levels, prevention from uncoupling of oxidative phosphorylation and protection from protein aggregation or DNA damage.

In this review, we will try to address the following issues:
i. What mediates the induction of a stress response during OFR production?

ii. Can HSPs function to protect cells from oxidative injury?
iii. If yes, what are the mechanisms for such a protective effect?

Our work in the heat shock system has been centered on human phagocytes. Monocytes-macrophages and polymorphonuclear cells are devoted to phagocytosis and during the phagocytic process, generate high amounts of OFR. Molecular oxygen is converted to superoxide anion (O_2^-) by the phagocyte-specific respiratory burst enzyme NADPH oxidase, either in association with phagocytosis or after stimulation with non-particulate stimuli such as phorbol esters (PMA). Production of OFR is part of the normal physiology of monocytes-macrophages; OFR play an important role in the host's antibacterial defenses, as examplified by chronic granulomatous disease (CGD), a condition in which NADPH oxidase is non-functional (Curnutte et al., 1974). On the other hand, OFR can participate to tissue injury, by causing direct damage or by amplifying inflammation (Fantone and Ward, 1982). An increased production of OFR has been suggested to play a role in a variety of experimental conditions and human diseases (Halliwell, 1987). The *extracellular* generation of free radicals by phagocytes has been involved in inflammation (rheumatoid arthritis, the adult respiratory distress syndrome, pulmonary inflammation associated with exposure to mineral dusts or to the antitumor agent bleomycin) and the *intracellular* generation of free radicals (via xanthine-xanthine oxidase) in ischemia and reperfusion injury (Fantone and Ward, 1982; Lunec et al., 1981).

The toxicity of OFR can be modulated by endogenous protective mechanisms such as free radical scavengers, by drugs and eventually by the cellular metabolic response to injury, i.e., the heat shock, or stress response. We have developed the concept that OFR may be directly deleterious to the cells producing them and that these cells thus require peculiar protective mechanisms from oxidative damage. The analysis of the regulation of free radical production also has important physiological and pathological implications. Interestingly, production of OFR by phagocytes can be regulated by agents also known to modulate the heat shock reponse, including the steroid hormone 1,25-dihydroxyvitamin D_3 (1,25-$(OH)_2D_3$) (Polla et al., 1986; Polla et al., 1987), cytokines such as tumor necrosis factor α (TNF) or interferon gamma (IFNγ) (upregulation of OFR production), or by drugs such as dexamethazone or lipocortins (downregulation).

Induction of a stress response during OFR production

Induction of HSPs during phagocytosis: We have first focused on the interactions between OFR and HSPs and on the role(s) these proteins may play during phagocytosis. Having established that exogenous hydrogen peroxide (H_2O_2) induces HSP synthesis in human monocytes-macrophages, we asked the question as to whether synthesis of HSPs would also be associated with endogenous generation of OFR during phagocytosis. As a model system, we used sheep erythrocytes as phagocytic stimulus and the human premonocytic line U937 cells. U937 cells are non-phagocytic and unable to produce superoxide in their undifferentiated state, but phagocytosis and a functional NADPH oxidase can be induced in these cells by incubation with the steroid hormone 1,25-$(OH)_2D_3$. In this model, we have shown that generation of OFR during erythrophagocytosis is associated, in 1,25-$(OH)_2D_3$-differentiated U937 cells, with the synthesis of the classical HSPs and of the 32kD oxidation-specific stress protein heme oxygenase (Clerget and Polla, 1990). In this particular case, the question arises as to whether the inducer of HSPs is represented by OFR, by the phagocytosed particles recognized as "non-self" (Forsdyke, 1985), or both. Our data on erythrophagocytosis suggest that in U937 cells, both phagocytosis and generation of OFR (probably · OH, via the Haber-Weiss reaction catalyzed in the presence of hemoglobin-derived iron) are required for HSP synthesis. We also observed the induction, during erythrophagocytosis, of a doublet of an apparent molecular weight of 21 kD, likely corresponding to ferritin subunits. Translational induction of ferritin by heat shock has been reported in avian reticulocytes (Atkinson et al., 1990). In the monocytes-macrophages however, ferritin does not appear to be induced by heat shock. Thus, both in the avian reticulocytes after heat shock and in the monocytes-macrophages during erythrophagocytosis, ferritin induction is likely mediated by the release of iron from iron-containing proteins, i.e., hemoglobin, denatured by heat shock or by OFR respectively.

Interestingly, the synthesis of the classical HSPs as well as of heme oxygenase were prevented by the free thiol N-(2-mercaptoethyl)-1,3-propanediamine, WR 1065 (dephosphorylated from the radioprotective agent WR 2721) (Clerget and Polla, 1990; Polla et al., 1990a). WR 1065 represents a pharmacological tool to manipulate production and toxicity of OFR. WR 1065 prevents in U937 cells H_2O_2-induced cell death and inhibits the rise in $[Ca^{2+}]_i$ associated in these cells with oxidative injury (Polla et al., 1990a). Under these conditions, i.e., in the presence of WR 1065, one might concieve that induction of HSPs is prevented either because the toxic events related to free radical generation during phagocytosis do not occur, or because the pharmacological protection renders endogenous protection mechanisms unnecessary.

Induction of HSPs by ischemia: Ischemia is associated with a decrease in ATP levels related to uncoupling of oxidative phosphorylation and leads to the accumulation of xanthine and hypoxanthine. These substrates are normally metabolized by xanthine dehydrogenase, but during ischemia and in the presence of calcium, this enzyme reverts to xanthine oxidase. During reperfusion and in the presence of oxygen, xanthine is then metabolized by xanthine oxidase, leading to the generation of high amounts of O_2^-. Induction of HSPs in the ischemic organ (brain, heart) has been demonstrated in several animal models (gerbil, dog) for ischemia and reperfusion injury (Dillmann et al., 1986). Our hypothesis is that accumulation of HSPs in the ischemic organs is secondary to the oxidative stress associated with reperfusion injury. We have demonstrated accumulation of transcripts for HSP70 in rodent kidneys after unilateral ischemia and reperfusion (Donati, Polla, Bonventre, unpublished). In this *in vivo* model however, preliminary data suggests that WR 2721 does not modulate the accumulation of HSP70 transcripts.

Potential roles of calcium and potassium in induction of HSPs: calcium, besides its widespread role as second messenger in many biological systems, participates to cell death in particular in the context of oxidative injury and has the ability to potentiate oxidation-induced cell damage (Malis and Bonventre, 1986). Although some reports have suggested that calcium plays a role as a second messenger in the cascade of stressful events leading to the activation of heat shock genes (Stevenson et al., 1986), others did not support a requirement for calcium in HSP induction. We reexamined this issue in the U937 cells and found that exposure of these cells to elevated temperatures induced the synthesis of HSPs whatever the concentrations of extracellular calcium. Furthermore, we found no rise in $[Ca^{2+}]_i$ in the U937 cells during exposure to heat shock (Kantengwa et al., 1990). In contrast, during oxidative injury, calcium enters the cells and calcium fluxes are involved in the complex processes associated with phagocytosis and OFR generation. Whereas calcium is neither required nor sufficient for induction of HSP synthesis after heat shock, its role in HSP induction during phagocytosis and oxidative injury remains elusive. Oncogenes, in particular *c-fos*, represent one possible link between calcium and the induction of the heat shock response during phagocytosis (Donati et al., in press). Indeed, *c-fos* gene expression increases in parallel with the induction of HSP70 under a variety of conditions, including ischemia (Higo et al., 1989). Furthermore, *c-fos* expression is increased during phagocytosis and this increased expression is calcium dependent. Finally, oxidative stress also has been shown to induce oncogene expression (Crawford et al., 1988).

Little is known about the potential role of potassium as second messenger of cellular injury. Although Hightower and White reported in 1981 that in rat brain slices varying potassium ion concentration in the perfusate 0 to 50 mM did not alter the stress response (Hightower and White, 1981), a role for potassium and potassium channels in the induction of a heat shock response is being reevaluated in our laboratory, in particular with respect to oxidative injury. In order to investigate which are, in isolated cells, the relevant ion fluxes possibly initiating a G protein-dependent signaling cascade participating to transcriptional activation of heat shock genes, cells are incubated with ouabain, potassium channel blockers such as quinine, or potassium channel activators such as cromakalim prior to exposure to heat shock (Okada and Azama, 1989). Separate activation of K^+ and Cl^- channels could occur during exposure to elevated temperatures, oxidative injury, or other inducers of the stress response.

Do HSPs have or not the potential to protect cells from oxidative injury?

In general, there is growing evidence that HSPs play an essential role in protecting cells from thermal and other injuries (Welch and Mizzen, 1988). The possibility that HSPs are part of the cell's protective mechanisms against oxidative injury is supported by several observations, including:

a) the *in vitro* and *in vivo* synthesis of HSPs by cells or tissues exposed to oxidative injuries;

b) the description, by Christman, of a mutant of *Salmonella typhimurium*, oxyR1, which is resistant to a variety of oxidizing agents and constitutively overexpresses HSPs (Christman et al., 1985);

c) the partial protection by heat shock from H_2O_2-induced cell death in the human premonocytic line U937 and from oligomycine-induced cell death in the DS7 cells (Polla et al., 1988; Polla and Bonventre, 1987). The DS7 is a glycolytic mutant of chinese hamster lung fibroblasts which is strictly restricted to aerobic metabolism, unable to utilize glucose and therefore dependent upon oxidative metabolism for its energy supply. This cell dies upon exposure to inhibitors of oxidative phosphorylation such as oligomycin. In these cells, we have been able to show that preexposure to heat partially prevents cell death secondary to inhibition of oxidative phosphorylation (Polla and Bonventre, 1987);

d) the protection by heat shock from glutamate, a model used *in vitro* for ischemic (oxidative) brain injury in neural cells (Bonventre and Polla, unpublished).

e) the protection against light-induced (oxidation-mediated) damage in the rat retina by *in vivo* hyperthermia (Barbe et al., 1988);

f) the identity of the 32 kD oxidation-specific stress protein, P32, to an enzyme with known scavenging potential, heme oxygenase (Keyse and Tyrrell, 1987; Keyse and Tyrrell, 1989).

With respect to the *in vitro* cellular models mentioned under c) (i.e., the U937 cells stressed by the addition of exogenous H_2O_2 and the DS7 cells stressed by the addition of oligomycin) we have however been confronted with experimental problems and variability in the results. In the U937 cells, the toxicity of H_2O_2 greatly varies depending upon the culture systems; furthermore, in this rapidly growing cell line, true protection is difficult to distinguish from selection of resistant cells. In the DS7 cells, there is an interexperiment variability in sensitivity to oligomycin, suggesting rapid selection of cells able to switch to anaerobic glycolysis. We have therefore worked at establishing another model for oxidative injury in the U937 cells, using a drug, bleomycin

284

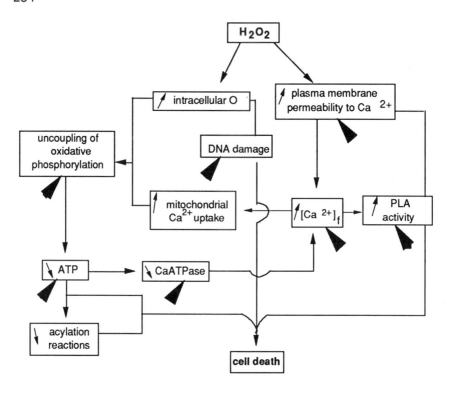

Figure 1. Cellular effects of oxygen free radicals. Cell death may be the consequence of increased intracellular concentration of free calcium ($[Ca^{2+}]_f$), uncoupling of oxidative phosphorylation and/or DNA damage.
Potential targets for protection by heat shock or heat shock proteins are indicated by the bold arrows.

(BLM), easier to titrate and to reproducibly manipulate than H_2O_2. BLM is a chemotherapeutic agent with high affinity for DNA, acting via free radicals generated through oxidation-reduction cycles involving iron-BLM complexes. Whereas SOD has previously been shown to inhibit BLM-induced DNA strand breaks (Galvan et al., 1989), we have investigated the effects of heat shock on subsequent BLM toxicity in U937 cells. Preexposure of U937 cells to a range of temperatures (43° to 45°C for 20 min) consistently prevents the dose-dependent decrease in ^{3}H-thymidine incorporation induced by BLM in these cells. Maximum protection occurs at 44.5°C, the temperature at which HSPs are most induced in these cells (Polla et al., 1990b). The observation that 1,25-$(OH)_2D_3$, which decreases DNA synthesis and increases HSP synthesis in U937 cells, potentiates these protective effects of heat shock (Polla et al., 1990b) is one more argument in favor of a protective role of HSPs against oxidative damage.

What are the mechanisms for protection from oxidative injury by heat shock?

Figure 1 illustrates both the cellular targets for injury mediated by OFR and the potential *points d'impact* for protection by HSPs. After exposure to H_2O_2, cell membranes may undergo lipid peroxidation, proteins may be degraded, enzymes inactivated, and DNA damaged. DNA strand breaks induced by radicals are a major cause of cell death following irradiation or treatment with agents such as bleomycin, although DNA damage, alternatively to cell death, may also lead to carcinogenesis (Cerutti, 1989).

The different levels where HS, or HSPs, could interfere and protect cells from H_2O_2 or other types of oxidative injuries are outlined below.

Scavenging enzymes and metal chelators: Free radical scavengers such as SOD, glutathione peroxidase or catalase, as well as non enzymatic systems such as selenium, vitamin E or iron chelators are the classical mechanisms for protection from oxidative injury. There still is controversy as to whether or not SOD is a HSP: although it is not considered as a classical heat shock protein in the human system, inasmuch it is not induced by heat, in bacteria however (*Escherichia coli*) and eventually in other cells, it is indeed a HSP (Privalle and Fridovich, 1987; Loven et al., 1985). In any case, both SOD and HSPs can provide protection against the lytic effects of subsequent exposure to TNF (Jäättelä et al., 1989). The protective effects of heat shock do however not appear to be mediated by an increase in SOD. Other stress proteins endoved with scavenging properties are metallothioneins (Schroeder and Cousins, 1990) and the 32 kD oxidation-specific stress protein heme oxygenase.

Interestingly, TNF, as OFR, also induces "autoprotective" mechanisms against its own toxicity, i.e., SOD and eventually HSPs (Wong et al., 1989; Kantengwa et al., in press). Mitochondria are the first targets for TNF toxicity (Matthews and Neale, 1987); the increased synthesis and activity of Mn-SOD, essential for cellular resistance to TNF, is taking place in this organelle.

Oxidative damage mediated either by phagocytes or by the superoxide radical-generating system hypoxanthine-xanthine oxidase are amplified by iron (Aruoma et al., 1989). Thus, since iron may potentiate free radical injury, iron chelators have been considered as part of the protective tools against oxidative injury. Desferrioxamine for example has already been used in clinical conditions associated with oxidative injury such as the adult respiratory distress syndrome. Ferritin, which is induced by iron and can be considered as a stress protein, also has the potential to prevent iron-induced free radical damage and thus represents another potential example of autoprotection.

The role HSPs may play as radical scavengers could be of particular relevance under conditions in which the classical scavenging systems are insufficient. Glutathione depletion for example potentiates HSP induction. Another such situation is represented by the unique induction of HSPs by Interleukine 1(IL-1) in isolated rat pancreatic islets. The interactions between IL-1 and HSPs have been investigated for many years. In 1983, Duff and Durum reported that the pyrogenic and mitogenic effects of Il-1 were linked (Duff and Durum, 1983). Il-1 however does not induce synthesis of HSPs (only phosphorylation, but not synthesis, of the low molecular weight HSPs [Kaur et al., 1989]) except in isolated rat islets of Langerhans (Helqvist et al., 1989; Helqvist, Polla, Johannesen and Nerup, submitted). Compared to other cell types, islet cells contain small amounts of the classical scavengers such as SOD or glutathione peroxidase. Thus, the selective Il-1-mediated cytotoxicity toward β cells may relate to the insufficient capability of these cells to produce classical radical scavengers. Synthesis of HSPs by β cells exposed to IL-1 may represent an alternative, antioxidant, stress response.

Prevention of accumulation of intracellular free calcium: Although calcium does participate to DNA strand breaks and cell death during oxidative injury, it is not involved in induction of HSPs, thermal killing nor thermotolerance in the U937 cells (Kantengwa et al., 1990). Heat shock does protect U937 cells from subsequent exposure to H_2O_2, but does not prevent the rise in $[Ca^{2+}]_i$ induced by H_2O_2 in these cells (Polla et al., 1988). The protective effects of heat shock must thus be distal to calcium entry.

Protection of oxidative phosphorylation and maintenance of ATP levels: In the DS7 cells, preexposure to elevated temperatures decreases the toxic effects of the ATPase inhibitor oligomycine but does not prevent the drop in ATP induced in these cells by the drug (Polla and Bonventre, 1987). In *Saccharomyces cerevisiae* however, HSP induction protects mitochondrial ATPase activity (Patriarca and Maresca, 1990). This effect of heat appears to be operative only at the F_o subunit of ATPase, and HSP synthesis does not affect electron transport efficiency. These authors also found a strong correlation between the importance of the heat shock response and the capacity of the cells to maintain a cellular respiration coupled to ATP synthesis. The specific role the mitochondrial chaperonin HSP65 may play in prevention from uncoupling of oxidative phosphorylation and eventually in human diseases associated with defects in the respiratory chain has not been investigated yet. Interactions between heat shock and cellular respiration certainly deserve further investigations.

Prevention of DNA damage induced by OFR: In unstressed cells, HSPs exert important functions such as subcellular protein transport, import-export of other proteins ("chaperoning") and prevention from uncorrect protein folding (Young, 1990). Whether these functions mediate all the protective effects of HSPs remains unclear. We hypothesize that as they bind to and protect proteins, HSPs may as well bind to DNA in a similar manner. The fact that HSPs move inside the nucleus immediately after stress could be related to this hypothesis.

Heat shock (HSPs) could interfere with the effects of OFR for example by preventing DNA strand breaks induced by BLM or by the superoxide radical-generating system hypoxanthine-xanthine oxidase (Galvan et al., 1989; Aruoma et al., 1989). Alternatively, heat shock (HSPs) may protect cells not by preventing damage but by participating in DNA repair mechanisms such as poly (ADP-ribose) polymerase (Cerutti, 1989). Heat shock may also prevent the activation of endonucleases responsible for DNA breakage after exposure to H_2O_2.

Inhibition of production of OFR by heat shock: We have previously reported that heat shock inhibits superoxide production by human neutrophils (Maridonneau-Parini et al., 1988). This inhibition however does not appear to be mediated by the HSPs themselves, inasmuch HSP synthesis and inhibition of NADPH oxidase activation are not temporally associated and this inhibition is also found in cytoplasts (i.e., anucleated neutrophils) in which transcritpional induction of HSP synthesis cannot occur (Maridonneau-Parini, Russo-Marie, Malawista and Polla, unpublished). Although the precise mechanism by which heat shock inhibits O_2^- production via NADPH oxidase activation remains unclear, this inhibition may provide another potential end-point for modulation of oxidative damage. Whether or not heat shock prevents generation of OFR by other mechanisms than NADPH oxidase activation (for example, xanthine-xanthine oxidase) remains to be elucidated.

Inhibition of phospholipase A_2: Phospholipase A_2 (PLA$_2$) may participate to propagation of alterations in membrane permeability. PLA$_2$ thus represents one more potential target for HSPs to protect cells from OFR. We found however no inhibition of prostaglandin E_2 production by neutrophils after heat shock, whereas under the same conditions, the enzyme NADPH oxidase was inhibited (Maridonneau-Parini et al., 1988).

Conclusions

Inflammatory cells, in particular those producing OFR, are characterized, as compared to many other cells, by a high level of HSP synthesis after exposure to heat or to other types of injury, in particular oxidative injury. One interesting question we are confronted with is whether HSPs induced during phagocytosis play a role in protection of the phagocyte from the oxidative stress it is itself generating. Indeed, cells producing OFR are themselves exposed to their toxicity. We hypothesize that these cells which are generating high levels of OFR and proinflammatory cytokines have to be particularly well equipped in protective mechanisms and that the high levels of HSP synthesis observed in monocytes-macrophages deserve such a function.

288

Aknowledgements

Supported in part by FNRS 3.960.0.87 and 32.28645.90 and the Fondation Centre de Recherches Médicales Carlos et Elsie de Reuter to BSP.

References

Aruoma, OI, Halliwell, B and Dizdaroglu, M, (1989) Iron ion-depedent modification in bases in DNA by superoxide radical-generating system hypoxanthine/xanthine oxidase. J. Biol. Chem., 264: 13024-13028.

Atkinson, BG, Blaker, TW, Tomlinson, J and Dean, RL, (1990) Ferritin is a translationally regulated heat shock protein of avian reticulocytes. J. Biol. Chem., 265: 14156-14162.

Barbe, MF, Tytell, M, Gower, DJ and Welch, WJ, (1988) Hyperthermia protects against light damage in the rat retina. Science, 241: 1817-1820.

Cerutti, PA, (1989) Mechanisms of action of oxidant carcinogens. Cancer Detection and Prevention, 14: 281-284.

Christman, MF, Morgan, RW, Jacobson, FS and Ames, BN, (1985) Positive control of a regulon for defenses against oxidative stress and some heat shock proteins in *Salmonella typhimurium*. Cell, 41: 753-762.

Clerget, M and Polla, BS, (1990) Erythrophagocytosis induces heat shock protein synthesis by human monocytes-macrophages. Proc. Natl. Acad. Sci. USA, 87: 1081-1085.

Crawford, D, Zbinden, I, Amstad, P and Cerutti, P, (1988) Oxidant stress induces the proto-oncogenes *c-fos* and *c-myc* in mouse epidermal cells. Oncogene, 3: 27-32.

Curnutte, JT, Whitten, DM and Babior, BM, (1974) Defective superoxide production by granulocytes from patients with chronic granulomatous disease. N. Engl. Med., 290: 593-597.

Dillmann, WH, Mehta, HB, Barrieux, A, Guth, BD, Neeley, WE and Ross, J Jr, (1986) Ischemia of the dog heart induces the appearance of a cardiac mRNA coding for a protein with migration characteristics similar to heat-shock/stress protein 71. Circulation Res., 59: 110-114.

Donati, YRA, Kantengwa, S and Polla, BS, Phagocytosis and heat shock response in human monocytes-macrophages. Pathobiol., in press.

Duff, GW and Durum, SK, (1983) The pyrogenic and mitogenic actions of interleukin-1 are related. Nature, 304: 449-451.

Fantone, JC and Ward, PA, (1982) Role of oxygen-derived free radicals and metabolites in leukocyte-dependent inflammatory reactions. Am. J. Pathol., 107: 397-418.

Forsdyke, DR, (1985) Heat shock proteins defend against intracellular pathogens: a non immunological basis for self/nonself discrimination? J. Theor. Biol., 115: 471-473.

Galvan, L, Huang, CH, Prestayko, AW, Stout, JT, Evans, JE and Crooke, ST, (1989) Inhibition of bleomycin-induced DNA breakage by superoxide dismutase. Cancer Res., 41: 5103-5106.

Halliwell, B, (1987) Oxidants and human disease: some new concepts. FASEB J., 1: 358-364.

Helqvist, S, Hansen, BS, Johannesen, J, Andersen, HU, Nielsen, JH and Nerup, J, (1989) Interleukin 1 induces new protein formation in isolated rat islets of Langerhans. Acta Endocrinol. (Copenh), 121: 136-140.

Hightower, LE and White, FP, (1981) Cellular response to stress: comparison of a family of 71-73-kilodalton proteins rapidly synthesized in rat tissue slices and canavanine-treated cells in culture. J. Cell Physiol., 108: 261-275.

Higo, H, Lee, JY, Satow, Y and Higo, K, (1989) Elevated expression of proto-oncogenes accompanies enhanced induction of heat-shock genes after exposure of rat embryos *in utero* to ionizing irradiation. Teratogen Carcinogen Mutagen, 9: 191-198.

Jäättelä, M, Saksela, K and Saksela, E, (1989) Heat-shock protects WEHI-164 target cells from the cytolysis by tumor necrosis factor α and β. Eur. J. Immunol., 19: 1413-1417.

Kantengwa, S, Capponi, AM, Bonventre, JV and Polla, BS, (1990) Calcium and the heat shock response in the human premonocytic line U937 : effects of 1,25-dihydroxyvitamin D3. Am. J. Physiol., 259 (Cell Physiol 28): C77-C83.

Kantengwa, S, Donati, RYA, Clerget, M, Maridonneau-Parini, I, Sinclair F, Mariéthoz, E, Rees, ADM, Slosman, DO and Polla, BS, Heat shock proteins: an autoprotective mechanism for inflammatory cells. Sem. Immunol, in press.

Kaur, P, Welch, WJ and Saklatvala, J, (1989) Interleukin 1 and tumor necrosis factor increase phosphorylation of the small heat shock protein. Effects in fibroblasts, Hep G2 and U937 cells. FEBS, 258: 269-273.

Keyse, SM and Tyrrell, RM, (1987) Both near ultraviolet radiation and the oxidizing agent hydrogen peroxide induce a 32-kDa stress protein in normal skin fibroblasts. J. Biol. Chem., 262: 14821-14825.

Keyse, SM and Tyrrell, RM, (1989) Heme oxygenase is the major 32-kDa stress protein induced in human skin fibroblasts by UVA radiation, hydrogen peroxide, and sodium arsenite. Proc. Natl. Acad. Sci. USA, 86: 99-103.

Loven, DP, Leeper, DB and Oberley, LW, (1985) Superoxide dismutase levels in chinese hamster ovary cells and ovarian carcinoma cells after hyperthermia or exposure to cycloheximide. Cancer Res., 45: 3029-3033.

Lunec, J, Halloran, SP, White, AG and Dormandy, TL, (1981) Free-radical oxidation (peroxidation) products in serum and synovial fluid in rheumatoid arthritis. J. Rheumatol., 8: 233-245.

Malis, CD and Bonventre, JV, (1986) Mechanism of calcium potentiation of oxygen free radical injury to renal mitochondria. A model for post-ischemic and toxic mitochondrial damage. J. Biol. Chem., 261: 14201-14208.

Maridonneau-Parini, I, Clerc, J and Polla, BS, (1988) Heat shock inhibits NADPH oxidase in human neutrophils. Biochem. Biophys. Res. Commun., 154: 179-186.

Matthews, N and Neale, ML, (1987) Studies of the mode of action of tumor necrosis factor on tumor cells *in vivo*. Lymphokine, 14: 323-352.

Okada, Y and Hazama, A, (1989) Volume-regulatory ion channels in epithelial cells. NIPS, 4: 238-242.

Patriarca, EJ and Maresca, B, (1990) Acquired thermotolerance following heat shock protein synthesis prevents impairment of mitochondrial ATPase activity at elevated temperatures in *Saccharomyces cerevisiae*. Exp. Cell Res., 190: 57-64.

Polla, BS, Healy, A, Amento, EP and Krane, SM, (1986) 1,25-dihydroxyvitamin D_3 maintains adherence of human monocytes and protects them from thermal injury. J. Clin. Invest., 77: 1332-1339.

Polla, BS and Bonventre, JV, (1987) Heat shock protects cells dependent on oxidative metabolism from inhibition of oxidative phosphorylation. Clin. Res., 35: 555A.

Polla, BS, Healy, AM, Wojno, WC and Krane, SM, (1987) Hormone 1α,25-dihydroxyvitamin D_3 modulates heat shock response in monocytes. Am. J. Physiol., 252 (Cell Physiol 21): C640-C649.

Polla, BS, Bonventre, JV and Krane, SM, (1988) 1,25-dihydroxyvitamin D_3 increases the toxicity of hydrogen peroxide in the human monocytic line U937 : role of calcium and heat shock. J. Cell Biol., 107: 373-380.

Polla, BS, Donati, YR, Kondo, M, Tochon-Danguy, HS and Bonjour, JP, (1990a) Protection from cellular oxidative injury and calcium intrusion by N-(2-mercaptoethyl)-1,3-propanediamine, WR 1065. Biochem. Pharmacol., 40: 1469-1475.

Polla, BS, Pittet, N and Slosman, DO, (1990b) Heat shock (HS) protects the human monocytic line U937 from bleomycin (BLM)-mediated oxidative injury. Eur. Resp., 3(suppl) 335s, 1355A.

Privalle, CT and Fridovich, I, (1987) Induction of superoxide dismutase in *Escherichia coli* by heat shock. Proc. Natl. Acad. Sci. USA, 84: 2723-2726.

Schroeder, JJ and Cousins, RJ, (1990) Interleukin 6 regulates metallothionein gene expression and zinc metabolism in hepatocyte monolayer cultures. Proc. Natl. Acad. Sci. USA, 87: 3137-3141.

Stevenson, MA and Calderwood, SK, Hahn, GM, (1986) Rapid increases in inositol triphosphate and intracellular Ca^{++} after heat shock. Biochem. Biophys. Res, Commun., 137: 826-833.

Welch, WJ and Mizzen, LA, (1988) Characterization of the thermotolerant cell. II. Effects on intracellular distribution of heat-shock protein 70, intermediate filaments, and small nuclear ribonucleoprotein complexes. J. Cell Biol., 106: 1117-1130.

Wong, GHW, Elwell, JH, Oberley, LW and Goeddel, DV, (1989) Manganous superoxide dismutase is essential for cellular resistance to cytotoxicity of tumor necrosis factor. Cell, 58: 923-931.

Young, DB, (1990) Chaperonins and the immune response. Sem. Cell Biol., 1: 27-35.

Induction of HSP70 Genes in the Mammalian Nervous System by Hyperthermia, Tissue Injury and Other Traumatic Events

I.R. Brown
Department of Zoology
University of Toronto, Scarborough Campus
Scarborough, Ontario M1C 1A4
Canada

Introduction

The induction of heat shock proteins has been studied in a wide range of organisms including *E. coli*, yeast, *Drosophila* and eukaryotic cells grown in culture (Pardue et al., 1989), however, comparatively little work has been carried out on intact thermoregulating animals. In this laboratory we are exploring whether the heat shock response is turned on in the mammalian nervous system following such traumatic conditions as fever-like temperatures and tissue wounding, and if so, which cell types show induction.

Early studies in our laboratory demonstrated that fever-like temperatures, produced by the injection of the psychotropic drug LSD, rapidly induced synthesis of HSP70 in the adult and fetal rabbit brain (Freedman et al., 1981; Brown, 1983; Cosgrove and Brown, 1983; Brown, 1985a; Brown, 1985b) concomitant with a transient disaggregation of brain polysomes (Brown et al., 1982). The transient inhibition of brain protein synthesis appears to result from a lesion at an early step in initiation of translation (Cosgrove et al., 1981; Brown et al., 1982; Cosgrove and Brown, 1984). A translational inhibitor has been partially purified which results in a decreased formation of ternary complexes when added to a brain cell-free protein synthesis system (Fleming and Brown, 1986; Fleming and Brown, 1987).

Regional and cell type differences in expression of *hsp70* genes in brain detected by *in situ* hybridization

Early experiments at the protein synthesis level indicated that fever-like temperatures induced the rapid synthesis of *hsp70* in the mammalian brain. We have addressed the question of whether there are regional and cell type differences in the expression of *hsp70* genes in the nervous system. If the ability of brain cells to survive heat shock and other traumas is related to their capacity to induce HSPs, an analysis of the pattern of expression of heat shock genes in the brain may further our understanding of the selective vulnerability of certain brain cells to various forms of trauma and neurogenetic diseases. Northern blot analysis of RNA isolated from rabbit brain 1 hr after elevation of body temperature by 2°-3°C revealed the massive induction of a 2.7 kilobase (kb) mRNA species while in control animals the presence of a constitutively expressed 2.5 kb mRNA was apparent (Brown et al., 1985; Sprang and Brown, 1987). *hsp70* nucleic acid probes have recently been reported which distinguish transcripts in brain which are derived from either inducible or constitutively expressed members of the *hsp70* multigene family (Miller et al., 1989; Brown and Rush, 1990; Nowak et al., 1990). Use of an inducible-specific riboprobe revealed that induction of the 2.7 kb brain transcript in hyperthermic rabbits is transient and parallels the rise and fall in body temperature (Brown and Rush, 1990).

In situ hybridization can be utilized to identify which cell types in brain tissue demonstrate constitutive and stress-inducible *hsp70* gene expression (Sprang and Brown, 1987). A labeled nucleic acid probe is hybridized to a brain tissue section under conditions in which the probe binds to target RNA with high specificity (Uhl, 1989). By subsequent autoradiography one can identify those cells in a complex tissue which are expressing the gene of interest.

In situ hybridization with *hsp70* riboprobes revealed striking regional differences in the expression of constitutive and stress-inducible heat shock genes in the rabbit brain 1 hr after drug-induced hyperthermia. Constitutive expression of an *hsp70* gene was observed in several neuronal enriched areas such as hippocampal regions CA1 to CA4 and in the granule and Purkinje layers of the cerebellum. One hour after hyperthermia, expression of an inducible *hsp70* gene was noted in certain neuronal enriched regions but not in others. For example, induction was noted in the granule cell layer of the cerebellum but not in the hippocampal neurons. A dramatic induction of heat shock mRNA was noted 1 hr after hyperthermia in fibre tracts throughout the rabbit forebrain, a pattern consistent with a strong glial response to heat shock. Strong induction was also detected in the choroid plexus which lines the ventricles of the brain, in the microvasculature and in the pia mater, a cellular layer which surrounds the cerebellum and cerebral cortex. The magnitude of the induction response appears to be greatly reduced in cell types which show a high level of constitutive expression of *hsp70* mRNA such as hippocampal neurons. The rabbit spinal cord was selected for investigation because of the ease of differentiating between white matter fibre tracts and neuronal enriched gray matter regions. In control animals the principle cell type which demonstrated high levels of constitutive expression of an *hsp70* gene was large motor neurons in the anterior horns of the gray matter. One hr after a fever-like heat shock, a massive induction of *hsp70* mRNA was noted in glial cells throughout white matter areas of the spinal cord, however little induction was observed in the large motor neurons which showed a high level of constitutive expression.

To extend the investigation of heat shock gene expression in the brain to a higher level of cellular resolution, *in situ* hybridization utilizing plastic-embedded tissue sections was carried out (Masing and Brown, 1989). Decreased section thickness compared with frozen tissue sections, enhanced tissue integrity and examination at increased magnification have facilitated analysis of cell types which are engaged in the expression of HSP70 genes in the rabbit brain after hyperthermia and other traumatic events. A prominent induction of HSP70 mRNA was observed in oligodendroglia in fibre tracts of the deep white matter of the rabbit cerebellum 1 hr after elevation of body temperature to fever-like temperatures.

In the isolated squid giant axon, *hsp70* is synthesized in adaxonal glial cells and rapidly exported into the axon (Tytell et al., 1986). Induction of *hsp70* mRNA in glial cells in fibre tracts of the mammalian brain and export of the resultant protein into adjacent axons could provide a 'fast response' mechanism to ensure the delivery of the heat shock protein to regions of the neuron which are distant from the cell body. In earlier studies we demonstrated that *hsp70* is induced in the cell bodies of retinal ganglion neurons and transported down the optic nerve (Clark and Brown, 1985). However, this is by slow axonal transport (Clark and Brown, 1985) specifically with slow component b which includes such transported molecules as actin microfilaments, spectrin, clathrin and calmodulin (Tytell and Barbe, 1987).

Constitutive expression of *hsp70* mRNA

In situ hybridization studies in both the rabbit and the rat brain have revealed an interesting pattern of constitutive expression of an HSP70 gene in neuronal enriched layers of the hippocampus. The role of the encoded protein in normal metabolic functions of neurons could include participation in neuronal mechanisms related to membrane cycling events associated with neurotransmission and to the translocation of proteins between complex neuronal intracellular compartments. Clathrin-uncoating ATPase isolated from bovine brain has been reported to be a constitutively expressed membrane of the HSP70 family (Ungewickell, 1985; Chappell et al., 1986). This abundant protein comprises 1% of soluble brain protein (Schlossman et al., 1984). Members of the HSP70 family may function in the maintenance of neuronal membrane cycling events and perhaps play a critical role in restoring membrane integrity following traumatic events.

Induction of *hsp70* mRNA at the site of tissue injury in the brain

Localized tissue injury induced expression of an *hsp70* gene in the mammalian nervous system. A small surgical cut was made in the rat cerebral cortex (Brown et al., 1989). By 2 hr postsurgery a dramatic and highly localized induction of *hsp70* mRNA was detected in cells proximal to the lesion site using in situ hybridization. High resolution analysis indicated that both neuronal and glial cells at the injury site respond by inducing *hsp70* mRNA. By 12 hr the intensity of the signal had diminished, and by 24 hr only a few cells along the walls of the cut demonstrated a high level of *hsp70* mRNA. This study suggests that induction of *hsp70* mRNA is a physiologically relevant response which is activated at an early stage following tissue injury in the nervous system. The pattern of induction of *hsp70* mRNA serves as a marker to identify a population of reactive cells that respond rapidly to trauma.

Immunological detection of HSP70 in brain tissue

The availability of antibodies directed against HSP70 has permitted us to detect the induction of this protein in the rat hippocampus after neonatal hypoxia-ischemia which is produced by unilateral common carotid artery ligation combined with hypoxia (Dwyer et al., 1989). Induction was noted in the ipsilateral but not contralateral hippocampus. Immunocytochemical studies have revealed a neuronal pattern of induction of HSP70 following either ischemia (Vass et al., 1988) or kainic acid-induced seizures (Gonzalez et al., 1989; Vass et al., 1989). Accumulation of HSP70 was minimal in hippocampal CA1 neurons, which die after brief ischemia, but was pronounced in dentate granule cells and CA3 neurons, which survive. In the gerbil the peak of CA3 immunoreactivity did not occur until 48 hr postischemia, suggesting that *hsp70* induction is a response to delayed hippocampal pathophysiology rather than a direct response to the initial ischemic insult. Results suggest that HSP70 immunocytochemistry may serve as a convenient marker for neuronal circuitry involved in excitotic mechanisms after ischemia and other stresses (Vass et al., 1988).

In agreement with our *in situ* hybridization studies, immuno-cytochemical observations have revealed a strong induction of HSP70 in glial cells either in vivo or in vitro following temperature elevation (Nishimura et al., 1988; 1990; Marini et al., 1990).

Different types of neural trauma induce characteristic cellular responses in the mammalian brain with regard to the type of brain cell that responds by inducing *hsp70* mRNA and the timing of the induction response. Fever-like temperatures cause a dramatic induction of *hsp70* mRNA within 1 hr in glial cells in fibre tracts of the rabbit forebrain, spinal cord and cerebellum. In the case of tissue injury to the brain, a rapid and highly localized induction of *hsp70* mRNA is noted in both neuronal and glial cells which are proximal to the injury site in the cerebral cortex. A neuronal pattern of induction of *hsp70* is observed after ischemia and the time course of induction in the gerbil is slow compared to the rapid induction which is seen in response to hyperthermia or tissue injury.

Tissue protective effects of heat shock in the nervous system

Brief heat shock has been found to confer tissue protection in the nervous system against subsequent traumatic events. For example, if rat embryos are exposed to heat shock at 43°C at a specific stage of neural development, major regions of the brain fail to develop. A brief, nonteratogenic heat shock at 42°C administered prior to the defect-inducing temperature shock confers complete tissue protection against the teratogenic effects (Walsh et al., 1987, 1989). Protection against light-induced degeneration of rat retinal photoreceptors is also confered by prior whole body hyperthermia (Barbe et al., 1988).

In collaboration with Tytell's group, we have been determining which cell types in the retina induce *hsp70* mRNA in response to heat shock and what is the time course of this induction in relation to the window of the tissue protective effects of the heat shock. *In situ* hybridization revealed a massive induction of *hsp70* mRNA in retinal photoreceptors at 4 and 18 hr, a time which matches the window of the tissue protective effects of the heat shock. *In situ* hybridization with plastic-embedded retinal tissue permitted a much improved level of cellular resolution compared to frozen sections cut on a cryostat.

The pattern of induction of *hsp70* mRNA following various traumatic conditions has proved to be a useful early marker of cellular perturbation in the nervous system. The ability of brain cells to survive fever-like temperatures and other stresses may be related to their capacity to induce HSP70. It would be interesting to explore whether this induction response is altered in selected neural cells during aging or neurodegenerative diseases. Elevation of the levels of HSP70 in selected neural cell types may open up new avenues for enhancing cellular repair in the nervous system and reducing loss of neurons which occurs after injury.

Conclusions

Our studies have demonstrated that *hsp70* genes are induced in the mammalian brain following traumatic events such as hyperthermia and tissue injury. Different types of neural trauma have been found to induce characteristic cellular responses in the mammalian brain with regard to i) the type of brain cell that responds by inducing *hsp70* mRNA and ii) the timing of the induction response. The pattern of induction of *hsp70* mRNA is a useful early marker of cellular perturbation in the mammalian nervous system and may identify previously unrecognized areas of vulnerability. The rapid induction of an *hsp70* gene at the site of tissue injury in the mammalian brain and induction of *hsp70* mRNA in specific populations of neural cells following hyperthermia and other traumatic events, suggests the *hsp70* may play a role in reactive and perhaps protective mechanisms in the nervous system.

296

Acknowledgments

I thank Sheila Rush, Julie Silver and Pat Manzerra for helpful comments. This work was supported by grants from the Medical Research Council of Canada.

References

Barbe, MF, Tytell, M, Gower, DJ and Welch, WJ, (1988) Hyperthermia protects against light damage in the rat retina. Science, 241: 1817-1820.

Brown, IR, (1983) Hyperthermia induces the synthesis of a heat shock protein by polysomes isolated from the fetal and neonatal mammalian brain. J. Neurochem., 40: 1490-1493.

Brown, IR, (1985a) Effect of hyperthermia and LSD on gene expression in the mammalian brain and other organs. In Atkinson, BG, Walden, CB, eds. Changes in Eukaryotic Gene Expression in Response to Environmental Stress. Academic Press, Inc., New York, pp 211-225.

Brown, IR, (1985b) Modification of gene expression in the mammalian brain after hyperthermia. In Zomzely-Neurath, C, Walker, WA, eds., Gene expression in brain. John Wiley and Sons, Inc., New York, pp 157-171.

Brown, IR and Rush, SJ, (1990) Expression of heat shock genes (hsp70) in the mammalian brain: distinguishing constitutively expressed and hyperthermia-inducible species. J. Neurosci. Res., 25: 14-19.

Brown, IR, Heikkila, JJ and Cosgrove, JW, (1982) Analysis of protein synthesis in the mammalian brain using LSD and hyperthermia as experimental probes. In Brown, IR, ed., Molecular Approaches to Neurobiology. Academic Press, Inc., New York, pp 221-253.

Brown, IR, Lowe, DG and Moran, LA, (1985) Expression of a heat shock gene in fetal and maternal rabbit brain. Neurochem. Res., 10: 1277-1284.

Brown, IR, Rush, SJ and Ivy, GO, (1989) Induction of a heat shock gene at the site of tissue injury in the rat brain. Neuron, 2: 1559-1564.

Chappell, TG, Welch, WJ, Schlossman, DM, Palter, KB, Schlesinger, MJ and Rothman, JE, (1986) Uncoating ATPase is a member of the 70 kilodalton family of stress proteins. Cell, 45: 3-13.

Clark, BD and Brown, IR, (1985) Axonal transport of a heat shock protein in the rabbit visual system. Proc. Natl. Acad. Sci. USA, 82: 1281-1285.

Cosgrove, JW and Brown, IR, (1983) Heat shock protein in the mammalian brain and other organs following a physiologically relevant increase in body temperature induced by LSD. Proc. Natl. Acad. Sci. USA, 80: 569-573.

Cosgrove, JW and Brown, IR, (1984) Effect of intravenous administration of d-lysergic acid diethylamide on initiation of protein synthesis in a cell-free system derived from brain. J. Neurochem., 42: 1420-1426.

Cosgrove, JW, Clark, BD and Brown, IR, (1981) Effect of intravenous administration of d-lysergic acid diethylamide on subsequent protein synthesis in a cell-free system derived from brain. J. Neurochem., 36: 1037-1045.

Dwyer, BE, Nishimura, RN and Brown, IR, (1989) Synthesis of the major inducible heat shock protein in rat hippocampus after neonatal hypoxia-ischemia. Exper. Neurol., 104: 28-31.

Fleming, SW and Brown, IR, (1986) Characterization of a translational inhibitor isolated from rabbit brain following intravenous administration of d-lysergic acid diethylamide. J. Neurochem., 46: 1436-1443.

Fleming, SW and Brown, IR, (1987) Effect on in vitro brain protein synthesis of a translational inhibitor isolated from rabbit brain following intravenous administration of LSD. Neurochem. Res., 12: 323-329.

Freedman, MS, Clark, BD, Cruz, TF, Gurd, JW and Brown, IR, (1981) Selective effects of LSD and hyperthermia on the synthesis of synaptic proteins and glycoproteins. Brain Res., 207: 129-145.

González, MF, Shiraishi, K, Hisanaga, K, Sagar, SM, Mandabach, M and Sharp, FR, (1989) Heat shock proteins as markers of neural injury. Mol. Brain. Res., 6: 93-100.

Marini, AM, Kozuka, M, Lipsky, RH and Nowak, TS, (1990) 70-kilodalton heat shock protein induction in cerebellar astrocytes and cerebellar granule cells in vitro: comparison with immunocytochemical localization after hyperthermia in vivo. J. Neurochem., 54: 1509-1516.

Masing, TE and Brown, IR, (1989) Cellular localization of heat shock gene expression in rabbit cerebellum by in situ hybridization with plastic-embedded tissue. Neurochem. Res., 14: 725-731.

Miller, EK, Raese, JD and Morrison-Bogorad, MR, (1989) The family of heat shock protein 70 mRNAs are differentially induced in rat cerebellum, cortex, and non-neuronal tissues. Soc. Neurosci. Abstr., 15: 1127.

Nishimura, RN, Dwyer, BE, Welch, W, Cole, R, de Vellis, J and Liotta, K, (1988) The induction of the major heat-stress protein in purified rat glial cells. J. Neurosci. Res., 20: 12-18.

Nishimura, RN, Dwyer, BE, Clegg, K, Cole, R and de Vellis, J, Comparison of the heat shock response in cultured cortical neurons and astrocytes. Molec Brain Res., in press.

Nowak, TS, Bond, U and Schlesinger, MJ, (1990) Heat shock RNA levels in brain and other tissues after hyperthermia and transient ischemia. J. Neurochem., 54: 451-458.

Pardue, ML, Feramisco, JR and Lindquist, S, (1989) Stress-induced proteins. UCLA Symp Mol Cell Biol New Series v 96, Alan R Liss, Inc., New York, pp 1-294.

Schlossman, D, Schmid, S, Braell, W and Rothman, J, (1984) An enzyme that removes clathrin coats: Purification of an uncoating ATPase. J. Cell. Biol., 99: 723-733.

Sprang, GK and Brown, IR, (1987) Selective induction of a heat shock gene in fibre tracts and cerebellar neurons of the rabbit brain detected by in situ hybridization. Mol. Brain. Res., 3: 89-93.

Tytell, M and Barbe, MF, (1987) Synthesis and axonal transport of heat shock proteins. In Smith, RS, Bisby, MA, eds., Axonal Transport. Neurology and Neurobiology vol 25, Alan R Liss, Inc., New York, pp 473-492.

Tytell, M, Greenberg, SG and Lasek, RJ, (1986) Heat shock-like protein is transferred from glial to axon. Brain Res., 363: 161-164.

Uhl, GR, (1986) In situ hybridization in brain. Plenum Press, Inc., New York, pp 1-300.

Ungewickell, E, (1985) The 70-kDa mammalian heat shock proteins are structurally and functionally related to the uncoating protein that releases clathrin triskelia from coated vesicles. EMBO J., 4: 3385-3391.

Vass, K, Welch, WJ and Nowak, TS, (1988) Localization of 70 kDa stress protein induction in gerbil brain after ischemia. Acta Neuropath. (Berl), 77: 128-135.

Vass, K, Berger, ML, Nowak, TS, Welch, WJ and Lassmann, H, (1989) Induction of stress protein hsp70 in nerve cells after status epilepticus in the rat. Neurosci. Lett., 100: 259-264.

Walsh, DA, Klein, NW, Hightower, LE and Edwards, MJ, (1987) Heat shock and thermotolerance during early rat embryo development. Teratology, 36: 181-191.

Walsh, DA, Li, K, Speirs, J, Crowther, CE and Edwards, MJ, (1989) Regulation of the inducible heat-shock 71 genes in early neural development of cultured rat embryos. Teratology, 40: 321-334.

Ubiquitin and the Lysosomal System: Molecular Pathological and Experimental Findings

R.J. Mayer, J. Lowe*, M. Landon, H. McDermott*, J. Tuckwell,
F. Doherty, L. Laszlo
Departments of Biochemistry and Pathology*, University of
Nottingham Medical School, Queens Medical Centre,
Nottingham, NG7 2UH
UK

Introduction

Considerably advances in our understanding of the molecular pathology of the major human neurodegenerative diseases have been made by the use of ubiquitin immunocytochemistry. The technique demonstrates that filamentous inclusions and vacuoles contain ubiquitin-protein conjugates. The molecular structure of the filaments and the morphological type of vacuoles is not completely understood but there is evidence that some of the filamentous inclusions contain intermediate filaments and the perinuclear distribution of the vacuoles resembles the distribution of intraneuronal lysosomes. Intermediate filaments and lysosomes are involved in the sequestration and degradation of viral membrane proteins in tissue culture cells. Immunogold electron microscopical and biochemical evidence indicates that ubiquitin-protein conjugates are normally considerably enriched in the lysosomes of fibroblasts relative to all other organelles. Immunogold electron microscopy shows a similar enrichment of ubiquitin-protein conjugates in the dense lysosomes of granulocytic precursor cells in the bone marrow. Filamentous inclusions showing several of the features seen in inclusions in the neurodegenerative diseases are seen in Epstein-Barr virus transformed lymphoblastoid cells. Immunohistochemistry shows that the inclusions contain vimentin intermediate filaments, the latent membrane transforming protein of the virus, ubiquitin-protein conjugates, and heat-shock protein 70 (HSP70). Immunogold electron microscopy demonstrates that the latent membrane protein, ubiquitin-protein conjugates and HSP70 are in lysosomes entwined in a intermediate filament cage centred on the microtubule organising centre. The implications of the combined observations for our understanding of the cell stress response in degenerating neurones and in virally infected cells are discussed.

The histopathological examination of tissues not only provides insights into the pathogenesis of disease but also leads to increased knowledge about normal molecular cell biological processes. We have used this principle to make significant advances in our understanding of the chronic human neurodegenerative diseases and to show that ubiquitin may have a key role in the uptake of proteins into the lysososmal system.

Molecular pathological findings

Ubiquitin and filamentous inclusions: Ubiquitin immunocytochemistry, with polyclonal antisera which primarily detect ubiquitin-protein conjugates, has shown that the ostensibly unrelated major human neurodegenerative diseases, Alzheimer's disease, Parkinson's disease, and motor neurone disease (amyotrophic lateral sclerosis), belong to a family of ubiquitin-filament disorders (Lowe et al., 1988). Furthermore ubiquitin immunocytochemistry has shown that diffuse Lewy body disease, previously considered to be very rare, is in fact a major cause of dementia globally and accounts for up to one third of all cases (Lennox et al., 1989; Dickson et al., 1989; Hansen et al., 1990). The causes of dementia can be currently seen as Alzheimer's disease (accounting for approximately 50% of cases), diffuse Lewy body disease (accounting for up to 33% of cases) with the remaining cases being mainly associated with multi-infarct dementia together with a number of rarer causes.

We were stimulated to look for intraneuronal ubiquitin-protein conjugates in filamentous inclusions in the different neurodegenerative diseases for two reasons. The first reason was the demonstration in several laboratories that neurofibrillary tangles in Alzheimer's disease were ubiquitinated (Mori et al., 1987; Perry et al., 1987; Cole and Timiras, 1987). The second reason was our observation that reversible vimentin intermediate filament collapse occurs in tissue culture cells when viral membrane proteins are experimentally inserted into the cell surface (Mayer et al., 1989) before the transmembranous proteins are degraded in the lysosomal system (Earl et al., 1987). There was no *a priori* reason to suppose that there might be a connection between ubiquitin, intermediate filaments, the lysosomal system and the filamentous inclusions in the neurodegenerative diseases: indeed observations suggesting the presence of neurofilaments in neurofibrillary tangles in Alzheimer's disease are contentious (Ksiezak-Reding et al., 1987; Nukina et al., 1987). However, there is of course much evidence that ubiquitin is covalently conjugated to proteins as a signal for extralysosomal proteolysis (Hershko, 1988) and intermediate filaments appear to be the element of the cytoskeleton most sensitive to cell stress (Collier and Schlesinger, 1986).

Although the major human neurodegenerative diseases can now be grouped as a family of ubiquitin-filament diseases, it cannot be assumed that the molecular mechanisms involved in the biogenesis of the filamentous inclusions in the different diseases are identical. The molecular components of neurofibrillary tangles in Alzheimer's disease

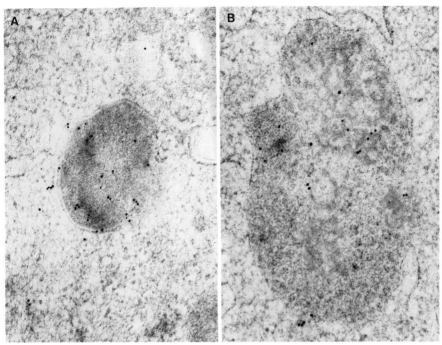

Figure 1. Immunogold electron microscopy of (A) fibroblasts and (B) fibroblasts treated with the thiol cathepsin inhibitor E-64. Fixed sections were immunostained with rabbit anti-ubiquitin protein-conjugate antiserum (DAKO, Glostrup, Denmark), followed by biotinylated anti-rabbit IgG antibody and streptavidin-gold (10nm gold particles). Magnifications; (A) 96,000; (B) 108,000.

e.g., the molecular progenitor of the paired helical filaments and tau protein, are not shared with Lewy bodies in diffuse Lewy body disease. Conversely, Lewy bodies definitely appear to contain neurofilaments (Forno et al., 1986). We have recently shown that a ubiquitin carboxyl-terminal hydrolase is found routinely in Lewy bodies but only in a minority of neurofibrillary tangles (Lowe et al., 1990a). We have also shown that the two types of filamentous inclusions can be further distinguished by the presence of αB crystallin which is again only found in a proportion of Lewy bodies and not in neurofibrillary tangles (Lowe et al., 1990b). The αB crystallin gene has a heat-shock promoter (de Jong et al., 1989). through which the elevated expression of the protein may be driven in some neurones containing Lewy bodies.

The combined observations indicate that either cell stress proteins and other proteins are found in both types of filamentous inclusions at different stages of inclusion biogenesis or that the molecular mechanisms involved in the biogenesis of the inclusions and the objectives of inclusion formation are different. We favour the latter notions and feel that the intermediate filament (neurofilament) containing Lewy body may have some special relationship to cell stress mediated events. The formation of intermediate filament containing inclusions may be part of a cell stress gene-mediated cytoprotective mechanism in neurones (Lowe and Mayer, 1990). This is supported by our findings that the brain specific ubiquitin carboxyl-terminal hydrolase is present in Rosenthal fibres of cerebellar astrocytomas, which contain intermediate filaments (glial filaments and vimentin filaments), but not in filamentous inclusions of unknown protein composition in brain and spinal cord in motor neurone disease (Lowe et al., 1990a). In addition αB crystallin is also found in Rosenthal fibres of cerebellar astrocytomas and in cytokeratin containing Mallory bodies in liver.

There certainly appears to be a molecular cell biological relationship between intermediate filament inclusions, ubiquitin and other cell stress proteins.

Ubiquitin and intraneuronal vacuoles: Some hippocampal neurones in Alzheimer's disease exhibit areas of granulovacuolar degeneration. We have shown by ubiquitin immunocytochemistry that ubiquitin-protein conjugates are present in these perinuclear granulovacuoles (Lowe et al., 1988); presumably proteins in the granular material are ubiquitinated. Recently we have also observed that ubiquitin-protein conjugates occur in perinuclear membrane bound vacuoles in neurones in scrapie infected mouse brain (Lowe et al., 1990c). These vacuoles are probably identical to the giant autophagic vacuoles seen in mouse brain in a model of human Creutzfeldt-Jakob disease (Boellaard et al., 1989).

The precise morphological identity of areas of granulovacuolar degeneration in hippocampal neurones in Alzheimer's disease has not been rigorously defined at the ultrastructural level. The granular material may represent secretory proteins in secretory vesicles; alternatively the ubiquitinated granular material may be in lysosomes. Consideration of the latter notion led to the experimentation and findings indicated below.

Conundrum

Molecular pathological findings are intrinsically limited by the availability and quality of autopsy material. Paraffin embedded material is an ideal substrate for histology and immunocytochemistry. However, the limitations of the light microscope do not allow clear molecular cell biological interpretation of the immunocytochemical observations. These observations are:

1. Intraneuronal vacuoles (possibly related to lysosomes) contain ubiquitin-protein conjugates.
2. Filamentous inclusions contain ubiquitinated proteins.
3. Inclusions containing intermediate filaments are immunoreactive for an enzyme of ubiquitin mediated metabolism (a ubiquitin carboxyl-terminal hydrolase) and for αB crystallin which is highly homologous to the small heat shock proteins of *Drosophila* (Ingolia and Craig, 1982).

The elements of the conundrum therefore are:
1. Do ubiquitinated proteins normally accumulate in lysosomes for catabolism?
2. Can some relationship be demonstrated between intermediate filaments and ubiquitinated proteins at the ultrastructural level in a model system?
3. Are heat shock proteins associated with intermediate filament inclusions in such a model system?
4. Could all the elements indicated above i.e., intermediate filaments, ubiquitinated proteins and lysosomes be associated with inclusions in such a model system?

Some of the answers to these questions are indicated below!

Ubiquitin and the lysosome system: The first question is whether ubiquitinated protein conjugates can ever gain access to a membrane limited compartment like the lysosomes since all of the published work (Hershko, 1988) has indicated that the conjugation of ubiquitin to target proteins serves to mark the proteins for cytosolic extra-lysosomal proteolysis. We tested this notion in conditions where lysosomal proteolysis is severely compromised by the thiol protease inhibitor E-64 which gains access to the lysosomal system by cellular pinocytosis. We discovered that ubiquitinated protein conjugates could be demonstrated to accumulate in the compromised lysosomal system of such fibroblasts by immunocytochemistry and also by subcellular fractionation coupled with Western blotting analysis (Doherty et al., 1989). Careful examination of Nycodenz gradients of homogenised cells by dot immunoblots appeared to show that ubiquitin-protein conjugates are enriched in lysosomes in the absence of the inhibitor i.e., in normal cells. This finding was confirmed by Western blotting of all fractions from the density gradient (Doherty et al., 1989). We have extended these observations by the use of immunogold electron microscopy in normal untreated fibroblasts (Laszlo et al., 1990). Quantitative analysis of the distribution of gold particles reveals that ubiquitin-protein conjugates are enriched some 12-fold in normal untreated fibroblasts relative to all other cytoplasmic organelles and, importantly, this ratio is not altered by E-64. The drug does not cause the non-specific accumulation of ubiquitinated proteins in the lysosomes: the drug simply causes a massive accumulation of proteins including ubiquitinated proteins in a volume-expanded lysosome system. The distribution of gold particles in lysosomes in normal and E-64 treated fibroblasts is shown in Figure 1. We have previously shown in E-64 treated cells that clusters of gold particles can be seen near to the invaginating membrane of what appear to be multivesicular bodies i.e., microautophagy is bringing about the accumulation of ubiquitinated proteins in lysosome-related multivesicular dense bodies (Laszlo et al., 1990). Multivesicular bodies capable of surface microinvagination are now being seen as part of the cell endosome-lysosome system involved in the sorting and degradation of some receptors (Hopkins et al., 1990) and viral proteins (Earl et al., 1987) as

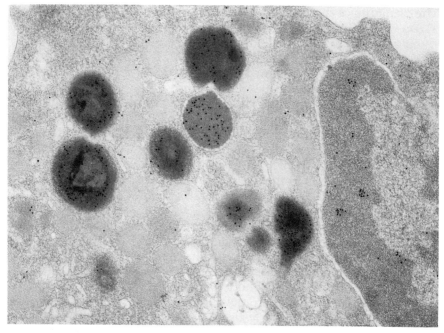

Figure 2. Granulocytic precursor cells from bone marrow are immunostained by immunogold electron microscopy as described in the legend to Figure 1. Magnification, 68,000.

well as in the ejection of some receptors (Davis et al., 1986) and viral membrane proteins (Earl et al., 1987) from the cell.

The clusters of ubiquitin-protein conjugates at points of microinvagination on the surface of multivesicular bodies indicate that protein ubiquitination, and possibly aggregation in the cytosol, may serve as one signal for the uptake of proteins into the lysosomal system by microautophagy. The considerable enrichment in the lysosomes (12-fold) of ubiquitinated proteins indicates that some specific uptake mechanism exists in the lysosomal membrane. However, since some membrane proteins are ubiquitinated on their extracellular domains e.g., the lymphocyte homing receptor (Seigelman et al., 1986), the possiblity cannot be excluded that membrane protein ubiquitination may occur in the lumen of the endoplasmic reticulum and all of its topological equivalents. Ubiquitinated proteins may be concentrated in the lysosomal system during the degradation of these membrane proteins together with ubiquitinated proteins derived from the cytosol.

These experimental findings, which indicate a role for protein ubiquitination in the uptake of proteins into the lysosomal system of fibroblasts, support the notion that the ubiquitinated materials seen in vacuoles in neurones in Alzheimer's disease and in scrapie infected mouse brain represent ubiquitin-protein conjugates in some form of

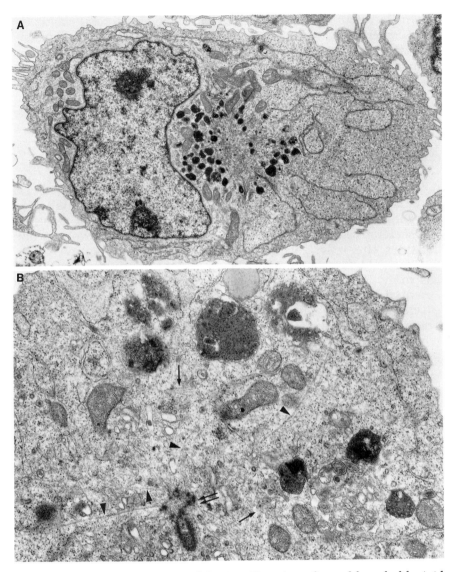

Figure 3. Electron microscopy of Epstein-Barr transformed lymphoblastoid cells. (A) The clustering of dense lysosomes and mitochondria around the micotubule organising centre; magnification, 14,400. (B) Higher magnification (40,600) showing the microtubule organising centre (double arrows), microtubules emanating from the centre (arrow head) and bundles of intermediate filaments (single arrow).

autophagic vacuoles (Lowe et al., 1988; Lowe et al., 1990c; Boellaard et al., 1989). The accumulation of ubiquitinated proteins or ubiquitin-protein fragments in the lysosomal system appears to be a general phenomenon in cells since we have now demonstrated ubiquitin-protein conjugates in lysosomes in granulocyte precursor cells from the bone marrow (Fig. 2) and in lymphoblastoid cells (see below).

A viral membrane protein together with ubiquitin-protein conjugates and HSP70 are found in lysosomes around the microtubule organisisng centre in an intermediate filament cage in Epstein-Barr virus transformed lymphoblastoid cells: The molecular and cell biological properties of the intermediate filament inclusions in some chronic human degenerative diseases (Lowe et al., 1988) together with their mechanism of biogenesis and functions will only be resolved with the help of model systems. Indeed it may not be possible to find the combination of cellular elements found in the diseases in a model system. There is currently no animal model system in which to study the inclusions seen in Alzheimer's disease!

A comprehensive survey of published works on virally infected cells reveals consistent reference to virally-induced cytoskeletal reorganisation, particularly affecting the intermediate filament cytoskeleton (Garcia-Barreno et al., 1988). Almost exclusively authors consider the rearrangements in terms of manipulation of the intermediate filament system by viral gene products to aid viral assembly: this notion may be partially correct by analogy with the essential role of the *E. coli* heat shock proteins in the replication and assembly of bacteriophage λ in bacteria (Alfano and McMacken, 1989). There is of course an alternative reason for intermediate filament reorganisation, namely to bring about the sequestration of viral proteins in a filamentous "cocoon" prior to degradation of the proteins or the ejection of the viral proteins from the cell (Mayer et al., 1989; Earl et al., 1987). Such a process might be linked to viral protein fragmentation by limited proteolysis for antigen presentation by the Class 1 histocompatibility molecules.

The ultrastructural examination of Epstein-Barr transformed lymphoblastoid cells (Fig. 3A) reveals extensive changes in the arrangement of the organelles with aggregation of the dense lysosomes around the microtubule organising centre. Higher magnification reveals that microtubules and bundles of intermediate filaments pervade the perinuclear constellation of organelles (Fig. 3B). Immunogold electron microscopy (Fig. 4B) shows that the latent membrane transforming protein (LMP) of the virus is found in lysosomes (the detailed ultrastructure of a dense secondary lysosome can be seen in Figure 4A). Immunohistochemistry shows that the intermediate filament cytoskeleton is also rearranged to form a filamentous inclusion centred on the microtubule organising centre (not shown here, but demonstrated in Laszlo et al., 1991) and surrounding and entwining the dense lysosomes. The lysosomes fuse together to form large autophagic vacuoles and, in a minority of cells, giant swollen membrane-bound vacuoles can be seen which contain LMP (not shown); LMP is also present in multivesicular bodies. The transmembranous protein would be expected to undergo lysosomal degradation and possibly ejection from the cell in vesicles following fusion of the giant membrane bound vacuoles and multivesicular bodies with the cell membrane: this process occurs during reticulocyte maturation to eject transferrin receptors in membrane bound vesicles from the cells (Davis et al., 1986).

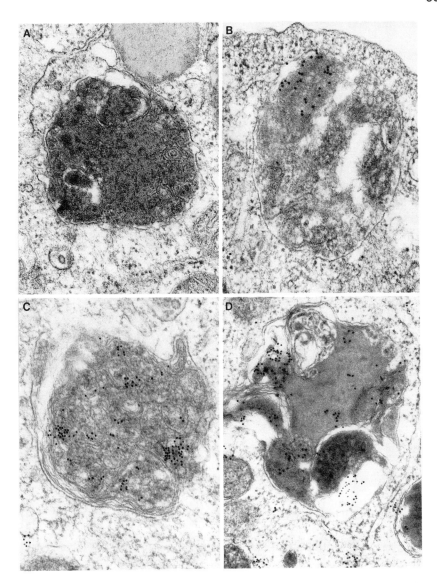

Figure 4. Immunogold electron microscopy of Epstein-Barr transformed lymphoblastoid cells showing lysosome related organelles. Cells were immunostained using 10nm gold particles as described in the legend to Figure 1. (A) Lysosome-related organelle not immunostained; magnification, 87,000. (B) Immunostained for the viral latent membrane transforming protein (LMP); magnification, 102,000. (C) Immunostained for ubiquitin-protein conjugates; magnification, 92,800. (D) Immunostained for HSP70; magnification, 92,800.

Immunogold electron microscopy also demonstrates that ubiquitin protein conjugates (Fig. 4C) and heat-shock protein 70 (HSP70; Fig 4D) are found in the expanded lysosomal system of the virally

308

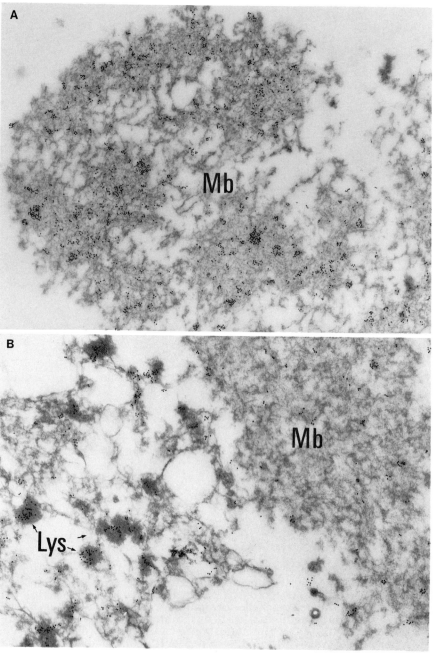

Figure 5. Mallory bodies partially purified from human liver by differential centrifugation and viscosity barrier centrifugation on 50% (v/v) Ficoll. Immunogold electron microscopy with 10nm gold particles performed as described in the legend to Figure 1. (A) Cytokeratin containing Mallory body immunostained for ubiquitin-protein conjugates; magnification, 52,000. (B) Immunostaining of HSP70 in dense lysosome-like structures associated with the Mallory body; magnification, 56,000.

transformed lymphoblastoid cells. Surface invagination of some of the membrane bound vacuoles can be seen; such a process will result in the formation of multivesicular bodies. These observations indicate that microautophagy by surface microinvagination of lysosome-related vesicles is active in Epstein-Barr transformed lymphoblastoid cells. We have no evidence that viral LMP is ubiquitinated although there is an increased accumulation of ubiquitinated proteins or ubiquitinated protein fragments in lymphoblastoid cells compared to lymphocytes (Laszlo et al., 1991).

We do not know the roles, if any, of protein-ubiquitination and HSP70 in the process by which LMP is directed into the lysosomal system. However, the fact that ubiquitin-protein conjugates are normally concentrated in lysosomes in fibroblasts (Doherty et al., 1989; Laszlo et al., 1990) and granulocytic precursor cells (Fig. 2) and that HSP70 has a function in protein sequestration into lysosomes (Chang et al., 1989) may indicate that protein ubiquitination and HSP70 are involved in events in LMP sequestration into the lysosomal system in lymphoblastoid cells. One intriguing possibility is that HSP70, and perhaps other heat-shock proteins, cooperate in a process, which also requires protein ubiquitination, driving the microinvagination of vacuolar membranes and generating the multivesicular bodies. Clearly such a molecular motor must exist in cells to facilitate the formation of multivesicular bodies.

The combined studies on the Epstein-Barr virus transformed cells show, perhaps suprisingly, that the majority of LMP is in the lysosomal system. The lysosomes are gathered around the microtubule organising centre which might be expected in view of the close association of lysosomes with microtubules (Hollenbeck and Swanson, 1990). Finally, the lysosome-related vacuolar system is surrounded and entwined in intermediate filaments. The role of intermediate filaments in the processes described above is not known. The LMP can be seen by immunofluorescence microscopy in patches in Daudi lymphoblastoid cells which do not contain intermediate filaments (Liebowitz et al., 1986); the patches correspond to the lysosome-related vacuoles. These data may suggest that the entry of LMP into the lysosomal system, the fusion of lysosomes and the formation of large vacuoles and multivesicular bodies centred on the microtubule organising centre does not require intermediate filaments and occurs before intermediate filament collapse. This interpretation would further suggest that intermediate filament collapse occurs after the organellar reorganisation which causes the lysosomes to accumulate around the microtubule organising centre. Prolonged, perhaps irreversible, intermediate filament collapse to enmesh reorganised organelles may be a late event in Epstein-Barr transformed cells: reversible intermediate filament collapse is the response of tissue culture cells to the insertion by fusion of a single bolus of Sendai transmembranous glycoproteins into the cell surface (Mayer et al., 1989; Earl et al., 1987). The LMP gene is presumably continuously switched on in lymphoblastoid cells creating a permanent membrane protein disposal problem for the cells. This problem appears to be solved initially by protein sequestration into the lysosomal system followed by organellar movements and intermediate filament reorganisation to cocoon the entire rearranged organellar assembly. Whether the intermediate filament reorganisation is permanent or could be reversed if the LMP gene was switched off remains to be elucidated.

Implications for intracellular inclusions in chronic human degenerative diseases

We clearly need to understand the reasons and purpose of the formation of filamentous inclusions in nerones, astrocytes and hepatocytes, as well as other cells1in chronic degenerative disease if we are to comprehend the molecular mechanisms of pathogenesis and indeed search for novel therapies to combat the diseases e.g., in the neurodegenerative disorders. The fundamental question concerns the reason for inclusion body formation. Our recent finding that there are major differences in these inclusions with respect to heat-shock related gene products indicates that we should not consider that there is a single class of filamentous inclusion. The inclusions which contain ubiquitin protein-conjugates, the ubiquitin carboxyl terminal hydrolase and the αBcrystallin (Lowe et al., 1988; Lowe et al., 1990a; Lowe et al., 1990b) have been previously shown to contain intermediate filaments. Currently, we can only comment on the possible reasons for the formation of this type of inclusion. Given that heat-shock gene activation and intermediate filament reorganisation occur in response to a variety of cell stresses we can argue that the formation of inclusions is consequent upon cells being chronically stressed. The objective of the response is to destroy, eject or cocoon foreign exogenous or functionless endogenous proteins or supramacromolecular assemblies. We have recently shown by immungold electron microscopy that partially purified preparations of cytokeratin-containing Mallory bodies also contain ubiquitin-protein conjugates (Fig. 5A) as previously demonstrated by immunohistochemistry (Lowe et al., 1988). The Mallory bodies are generally closely associated with lysosome-like dense bodies containing HSP70 (Fig. 5B). Again as in Epstein-Barr transformed lymphoblastoid cells an intermediate filament-containing inclusion is in close juxtaposition with elements of the lysosome system.

At present the molecular pathological and experimental findings overlap enough to suggest a working hypothesis for the formation and function of intermediate filaments in response to cell stress and injury (Fig. 6). However, there are sufficient differences in our findings (e.g., αB crystallin is not found in Epstein-Barr generated intermediate filament inclusions in lymphoblastoid cells and HSP70 is not detected in Lewy bodies) to suggest that some revision of the hypothesis will as usual be inevitable.

At some point in the progression of some chronic degenerative

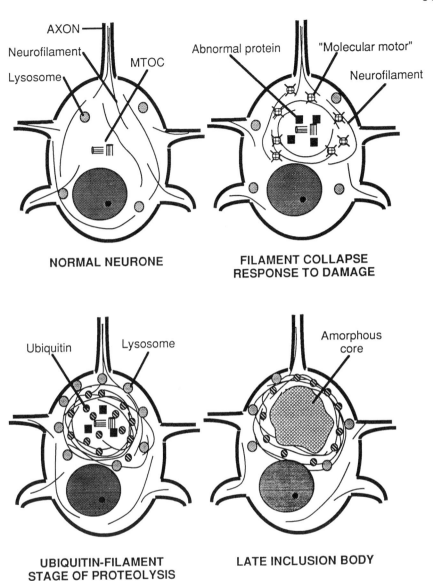

Fig. 6 Model of putative molecular events in the biogenesis of intraneuronal Lewy bodies and other intermediate filament containing inclusions in cells.

diseases, particularly certain neurodegenerative diseases, a process is activated which causes a reorganisation of the intermediate filament cytoskeleton. This process could follow organellar redistribution e.g., the microtubule dependent alteration in the distribution of lysosomes to mediate lysosomal fusions. The collection of proteins into the lysosomal system by microinvagination and/or the motor for the lysosomal fusion events may be assisted by protein ubiquitination and αB crystallin. The latter protein may be involved in large aggregate formation with other proteins prior to microautophagy. These events may occur in neurones poisoned by environmental toxins or as a response to latent viral infection. It is salutary that the events described in Epstein-Barr virus transformed lymphoblastoid cells might be seen because only a fraction of the large viral genome is activated in lymphoblastoid cells; the expression of the gene for a transmembranous glycoprotein (LMP) presumably triggers the changes in the distribution of the lysosomes and intermediate filaments. We might suppose that in lytic infections viruses would be able to switch off or overwhelm a mechanism for the sequestion and elimination of viral proteins. We must also not forget that events resulting in lysosomal fusion/microinvagination processes might occur without intermediate filament involvement, thus explaining the granulovacuoles seen in neurones in Alzheimer's disease (Lowe et al., 1988) and in scrapie (Lowe et al., 1990c).

Finally there are two intriguing independent concepts that emerge from our work with implications for both molecular pathology and experimental cell research.

1. The search for protein abnormalities in neurodegenerative diseases has identified the transmembranous β-amyloid precursor protein in Alzheimer's disease and the phosphatidyl-inositol membrane-anchored scrapie prion protein as proteins which appear in extracellular accumulations as truncated (Alzheimer's disease) or intact (scrapie) forms in the course of disease progression. In Epstein-Barr transformed lymphoblastoid cells only a limited repertoire of viral genes are expressed to include the multiple membrane spanning latent membrane transforming protein. Our initial observations on intermediate filament reorganisation and the lysosomal system were made after the fusion of reconstituted Sendai viral envelopes, containing transmembranous haemagglutinin and fusion proteins, into the cell surface (Mayer et al., 1989; Earl et al., 1987). The commonality in all these situations is, of course, one or two membrane proteins. This suggests that the molecular events leading to the major human neurodegenerative diseases and also occuring in some virally infected cells may be triggered by the abnormal processing or catabolism of a single membrane protein. With the benefit of hindsight it is not difficult to see that the cell-stress response, particulary of neurones, together with associated protein catabolic systems, must be of major importance for the inspection and degradation of membrane proteins: accumulation of a single membrane protein for a variety of reasons can severely compromise the system and result in disease progression and cell death.

2. Ubiquitin is only found in eukaryotic cells. There may be several reasons for this observation given the pleiotypic functions of ubiquitin in cellular homeostasis. However one major reason may be that protein ubiquitination serves to unify two major protein catabolic systems in eukaryotic cells acting, as a signal both for cytosolic protease action and for protein uptake into the lysosome system. Such a unifying function would explain why ubiquitin is only found in cells

with extensive intracellular membranes, including the nuclear envelope. Other heat-shock proteins e.g., HSP70 and crystallin may have evolved to assist in the protein-inspection and protein-gathering processes prior to degradation by either catabolic system.

Acknowledgements

We would like to thank the Parkinson's Disease Society of Great Britain (RJM, JL, ML), the Motor Neurone Disease Association (RJM, JL), and the Welcome Trust (RJM) for the support of these studies.

References

Alfano, C and McMacken, R, (1989) Heat shock protein-mediated disassembly of nucleoprotein structures is required for the initiation of bacteriopahge lambda DNA replication. J. Biol. Chem., 264: 10709-10718.

Boellaard, J, Schlote, W and Tateishi, J, (1989) Neuronal autophagy in experimental Creutzfeldt-Jacob's disease. Acta Neuropathol., 78: 410-418.

Chiang, H-L, Terlecky, SR, Plant, CP and Dice, JF, (1989) A role for a 70-kilodalton heat shock protein in lysosomal degradation of intracellular proteins. Science, 246: 382-385.

Cole, GM and Timiras, PS, (1987) Ubiquitin-protein conjugates in Alzheimer's lesions. Neurosci. Letts., 79: 207-212.

Collier, NC and Schlesinger, MJ, (1986) Heat shock proteins: the search for functions. J. Cell Biol., 103: 1495-1507.

Davis, JQ, Dansereau, D, Johnstone, RM and Bennett, V, (1986) Selective externalization of an ATP-binding protein structurally related to the clathrin-uncoating ATPase/heat shock protein in vesicles containing terminal transferrin receptors during reticulocytes maturation. J. Biol. Chem., 261: 15368-15371.

de Jong, WW, Hendriks, W, Mulders, JWM and Bloemendal, H, (1989) Evolution of eye lens crystallins: the stress connection. TIBS, 14: 365-368.

Dickson, DW, Crystal, H and Mattiace, LA, (1989) Diffuse Lewy-body disease: light and electron microscopy immunochemistry of senile plaques. Acta Neuropathol., 78: 572-84.

Doherty, FJ, Osborn, NU, Wassell, JA, Heggie, PE, Laszlo, L and Mayer, RJ, (1989) Ubiquitin-protein conjugates accumulate in the lysosomal system of fibroblasts treated with cysteine proteinase inhibitors. Biochem. J., 263: 47-55.

Earl, RT, Mangiapane, EH, Billett, EE and Mayer, RJ, (1987) A putative protein-sequestration site involving intermediate filaments for protein degradation by autophagy. Biochem. J., 241: 809-815.

Forno, LS, Sternberger, LA, Sternberger, NH, Strefling, AM, Swanson, K and Eng, LF, (1986) Reaction of Lewy bodies with antibodies to phosphorylated and non-phosphorylated neurophilaments. Neurosci. Lett., 64: 253-258.

Garcia-Barreno, B, Jorcano, JL, Aukenbauer, T, Lopez-Galindez, L and Melero, JA, (1988) Participation of cytoskeletal intermediate filaments in the infectious cycle of human respiratory syncytial virus. Virus Res., 9: 307-322.

Hansen, L, Salmon, D, Galasko, D and Masliah, E, (1990) The Lewy body variant of Alzheimer's disease: a clinical and pathological entity. Neurology, 40: 1-8.

Hershko, A, (1988) Ubiquitin-mediated protein degradation. J. Biol. Chem., 263: 15237-15240.

Hollenbeck, PJ and Swanson, JA, (1990) Radial extension of macxrophage tubular lysosomes supported by kinesin. Nature, 346: 864-866.

Hopkins, CR, Gibson, A, Shipman, M and Miller, K, (1990) Movement of internalized ligand-receptor complexes along a continuous endosomal reticulum. Nature, 346: 335-339.

Ingolia, D and Craig, EA, (1982) Four small Drosophila heat shock proteins are related to each other and to mammalian alpha-crystallin. Proc. Natl. Acad. Sci. USA, 79: 2360-2364.

Ksiezak-Reding, H, Dickson, DW, Davies, P and Yen, S, (1987) Recognition of tau epitopes by anti-neurofilament antibodies that bind to Alzheimer neurofibrillary tangles. Proc. Natl. Acad. Sci. USA, 331: 530-532.

Laszlo, L, Doherty, FJ, Osborn, NU and Mayer, RJ, (1990) Ubiquinated protein conjugates are specifically enriched in the lysosomal system of fibroblasts. FEBS Lett., 261: 365-368.

Laszlo, L, Tuckwell, J, Self, T, Lowe, J, Landon, M, Smith, S, Pike, I, Hawthorne, JN and Mayer, RJ, The latent membrane protein-1 in Epstein-Barr virus transformed lymphoblastoid cells is found with ubiquitin-protein conjugates and heat shock protein-70 in lysosomes oriented around the microtubule organising centre. J. Pathol., in press.

Lennox, G, Lowe, J, Landon, M, Byrne, EJ, Mayer, RJ and Godwin-Austen, RB, (1989) Diffuse Lewy-body disease:correlative neuropathology using anti-ubiquitin immunochemistry. J. Neurol. Neurosurg. Psychiat., 52: 1236-2347.

Liebowitz, D, Wang, D and Kieff, E, (1986) Orientation and patching of the latent infection membrane protein encoded by epstein-Barr virus. J. Virol., 58: 233-237.

Lowe, J, Blanchard, A, Morrell, K, Lennox, G, Reynolds, L, Billett, M, Landon, M and Mayer, RJ, (1988) Ubiquitin is a common factor in intermediate filament inclusion bodies of diverse type in man, including those of Parkinson's disease, Pick's disease, and Alzheimer's,as well as Rosenthal fibres in cerebellar astrocytomas, cytoplamsic bodies in muscle, and mallory bodies in alcoholic disease. J. Pathol., 155: 9-15.

Lowe, J and Mayer, RJ, (1990) Ubiquitin, cell stress and diseases of the nervous system. Neuropath. App. Neurobiol., 16: 281-291.

Lowe, J, McDermott, H, Landon, M, Mayer, RJ and Wilkinson, KD, (1990a) Ubiquitin carboxyl-terminal hydrolase (PGP-9.5) is selectively present in ubiquinated inclusion bodies characteristic of human neurodegenerative diseases. J. Pathol., 161: 153-160.

Lowe, J, Landon, M, Pike, I, Spendlove, I, McDermott, H and Mayer, RJ, (1990b) Dementia with β-amyloid deposition: involvement of αβ–crystallin supports two main diseases. Lancet, 336: 515-516.

Lowe, J, McDermott, H, Kenward, N, Landon, M, Mayer, RJ, Bruce, M, McBride, P, Somerville, RA and Hope, J, Ubiquitin conjugate immunoreactivity in the brains of scrapie infected mice. J. Pathol., 161. in press.

Mayer, RJ, Landon, M, Lennox, G, Lowe, J and Doherty, FJ, (1989) Intermediate filaments, ubiquitin, protein catabolism and chronic degenerative disease. In Intracellular Proteolysis (eds. Nobuhiki Katunuma and Eiki Kominami) Japan Scientific Societies Press, pp. 460-467.

Mori, H, Kondo, J and Ihara, Y, (1987) Ubiquitin is a component of paired helical filaments in Alzheimer's disease. Science, 235: 1641-1644.

Nukina, N, Kosik, KS and Selkoe, DJ, (1987) Recognition of Alzheimer paired helical filaments by monoclonal neurofilaments antibodies is due to crossreaction with tau protein. Proc. Natl. Acad. Sci. USA, 84: 3415-3419.

Perry, G, Friedman, R, Shaw, G and Chau, V, (1987) Ubiquitin is detected in neurofibrillary tangles and senile plaques neurites of Alzheimer's disease brains. Proc. Natl. Acad. Sci. USA, 84: 3033-3036.

Seigelman, M, Bond, MW, Gallatin, WM, St. John, T, Smith, HT, Fried, VA and Weissman, IL, (1986) Cell surface molecule associated with lymphocyte homing receptor is a ubiquitinated branch-chain glycoprotein. Science, 231: 823-829.

Subject index